UNIMOLECULAR REACTIONS

A Concise Introduction

The statistical theory of unimolecular reactions is now universally known as RRKM theory. This textbook covers the basics necessary for the understanding of RRKM theory in its original, angular-momentum conserved, phase-space, and simplified variational incarnations. Discussed are applications of RRKM theory to specific reaction systems, including thermal, chemical-activation, and multi-channel systems.

After a review of the Kassel quantum model and the theory of Slater, the specific-energy RRKM rate constant $k(E)$ is derived. Experimental evidence for (and some against) the RRKM form of $k(E)$ is given. The argument is then extended to the angular-momentum-dependent rate constant $k(E, J)$, to non-classical effects (tunneling and non-adiabatic transitions), the general problem of conservation of angular momentum, and to discussion of exit-channel effects. A chapter is devoted to the counting of quantum states, making extensive use of the Laplace-transform technique and the method of steepest descents. The chapter on thermal systems discusses both analytical and numerical solutions based on the master-equation approach, including activation by black-body radiation of interest in the biomedical field.

The book will be of primary interest to advanced undergraduate and graduate students studying chemical dynamics, chemical kinetics, and theoretical chemistry, as well as to interested researchers in other disciplines wishing to be acquainted with the essentials of RRKM theory.

UNIMOLECULAR REACTIONS

A Concise Introduction

WENDELL FORST

Université Bordeaux I, Talence, France

CAMBRIDGE UNIVERSITY PRESS

PUBLISHED BY THE PRESS SYNDICATE OF THE UNIVERSITY OF CAMBRIDGE
The Pitt Building, Trumpington Street, Cambridge, United Kingdom

CAMBRIDGE UNIVERSITY PRESS
The Edinburgh Building, Cambridge CB2 2RU, UK
40 West 20th Street, New York, NY 10011-4211, USA
477 Williamstown Road, Port Melbourne, VIC 3207, Australia
Ruiz de Alarcón 13, 28014 Madrid, Spain
Dock House, The Waterfront, Cape Town 8001, South Africa

http://www.cambridge.org

First published 2003

Printed in the United Kingdom at the University Press, Cambridge

Typeface Times 11/14 pt   *System* LATEX 2_ε [TB]

A catalogue record for this book is available from the British Library

Library of Congress Cataloguing in Publication data
Forst, Wendell.
Unimolecular reactions: a concise introduction / Wendell Forst.
p.   cm.
Includes bibliographical references and indexes.
ISBN 0 521 82190 8 – ISBN 0 521 52922 0 (pbk.)
1. Unimolecular reactions.   2. Statistical mechanics.   I. Title.
QD502.F68   2003   541.3′93–dc21   2003041049

ISBN 0 521 82190 8 hardback
ISBN 0 521 52922 0 paperback

Contents

Preface		*page* ix
List of symbols		xi
	Introduction: a brief historical perspective	1
1	The RRK model	4
	1.1 Isolated molecule	4
	1.2 Molecule in a thermal system	10
	References	14
	Problems	15
2	An outline of the theory of Slater	16
	2.1 Isolated molecule	16
	2.2 Thermal system	18
	References	24
	Problems	25
3	The RRKM rate constant $k(E)$	26
	3.1 Classical rate constant	26
	3.2 Semi-classical rate constant	32
	3.3 Implementation of the RRKM rate constant	37
	3.4 Rotational effects	41
	3.5 Experimental evidence	47
	References	52
	Problems	53
4	Calculation of energy-level densities	56
	4.1 General considerations	56
	4.2 Semi-classical degrees of freedom	64
	4.3 Hindered rotors	69
	4.4 Calculation of the inverse Laplace transform	78
	4.5 Quantized systems	83

	4.6	An overview	96
		References	99
		Problems	100
5	**Unimolecular reactions in a thermal system**		**105**
	5.1	Introduction	105
	5.2	The master equation	107
	5.3	Formal solution of the master equation	115
	5.4	Steady-state solutions	122
	5.5	Numerical steady-state solutions	136
	5.6	Thermal rate as a Laplace transform	143
	5.7	Non-collisional input	147
	5.8	Rotational effects	153
	5.9	Multichannel reactions	160
	5.10	Activation by black-body radiation	163
		References	165
		Problems	168
6	**Non-classical effects near barrier maximum**		**171**
	6.1	Tunneling	171
	6.2	Non-adiabatic transition	181
		References	187
		Problems	188
7	**A variational transition-state theory**		**189**
	7.1	General aspects	189
	7.2	Implementation	194
		References	205
		Problems	205
8	**Unimolecular decomposition under a central potential**		**207**
	8.1	Unimolecular decomposition from microscopic reversibility	207
	8.2	The potential	212
	8.3	Angular momentum	215
	8.4	$\Gamma(y, J)$ for simple cases	220
	8.5	$\Gamma(y, J)$ for the general case	222
	8.6	$\Gamma(y, J)$ for more complex cases	229
	8.7	Reduction of convolution integrals	237
	8.8	The J-conserved partition function	239
	8.9	Rate constant by phase-space theory	243
	8.10	Distribution of energy between fragments	249
		References	258
		Problems	260

9   Non-central-potential and exit-channel effects   265
    9.1   Triatomic systems   265
    9.2   Polyatomic systems   271
            References   273
            Problems   274
    Appendix 1   The sum of states, and the partition function as a
    semi-classical phase-space volume   276
    Appendix 2   Summary of the properties of the Laplace transform   284
    Appendix 3   Review of properties of the delta function   287
    Appendix 4   Potentials   288
    Appendix 5   Analytical solution at the low-pressure limit   293
    Appendix 6   Gamma function $\Gamma(n) = (n-1)!$ 298
    Answers to selected problems   300
    Author index   310
    Subject index   314

Preface

This textbook is an outgrowth of my *Theory of Unimolecular Reactions*, originally published in 1973 (now out of print), which was rather well received at the time. New developments of the subject since then seem sufficiently numerous and important to justify writing an entirely new text rather than merely updating the old one. Since in the meantime the subject matter has grown considerably, the focus of the present text is narrower, for instance excluding reactions of ionic species *per se* and isotope effects.

The aim of the presentation is a fundamental exposition intended as a graduate-level introduction to this particular field of chemical dynamics. It is primarily concerned with the basics of the theory, and therefore does not deal in much detail with interpretation of experimental results on specific reaction systems, except insofar as they illustrate a given point of theory, given that there are fairly recent publications that discuss such topics. (See R. G. Gilbert and S. C. Smith, *Theory of Unimolecular and Recombination Reactions* (Blackwell Scientific, 1990); K. A. Holbrook, M. J. Pilling and S. H. Robertson, *Unimolecular Reactions*, 2nd Ed. (Wiley, 1996), concerned primarily with thermal reactions; and T. Baer and W. L. Hase, *Unimolecular Reaction Dynamics* (Oxford University Press, 1996), which deals exlusively with state- or energy-selected systems.)

While, for reasons of presenting a coherent argument, some sections of the present work cover more or less the same ground as in these publications, though often from a different point of view, the major emphasis is on detailed treatment of topics not covered elsewhere. There are numerous problems illustrating various ramifications of the arguments developed in the main text, with which the reader can test his or her comprehension. Answers are provided for selected problems.

The subject matter is organized into nine chapters and six appendices. Equations, figures, and problems in each chapter are numbered from 1 onwards, preceded by the chapter number, separated by a full stop; thus, for example, "eq. (5.25)" refers to equation 25 of Chapter 5.

I am grateful to Professor J.-C. Rayez (Bordeaux) for support, to Professor J. C. Lorquet (Liège) for comments, and to Dr David Husain (Cambridge) who was instrumental in bringing this work to print.

Symbols

$\mathbb{M}$	matrix
$\mathbf{N}$	vector
$\mathscr{L}$	Laplace transform
A, B	rotational constants
E	energy
$G(E)$	sum of states at E
J, K	rotational quantum number
L	orbital quantum number
$N(E)$	density of states at E
Q	partition function
T	temperature
V	potential
k	Boltzmann's constant
k_{uni} k_∞ k_0	canonical (thermal) rate constant
$k(E)$	microcanonical (specific-energy) rate constant
log	logarithm base 10
ln	natural logarithm
t	time
μ	reduced mass
ω	collision frequency (except in Chapter 8)
$\Gamma(n)$	gamma function

Other symbols have a specific definition in each chapter.

Introduction: a brief historical perspective

The statistical theory of unimolecular reactions, now universally known as the "RRKM" theory, has its origins around 1925 when Oscar K. Rice, then a postdoc at Berkeley, became interested in interpreting experimental results on the decomposition of azomethane obtained by Herman C. Ramsperger, another postdoc in the same department. Rice later moved to Cal Tech, where he met Louis S. Kassel, who was working on the same problem. Thus was born what became known as the "RRK" theory.

While building on earlier treatments of unimolecular reactions by Lindemann and Hinshelwood [1, pp. 108ff.], RRK were astute to avoid all dynamical considerations by recognizing that the dissociation of a sufficiently large and suitably energized molecule is basically statistical. They viewed the reacting molecule as an assembly of harmonic oscillators that freely exchange energy; decomposition occurs when by chance a particular critical oscillator accumulates the minimum amount of energy necessary for reaction. RR worked out the problem in terms of classical statistical mechanics [2], whereas Kassel used the discrete combinatorial approach [3].

Since the process of activation (mainly collisional activation at the time) is unlikely to deposit energy directly in the critical oscillator, it can be expected that it will take a while for the requisite amount of energy to find its way into the critical oscillator, time that is longer the larger the number of oscillators in the molecule ("large molecules live longer"). This accounts for the time-lag between activation and decomposition postulated by Lindemann and Hinshelwood and links it to the size of the molecule. However, unlike in Hinshelwood's theory, the rate constant for decomposition (i.e. the inverse of the lifetime) becomes an increasing function of energy. A few years later, Rice [4] (with Gershinowitz) added the qualitative notion of an "activated complex" (i.e. transition state) as introduced by Eyring [5] in the "absolute reaction-rate theory" which dominated the thinking in the decade before World War II.

1

It is ironic that the basic idea of the theory survived the ravages of time much better than did the experimental results it sought to explain, for it is now known that the decomposition of azomethane is not a very good reaction on which to test unimolecular theory, afflicted as the reaction is by a free-radical chain [6]. In any event, the state of the technology at the time was such that it was not possible to carry out meaningful experimental tests of the theory, and hence for a time there were no new theoretical developments.

After a hiatus of more than ten years, Rice (who was by then at the University of North Carolina, Chapel Hill) took up the subject of unimolecular reactions again in 1949 when a new postdoc came to Chapel Hill in the person of R. A. Marcus. He introduced the notions of active and adiabatic (essentially rotational) degrees of freedom, as well as quantum considerations, and in general put the notion of the transition state in unimolecular reactions on a firm footing [7]. The circumstances were obviously ripe for a RRK-type unimolecular rate theory incorporating the transition-state concept, for almost at the same time Rosenstock, Wallenstein, Wahrhaftig, and Eyring [8] put forward a theory of mass spectra (known to mass spectrometrists as the QET, or the quasi-equilibrium theory), which, in slightly different language, invoked the same general idea as RRK, a fact not appreciated until many years later [9].

Marcus' improvements did not have an immediate impact since at about the same time a rival, mathematically very elegant, theory due to N. B. Slater [10] came into vogue. Unlike RRK or QET, this theory addressed the problem of the dynamics of reaction: a unimolecular decomposition was seen as the result of a collection of harmonic oscillators coming sufficiently into phase such that the bond extension of a particular oscillator exceeded a critical value and the oscillator dissociated. The time taken to reach this value could be calculated from detailed vibrational analysis of the normal modes. By then the advent of gas chromatography had greatly increased the sensitivity and selectivity of analysis, which encouraged the experimentalists to test Slater's theory. It eventually appeared that his theory predicted thermal rate constants at low pressures that were too small, which was ultimately traced to the unphysical assumption that a decomposing (and therefore highly excited) molecule could be treated as a superposition of normal modes, whereas such a molecule is in reality considerably anharmonic and therefore is able to share energy much more freely [11].

Largely as a result of the considerable amount of experimental data produced in the laboratory of B. S. Rabinovitch (BSR) from about 1960 onwards, much of it by the then-new technique of chemical activation, it became clear that the purely statistical treatment of "RRKM" (as the theory, with Marcus' additional improvements [12], became known) was the more physically reasonable. Also at this time computers started to appear, and BSR's laboratory developed most of the early computational routines necessary for practical numerical applications of the theory. As more and more experimental data on a large variety of unimolecular reactions appeared, it became evident that the main RRKM premise, namely complete randomization of energy among internal degrees of freedom prior to dissociation, is violated only in very special circumstances. With the wide-ranging applicability of the RRKM theory thus demonstrated, there appeared in due course a couple of monographs [9, 13] summarizing developments roughly up to 1970.

References

[1] K. J. Laidler, *Theories of Chemical Reaction Rates* (McGraw-Hill, New York, 1969).

[2] O. K. Rice and H. C. Ramsperger, *J. Am. Chem. Soc.* **49**, 1617 (1927).

[3] L. S. Kassel, *J. Phys. Chem.* **32**, 225, 1065 (1928); *Proc. Nat. Acad. Sci. USA* **14**, 23 (1928).

[4] O. K. Rice and H. Gershinowitz, *J. Chem. Phys.* **3**, 479 (1935).

[5] H. Eyring, *J. Chem. Phys.* **3**, 107 (1935). For a historical review of transition-state theory see K. J. Laidler and M. C. King, *J. Phys. Chem.* **87**, 2657 (1983).

[6] O. K. Rice and W. Forst, *Can. J. Chem.* **41**, 562 (1963).

[7] R. A. Marcus and O. K. Rice, *J. Phys. Coll. Chem.* **55**, 894 (1951); R. A. Marcus, *J. Chem. Phys.* **20**, 352, 355, 359, 364 (1952).

[8] H. M. Rosenstock, M. B. Wallenstein, A. L. Wahrhaftig and H. Eyring, *Proc. Nat. Acad. Sci. USA* **38**, 667 (1952).

[9] W. Forst, *Theory of Unimolecular Reactions* (Academic Press, New York, 1973).

[10] N. B. Slater, *Theory of Unimolecular Reactions* (Cornell University Press, Ithaca, 1959).

[11] E. K. Gill and K. J. Laidler, *Proc. Roy. Soc. (London)* **250**A, 121 (1959).

[12] G. M. Wieder and R. A. Marcus, *J. Chem. Phys.* **37**, 1835 (1962); see also O. K. Rice, *Statistical Mechanics, Thermodynamics and Kinetics* (W. H. Freeman, San Francisco, 1967), Ch. 20.

[13] P. J. Robinson and K. A. Holbrook, *Unimolecular Reactions* (Wiley, New York, 1972).

1

The RRK model

"RRK" was historically the first successful model of a unimolecular reaction, being put forward in 1926 by Rice and Ramsperger [1], and, independently and almost simultaneously, by Kassel [2]; hence the abbreviation "RRK." It conveys all the essential features of a unimolecular reaction, to the point where it is almost semi-quantitatively correct. Although nowadays it has been supplanted by more elaborate versions (known as the RRKM theory and variants thereof), the simplicity of the RRK version has the virtue of bringing into sharp focus the basic assumptions of the theory which perhaps tend to get obscured in the RRKM version. Therefore the exposition of RRK theory has instructional value, in that it lays the foundations on which to build a more elaborate theory.

The presentation below of the RRK theory follows the Kassel discrete-quantum version (sometimes abbreviated to "QRRK") which has been found useful in modeling multiphoton dissociation (see, e.g., [3–6]). The Rice–Ramsperger version forms the basis of the development given in Chapter 3.

1.1 Isolated molecule

1.1.1 The model

Consider an elementary unimolecular process in which a single isolated molecule dissociates, by simple bond rupture, into two fragments. Let us represent the stable (bound) reactant molecule by a collection of harmonic oscillators, say s in number. The reaction, i.e. the bond rupture, will correspond to one particular oscillator – the critical oscillator – dissociating as soon as it has accumulated the necessary energy. The critical oscillator is therefore considered to be a harmonic oscillator with a cut-off (cf. Fig. 2.1).

The process by which oscillators, and the critical oscillator in particular, accumulate energy is left unspecified.

At the simplest level, we shall assume that all the oscillators are of the same frequency v; this is the same as if the decomposing molecule were composed of one s-fold-degenerate oscillator of frequency v. The frequency restriction will be relaxed later.

1.1.2 The probability of dissociation

A suitable excitation process will cause the oscillators to take up energy in the form of quanta, i.e. multiples of hv (h is Planck's constant). Suppose that the excitation energy amounts to n quanta; the question of how the n quanta are distributed among the s oscillators now arises. The answer is that, if there is no information (and such was certainly the case around 1930), the only reasonable way to proceed is to assume that *all* ways to distribute the n quanta among the s oscillators are equally probable.

Formulated in this way, the question is identical to a problem in elementary probability, namely the total number of ways n indistinguishable balls can be distributed among s indistinguishable boxes (indistinguishable because obviously neither the quanta nor the oscillators can be distinguished one from the other). The answer can be found as follows: to delimit the s boxes, we need $s - 1$ partitions, represented below by a vertical stroke (|), whereas a ball will be represented by a circle (○). Imagine the balls and the partitions laid out in a linear array, for example

$$○ ○ ○ \mid ○ ○ \mid ○ ○ ○ ○ \mid ○ \mid ○ ○ ○ ○ ○ ○ \mid ○ \mid ○ ○ \mid ○ ○ ○ \cdots \text{etc.}$$

The number of permutations of $s - 1$ partitions and n balls, i.e. of $n + s - 1$ objects, is $(n + s - 1)!$ if the objects are distinguishable; since they are not, we have to divide by the number of permutations of n balls, multiplied by the number of permutations of $s - 1$ partitions, i.e. by $n!(s - 1)!$, so that W_n, the number of ways to put n balls (quanta) into s boxes (oscillators) is

$$W_n = \frac{(n + s - 1)!}{n!(s - 1)!} \tag{1.1}$$

With respect to the problem at hand, Eq. (1.1) represents the number of *states* of an s-fold-degenerate oscillator containing n quanta, or the degeneracy of state n for short.

By hypothesis the critical oscillator will dissociate as soon as it contains m quanta ($m \leq n$), which means that the remaining $s - 1$ oscillators will contain $n - m$ quanta. By an argument similar to the above, we see immediately that the number of ways to put $n - m$ quanta into $s - 1$ oscillators is

$$W_{n-m} = \frac{(n - m + s - 2)!}{(n - m)!(s - 2)!} \tag{1.2}$$

Thus the *probability* that, if s oscillators contain n quanta, one particular (critical) oscillator shall contain (exactly) m quanta is

$$p_{n,m} = \frac{W_{n-m}}{W_n} = \frac{(s-1)(n-m+s-2)!n!}{(n-m)!(n+s-1)!} \tag{1.3}$$

which makes use of the relation $(s-1)! = (s-1)(s-2)!$.

The critical oscillator obviously will dissociate whenever it contains a number of quanta that exceeds the critical threshold of m quanta, i.e. for any number of quanta between m and n. Hence the distribution of Eq. (1.2) has to be summed for all $m \leq n$:

$$W_{n\geq m} = \sum_{i=m}^{n} W_{n-i} = \sum_{i=m}^{n} \frac{(n-i+s-2)!}{(n-i)!(s-2)!} \tag{1.4}$$

Let $n - i = y$, hence $0 \leq y \leq n - m$, so that Eq. (1.4) becomes

$$W_{n\geq m} = \sum_{y=0}^{n-m} \frac{(y+s-2)!}{y!(s-2)!} \tag{1.5}$$

This summation can be evaluated by an argument similar to that used in connection with Eq. (1.1) by the device of adding one extra partition that provides an additional box into which to put the $n - m - y$ balls left over after y of them ($0 \leq y \leq n - m$) have been placed into the first $s - 1$ boxes. In this way we obtain

$$\sum_{y=0}^{n-m} \frac{(y+s-2)!}{y!(s-2)!} = \frac{(n-m+s-1)!}{(n-m)!(s-1)!} \tag{1.6}$$

and hence the *probability* that, if s oscillators contain a total of n quanta, a particular (critical) one shall contain *at least* m quanta is

$$p_{n\geq m} = \frac{W_{n\geq m}}{W_n} = \frac{(n-m+s-1)!n!}{(n-m)!(n+s-1)!} \tag{1.7}$$

1.1.3 The rate constant and lifetime

In chemical kinetics, the information desired is the rate constant. Before proceeding further it is of interest to consider briefly the wider implication of what in the present instance is the unimolecular rate constant k.

If isolated reactant molecule M decomposes (i.e. disappears) in a unimolecular reaction with rate constant k, this means that

$$-\frac{1}{[M(t)]}\frac{d[M(t)]}{dt} = k \tag{1.8}$$

where square brackets signify concentrations at time t. This is in fact the definition of a first-order rate constant. The ratio $d[M(t)]/[M(t)]$ is the fraction of molecules decomposing, that is, the probability of decomposition P, so that Eq. (1.8) can be written

$$-\frac{d[M(t)]}{[M(t)]} = k\,dt = P \tag{1.9}$$

on the assumption that P depends only on the length of the interval dt that M is under observation but is otherwise independent of time. Therefore k represents the probability of decomposition per unit time, likewise independent of time, i.e. independent of the time interval during which the decay of $M(t)$ is measured. Consequently Eq. (1.9) can be integrated directly to yield, with $[M(t = 0)]$ the initial concentration of M,

$$[M(t)] = [M(t = 0)]\exp(-kt) \tag{1.10}$$

which means that M will decay with a simple exponential time dependence.

Now the probability of M *not* decomposing in dt, i.e. the probability of survival in dt, is $1 - P = 1 - k\,dt$, and the probability of surviving n such intervals is $(1 - k\,dt)^n$, given that, by hypothesis, P is independent of time. If we let $n\,dt = t$, then the probability of survival in $(0, t)$ is

$$P(t) = \lim_{n\to\infty}(1 - k\,dt)^n = \lim_{n\to\infty}\left(1 - k\frac{t}{n}\right)^n = \exp(-kt) \tag{1.11}$$

In other words, $P(t) = \exp(-kt)$ is the probability of lifetime t. A lifetime that obeys Eq. (1.11) is said to be random, for reasons that will appear shortly. Normalized to unity, $P(t)$ becomes

$$P'(t) = k\exp(-kt) = -\frac{dP(t)}{dt} \tag{1.12}$$

which can be shown (Problem 1.1) to be the rate referred to the number of molecules present initially, rather than at time t, as in Eq. (1.8). A further ramification is the subject of Problem 1.2. The average lifetime $\langle t \rangle$, using the usual definition of an average, turns out to be the reciprocal of k:

$$\langle t \rangle = \int_0^\infty t P'(t)\,dt = \frac{1}{k} \tag{1.13}$$

1.1.4 The Kassel rate constant

Returning to the problem on hand, with Eq. (1.7) for the probability, and in view of the consequence of Eq. (1.9), according to which the rate constant k represents the probability of decomposition per unit time, we can now write for the rate constant

$$k_{n \geq m} = A p_{n \geq m} = A \frac{(n - m + s - 1)! \, n!}{(n - m)! (n + s - 1)!} \qquad (1.14)$$

where, since the probability $p_{n \geq m}$ is dimensionless, there appears a proportionality constant A of dimension s^{-1}. This is also the dimension of frequency, so that A can be interpreted as the characteristic frequency with which oscillators exchange quanta of energy. Since most oscillator frequencies are about $10^{13} \, s^{-1}$, we can expect A to be of the same order of magnitude.

It is worthwhile to stop here and ponder the meaning of Eqs. (1.7) and (1.14). The derivation is based on the assumption that there exists a large collection of reactant molecules with a distribution of quanta such that every possible distribution is represented with equal probability. In other words, $p_{n \geq m}$ measures the probability that a molecule chosen at random will have the specified property, which is reflected in the lifetime represented by Eq. (1.11). The implication is that $p_{n \geq m}$, and hence the rate constant $k_{n \geq m}$, measures an *average* property of decomposing molecules.

The second assumption is that $p_{n \geq m}$ and the underlying distribution are achieved whatever the excitation process. This means that, if, at the moment of excitation, the distribution is different (e.g. most molecules contain all n quanta in only a few of the s oscillators), then by the time the collection of molecules actually dissociates all possible distributions are equally represented. The implication is that (1) there is some mechanism for the redistribution of quanta among the oscillators, and (2) there is a time-lag between excitation and dissociation that allows the redistribution to take place. Thus $1/k_{n \geq m}$ may be interpreted as the average time-lag, i.e. average lifetime of an excited molecule (Eq. (1.13)).

The putative existence of a mechanism for the redistribution of quanta (or randomization as it is often called) means that the oscillators cannot be strictly harmonic since harmonic oscillators are unable to exchange quanta. The oscillators must therefore be considered perturbed (e.g. anharmonic), sufficiently so that a redistribution of quanta may take place, but not enough to invalidate the calculation that leads to the probability of dissociation $p_{n \geq m}$. The exact mechanism of the perturbation need not be specified since it does not appear anywhere explicitly.

1.1.5 The energy dependence

Equation (1.14) is not always useful for calculations because the factorials quickly become very large even for relatively small values of n. If the factorials are developed and common terms dropped, we get

$$p_{n \geq m} = \frac{(n - m + 1)(n - m + 2) \ldots (n - m + s - 1)}{(n + 1)(n + 2) \ldots (n + s - 1)} \qquad (1.15)$$

Assuming that $s - 1 \ll n - m$, which in a large molecule corresponds to the classical limit, Eq. (1.15) reduces to a simpler, but less realistic, version of Eq. (1.7):

$$p_{n \geq m} \approx \left(\frac{n - m}{n} \right)^{s-1} \tag{1.16}$$

Problems 1.3 and 1.4 explore an alternative approach using Stirling's approximation to the factorials. If it is more convenient to work in terms of energies rather than quanta, let $E = nh\nu$ and $E_0 = mh\nu$, where E is now the excitation energy and E_0 is the critical energy for dissociation. On combining Eqs. (1.15) and (1.16) we have

$$k(E) = A \left(1 - \frac{E_0}{E} \right)^{s-1} \tag{1.17}$$

where $k(E)$ is now the unimolecular rate constant (s^{-1}) in terms of energy. Equation (1.17) is known as the Kassel form of the unimolecular rate constant. A more elaborate version of $k(E)$ is the subject of Chapter 3.

Consider now the physical implications of Eq. (1.17). Since the rate constant $k(E)$ is by definition positive, Eq. (1.17) implies that $k(E) = 0$ for $E < E_0$. At the other extreme, as $E \to \infty$, $k(E) \to A$, which means that the rate constant cannot exceed the rate at which oscillators exchange energy. At intermediate energies, $k(E)$ will be an increasing function of energy, the rate constant with the *smallest s* being numerically largest (Fig. 1.1). Hence, at the same excitation energy, the average lifetime (given by $1/k(E)$) of a small molecule (small s) is shorter than the average lifetime of a large molecule (large s). Physically this means that, in

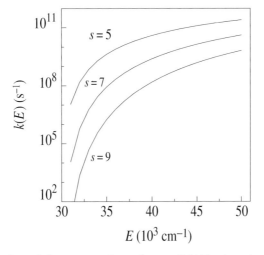

Fig. 1.1. An illustration of the energy dependence of $k(E)$ given by Eq. (1.17) at three values of the parameter s, assuming that $E_0 = 30\,000\,\text{cm}^{-1}$ and $A = 10^{13}\,\text{s}^{-1}$.

a large molecule, it takes longer for the energy to "find" its way into the critical oscillator, since there are many more ways to distribute a given energy. In short, "a large molecule lives longer." These are the principal characteristics of the RRK rate constant which are also found in every more elaborate version of unimolecular rate theory.

Note in particular that the rate constant is a function only of energy E and of the complexity of the molecule as represented by the parameter s. Implicit is the notion that the development is valid only insofar as the oscillators exchange energy at a rate $(A \, \mathrm{s}^{-1})$ that is faster than the rate at which the molecule actually dissociates. No reference is made anywhere to the dynamics of the dissociation process, which is avoided by the use of probabilistic arguments.

1.1.6 Non-degenerate oscillators

In any real molecule, oscillators are not all of the same frequency, and the problem of how to modify the preceding development then arises. The simplest way is to replace the individual frequencies $v_i (i = 1, \ldots, s)$ by some kind of average. Which average? It appears that the answer depends to some extent on the use one wishes to make of the RRK formulation. In infrared laser multiphoton dissociation, for example $SF_6 \rightarrow SF_5 + F$ [3, 4], it is useful to work with quanta rather than energies. In such cases some [5] argue in favor of the arithmetic-average frequency $\langle v \rangle = \Sigma_i v_i / s$ as most appropriate, whereas others [6] favor the geometric average

$$\langle v \rangle = \left(\prod_{i=1}^{s} v_i \right)^{1/s} \tag{1.18}$$

where s is defined, as before, as the total number of oscillators. In either case, if the total energy is E, the corresponding (average) quantum is then $\langle n \rangle = E/(h\langle v \rangle)$. Thus the essence of the previous development is preserved, with v and n merely replaced by averages.

1.2 Molecule in a thermal system

1.2.1 Lifetimes and collisions

So far we have considered only the somewhat artificial case of the decomposition of an isolated molecule. In a more realistic situation such as a thermal reaction, molecules undergo collisions that affect their survival probability.

Collisions being random, if the collision frequency is ω, the probability of collision in $t, t + \mathrm{d}t$ is $\omega \, \mathrm{d}t$, and the probability of *no* collision in $0, t$ is $\exp(-\omega t)$, using the same argument as that which led to Eq. (1.11). If decomposition is also random, the probability of survival is $P(t) = \exp(-kt)$ (Eq. (1.11)). The probability of dissociation actually taking place in $t, t + dt$ is composed of three factors: (probability of no collision up to t) × (probability of no dissociation up to t) × (probability of dissociation in $t, t + dt) = \exp[-(\omega + k)t] k \, dt$, and the actual proportion of molecules dissociating randomly will be

$$\int_0^\infty \exp[-(\omega + k)t] k \, dt = \frac{k}{\omega + k} \tag{1.19}$$

In a thermal system reacting molecules can accumulate energy only by collisions, so that the probability of reaction per unit time, i.e. the thermal rate constant k_{uni}, will be given by the number of molecules colliding per unit time, i.e. the collision frequency ω, times the probability of a molecule actually dissociating (Eq. (1.17)); thus, rearranging the terms,

$$k_{uni} = \omega \int_0^\infty \exp(-\omega t) k \exp(-kt) \, dt = -\int_0^\infty \omega \exp(-\omega t) \frac{dP(t)}{dt} \, dt \tag{1.20}$$

where $dP(t)/dt$ is given by Eq. (1.12). In this form $k_{uni}(s^{-1})$ can be considered quite generally as $(-\omega)$ × (Laplace transform of $dP(t)/dt$), the transform parameter being t (Appendix 2). Equation (1.20) offers in principle the possibility of evaluating k_{uni} for dissociations that are not necessarily random if the form of $dP(t)/dt$ is known. One such possibility is mentioned in Chapter 2.

In a realistic situation, of course, the rate constant k will be a function of (at least) energy or quantum number, as in Eqs. (1.16) and (1.17), so that the above expression for k_{uni} remains to be integrated or summed over the distribution of energies or quantum numbers, as shown next, and in considerably more detail later in Chapter 5.

1.2.2 Thermal reactions

Since for many purposes RRK is more useful in the representation of quanta rather than energies, the following development will continue to be given in terms of the discrete version. For example, in the cited multiphoton pumping of SF_6 [4] the energy distribution so produced is close to thermal.

Equation (1.1), which expresses the degeneracy of state n of an s-fold-degenerate oscillator, permits us to write directly the corresponding partition function Q, with $x = h\nu/(kT)$ (k is Boltzmann's constant, T is temperature):

$$Q = \sum_{n=0}^\infty \frac{(n + s - 1)! \exp(-nx)}{(s - 1)! n!} \tag{1.21}$$

If the summation is written out explicitly, it will be recognized as the binomial expansion of $Q = [1 - \exp(-x)]^{-s}$. Hence the probability that an s-fold-degenerate oscillator in a thermal system shall have energy equivalent to n quanta, i.e. the Boltzmann (equilibrium) fraction of oscillators of energy $nh\nu$, is, with Q as the normalization constant,

$$B_{nh\nu} = [1 - \exp(-x)]^s \frac{(n + s - 1)! \exp(-nx)}{(s - 1)! n!} \tag{1.22}$$

Using Eqs. (1.7) and (1.14), the corresponding thermal rate constant is, after some simplification,

$$k_\infty = \sum_{n=m}^{\infty} k(n \geq m) \, B_{nh\nu}$$

$$= A[1 - \exp(-x)]^s \sum_{n=m}^{\infty} \frac{(n - m + s - 1)!}{(s - 1)!(n - m)!} \exp(-nx) \tag{1.23}$$

The summation is in fact the same as in Eq. (1.21) if n is replaced with the dummy variable $z = n - m, 0 \leq z \leq \infty$, so that

$$k_\infty = A \exp(-mx) = A \exp[-E_0/(\ell T)] \tag{1.24}$$

Thus the thermal rate constant k_∞ has the Arrhenius form. It is well to remember that this applies only when the energy distribution corresponds to the equilibrium Boltzmann distribution $B_{nh\nu}$, which is the case only at "infinite" pressure in a gaseous system with collisional activation; hence the subscript "∞."

1.2.3 The pressure dependence

When collisional activation operates at finite pressures, there will be competition between activation and reaction. If M is the reactant molecule of interest, represented by an s-fold-degenerate oscillator containing n quanta, and X is any chemically inert species, this competition can be represented by the simple reaction scheme originally suggested by Lindemann [7]:

$$M + X \underset{k_2}{\overset{k_1}{\rightleftharpoons}} M^* + X$$
$$M^* \xrightarrow{k_3} \text{products} \tag{1.25}$$

where M^* symbolizes an activated reactant molecule, i.e. one with sufficient quanta $(n \geq m)$ to react. The time-lag between activation and reaction alluded to before is the result of M^* having two choices in this mechanism: it can be deactivated by collision with rate constant k_2, or decompose with rate constant k_3. Clearly this time-lag will depend on the number of deactivating collisions per unit time, i.e. on the pressure. Assuming a steady state for [M*], we have

$$\frac{d[M^*]}{dt} = k_1[M][X] - (k_2[X] + k_3)[M^*] = 0 \tag{1.26}$$

from which

$$[M^*] = \frac{k_1[M][X]}{k_2[X] + k_3} \tag{1.27}$$

The rate of formation of product(s) is $k_3[M^*]$, so that the pressure-dependent unimolecular rate constant k_{uni} (s^{-1}) = rate/[M], i.e.

$$k_{\text{uni}} = \frac{k_1 k_3[X]}{k_2[X] + k_3} = \frac{(k_1/k_2)k_3 k_2[X]}{k_2[X] + k_3} \tag{1.28}$$

On the assumption that [M*] represents only a small fraction of all molecules M, the ratio k_1/k_2 will be equal to the equilibrium fraction [M*]/[M], which therefore can be represented by the Boltzmann distribution (Eq. (1.22)). In the absence of any definite information, the term $k_2[X]$ may be replaced by the collision frequency ω, which amounts to the so-called strong-collision approximation (more on this in Section 5.4.6). On substituting for k_3 from Eq. (1.14) we have, after some reduction,

$$k_{\text{uni}} = \omega A[1 - \exp(-x)]^s \sum_{n=m}^{\infty} \frac{\dfrac{(n-m+s-1)!\exp(-nx)}{(s-1)!(n-m)!}}{\omega + A\dfrac{(n-m+s-1)!n!}{(n-m)!(n+s-1)!}} \tag{1.29}$$

With the substitution $i = n = m$, k_{uni} becomes

$$k_{\text{uni}} = \omega A[1 - \exp(-x)]^s \sum_{i=0}^{\infty} \frac{\dfrac{(i+s-1)!}{i!(s-1)!}\exp[-(i+m)x]}{\omega + A\dfrac{(i+s-1)!(m+i)!}{i!(i+m+s-1)!}} \tag{1.30}$$

A simple interactive computer program for calculations using Eq. (1.30) is available [8]. If $\omega \to \infty$ in Eq. (1.29), we recover Eqs. (1.23) and (1.24) for k_∞. At the other extreme, when $\omega \to 0$, k_{uni} becomes the limiting low-pressure rate constant k_0:

$$k_0 = \omega[1 - \exp(-x)]^s \sum_{i=0}^{\infty} \frac{(i+m+s-1)!}{(m+i)!(s-1)!}\exp[-(i+m)x] \tag{1.31}$$

In this case the time-lag between two successive collisions is infinitely long so that *every* molecule having at least m quanta will dissociate. Problem 1.5 shows that k_0 can be reduced to a simple expression at high temperature.

Since at a given temperature k_∞ (Eq. (1.24)) is a constant independent of [X], and k_0 is proportional to [X] (via ω), k_{uni} will exhibit a fall-off with pressure or concentration, a characteristic confirmed by every actual unimolecular reaction.

If the oscillator is not s-fold degenerate and we have instead a collection of oscillators with s different frequencies, we can try to replace s in the above formulas by an "effective" s_{eff}. One suggestion [9] has been to use $s_{\text{eff}} \approx \langle E_{\text{v}} \rangle /(\ell T)$, where $\langle E_{\text{v}} \rangle$ is the average vibrational energy of the s oscillators; another suggestion [10] has been $s_{\text{eff}} \approx C_v/\ell$, where C_v is the constant-volume specific heat of the s oscillators. In either case the s_{eff} so defined are a function of temperature. A closer examination [11] shows that neither definition of s_{eff} yields the correct pressure dependence of k_{uni}. It seems [12], however, that better results are obtained on the whole if the average $\langle v \rangle$ as defined in Eq. (1.18) is used in Eq. (1.30) with $x = h\langle v \rangle /(\ell T), m = E_0/(h\langle v \rangle)$, although a more elaborate procedure for specifying s_{eff} has been suggested [13].

This still leaves E_0 and the constant A undefined, beyond the previous *a priori* estimate $A \approx 10^{13} \text{ s}^{-1}$. Both are obtainable from experiment if k_∞ is known in its Arrhenius form (Eq. (1.24)). On a more fundamental level A can be interpreted as the rate of decay of the critical oscillator [14].

As a final comment it can be pointed out that Kassel's $k(E)$ or k $(n \geq m)$ is obtained entirely by probabilistic argument devoid of all dynamical or quantum considerations, and the only quantum feature is the calculation of the oscillator degeneracy W_n (Eq. (1.1)).

References

[1] O. K. Rice and H. C. Ramsperger, *J. Am. Chem. Soc.* **49**, 1617 (1927).
[2] L. S. Kassel, *J. Phys. Chem.* **32**, 225 (1928); *J. Phys. Chem.* **32**, 1065 (1928); *Proc. Nat. Acad. Sci. USA* **14**, 23 (1928); *Kinetics of Homogeneous Gas Reactions* (Chemical Catalog Co., New York, 1932).
[3] M. J. Shultz and E. Yablonovitch, *J. Chem. Phys.* **68**, 3007 (1978). For an elementary account see N. Bloembergen and E. Yablonovitch, *Phys. Today*, May 1978, p. 23.
[4] J. G. Black, P. Kolodner, M. J. Shultz, E. Yablonovitch and N. Bloembergen, *Phys. Rev.* **A19**, 704 (1979).
[5] E. Thiele, J. Stone and M. F. Goodman, *Chem. Phys. Lett.* **76**, 579 (1980).
[6] J. R. Barker, *J. Chem. Phys.* **72**, 3686 (1980).
[7] F. A. Lindemann, *Trans. Faraday Soc.* **17**, 598 (1922).
[8] R. J. Hanrahan, *Comp. Chem.* **6**, 21 (1982).
[9] J. Troe and H. G. Wagner, *Ber. Bunsenges.* **71**, 937 (1967).
[10] D. M. Golden, R. K. Solly and S. W. Benson, *J. Phys. Chem.* **75**, 1333 (1971).
[11] G. B. Skinner and B. S. Rabinovitch, *J. Phys. Chem.* **76**, 2418 (1972).
[12] R. E. Weston, *Int. J. Chem. Kinet.* **18**, 1259 (1986).
[13] H. W. Schranz, S. Nordholm and N. D. Hamer, *Int. J. Chem. Kinet.* **14**, 543 (1982).
[14] M. Solc, *Z. physik. Chem. (Leipzig)* **234**, 185 (1967); *Z. physik. Chem. (Leipzig)* **236**, 213 (1967).

Problems

Problem 1.1. Equation (1.8) refers the rate to the concentration of molecules present at time t. Show that, if, instead, the rate is referred to the concentration of molecules present initially at $t = 0$, then

$$-\frac{1}{[M(t = 0)]}\frac{d[M(t)]}{dt} = k\exp(-kt) = P'(t)$$

Problem 1.2. Show that, if the rate constant k is time-dependent (written as $k(t)$),

$$P'(t) = k(t)\exp\left(-\int_0^t k(t)\,dt\right)$$

of which $P'(t)$ of the preceding problem is a special case.

Problem 1.3. The simple form of Stirling's approximation to the logarithm of the factorial is $\ln n! \approx n \ln n - n$. Using this approximation, assuming that $n - m \gg s - 1$, show that Eq. (1.16) can also be derived from Eq. (1.7).

Problem 1.4. Using Stirling's approximation show that, for $n \gg s$, we have approximately $(n + s - 1)!/n! \approx n^{s-1}$, so that Eq. (1.1) becomes, for an s-fold-degenerate oscillator,

$$W_n \approx \frac{n^{s-1}}{(s-1)!} = \frac{E^{s-1}}{(h\nu)^{s-1}(s-1)!}$$

$W_n/(h\nu)$ then yields what is known as the *classical* approximation to the density of states.

Problem 1.5. Expand the summation in Eq. (1.31) and, keeping only the first two terms, show that, at sufficiently high temperature, assuming that $m \gg s$,

$$k_0 \approx \omega\frac{b^{s-1}}{(s-1)!}\exp(-b)$$

where $b = mh\nu/(\Bbbk T) = E_0/(\Bbbk T)$.

2

An outline of the theory of Slater

This theory [1], of great mathematical elegance, was much in the vogue in the 1950s and 1960s. It is based on a molecular model that is mathematically tractable in a good deal of detail but turned out to be unrealistic for reacting molecules, as was eventually realized, and, although it was later supplanted by the more successful theory of Rice–Ramsperger–Kassel–Marcus (RRKM), Slater's theory influenced the thinking about unimolecular processes for many years after its demise. It is for this reason that it is worthwhile to present here a short review, which, because of brevity, will not have the mathematical rigor of the original presentation.

2.1 Isolated molecule

2.1.1 Slater's model

The model for a polyatomic molecule considered by Slater is an assembly of n point-atoms vibrating about their equilibrium positions. This is tantamount to a collection of n harmonic oscillators, such that the characteristic motions of the atoms in the molecule about their equilibrium configuration (so-called normal modes) are described by n coordinates $q_1, \ldots, q_n$. Consider a collection of such molecules at some finite pressure, so that collisions take place. Suppose that a given molecule dissociates when a bond breaks, and suppose that this bond corresponds to the oscillator described by internal coordinate q_1; this coordinate will therefore be designated the reaction coordinate.

The time dependence of q_1 is given in principle by the superposition of all n normal modes:

$$q_1(t) = \sum_{i=1}^{n} \alpha_{1i} \sqrt{\epsilon_i} \cos[2\pi(\nu_i t + \psi_i)] \tag{2.1}$$

where t is the time elapsed from the last collision, α_{1i} is the amplitude factor, ν_i

16

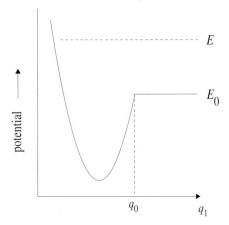

Fig. 2.1. The potential energy of a harmonic oscillator subject to cut-off. The oscillator dissociates when coordinate q_1 reaches the value $q_1 = q_0$, which will happen at any energy $E \geq E_0$.

the frequency of the ith mode, ϵ_i its energy and ψ_i its phase, the last two subject to change by collision. It is assumed that a molecule is capable of dissociation when q_1 attains some high value, say q_0, which means that the model for the breaking bond is akin to a harmonic oscillator with a cut-off (Fig. 2.1).

The condition for dissociation is thus

$$\sum_i \alpha_{1i} \sqrt{\epsilon_i} \geq q_0 \tag{2.2}$$

A molecule that satisfies this condition is said to be *energized* (Slater used the term "interesting"). However, an energized molecule will actually dissociate only if $q_1(t)$ reaches q_0 *before* the next collision, which will happen only if the n modes come sufficiently into phase at the right time. The reason is that a collision is likely to cause the energized molecule to lose some or all of its energy.

The minimum condition for oscillator 1 dissociating is $\Sigma_i \alpha_{1i} \sqrt{\epsilon_i} = q_0$. If the solution of this equation is ϵ_{i0}, the critical or minimum total energy required for the molecule to dissociate is $E_0 = \Sigma_i \epsilon_{i0} = q_0^2 / \Sigma_i \alpha_{1i}^2$ (Problem 2.1). Consequently, in Slater's theory the critical energy E_0, as well as the pre-exponential factor (Problem 2.3), are calculable from first principles, unlike the factor A in Kassel's theory (cf. Section 1.1.4). Note that a molecule with total energy $E > E_0$ is *not* considered capable of dissociation in Slater's theory, unless the much stricter condition given by Eq. (2.2) is satisfied.

These considerations show that Slater's theory, unlike Kassel's, is a non-statistical theory based on a detailed classical picture of the molecular dynamics of bond-breaking. Note in particular that the reaction coordinate is not assumed to be separable from the other degrees of freedom (cf. Chapter 3).

2.1.2 The specific dissociation probability

The key concept in Slater's theory is L, the specific dissociation probability (SDP) of an isolated molecule, defined as the average number of "up-zeros" of $F(t) = q_1(t) - q_0$. If $G(\tau; \Psi_1, \ldots, \Psi_n)$ is the number of zeros of $F(t)$ in $0 \leq t \leq \tau$, the long-time average SDP is

$$L = \frac{1}{2} \lim_{\tau \to \infty} \frac{G(\tau)}{\tau} \tag{2.3}$$

It can be shown [1] that, if the frequencies ν are linearly independent (i.e. incommensurable), the long-time average is the same as the phase-averaged SDP.

In other words, L is the average frequency with which $q_1(t)$ of an undisturbed molecule *rises* to q_0 (this excludes cases in which q *drops* to q_0, since this would represent a bond that "heals"). Clearly L depends on the detailed energies ϵ_i, and hence the *average* rate constant for molecules of total energy $E = \Sigma_i \epsilon_i$ is

$$k(E) = \frac{\int \ldots \int L \, d\epsilon_1 \ldots d\epsilon_n}{\int \ldots \int d\epsilon_1 \ldots d\epsilon_n} \tag{2.4}$$

where integration extends over the region $E \leq \Sigma_i \epsilon_i \leq E + dE$.

2.2 Thermal system

In a canonical (thermal) ensemble at temperature T the equilibrium fraction of energized molecules with total energy within dE at E, distributed such that the energy of the ith mode is within $d\epsilon_i$ at ϵ_i, with $E = \Sigma_i \epsilon_i$, is, from classical statistical mechanics (Problem 2.2),

$$f(E) \, d\epsilon_1 \ldots d\epsilon_n = \exp[-E/(\mathit{k}T)] \prod_{i=1}^{n} \frac{d\epsilon_i}{\mathit{k}T} \tag{2.5}$$

where k is Boltzmann's constant. If ω is the collision frequency and N the total concentration, then the number of molecules raised into the range $E, E + dE$ and distributed among the modes as above in Eq. (2.5) is $\omega N \exp[-E/(\mathit{k}T)]$ $\times \prod_i [d\epsilon_i/(\mathit{k}T)]$.

2.2.1 Arbitrary pressure

In the presence of reaction, a steady-state distribution will become established, which will be smaller than $f(E)$, say $g(E)$. The number of molecules lost by dissociation per unit time, with specified energies among the modes, is $N g(E) L \, d\epsilon_1 \ldots d\epsilon_n$. Assuming that every collision deactivates (the "strong-collision" assumption), the number of molecules de-energized by collisions is $\omega N g(E) \, d\epsilon_1 \ldots d\epsilon_n$. By virtue of the principle of detailed balancing, the gain and loss of molecules will balance for every detailed energy:

$$\omega N \exp[-E/(kT)] \prod_{i=1}^{n} \frac{d\epsilon_i}{kT} = N(L + \omega) \, g(E) \, d\epsilon_1 \ldots d\epsilon_n \qquad (2.6)$$

from which the steady-state distribution is

$$g(E) \, d\epsilon_1 \ldots d\epsilon_n = \frac{\omega}{L + \omega} \exp[-E/(kT)] \prod_{i=1}^{n} \frac{d\epsilon_i}{kT} \qquad (2.7)$$

The apparent first-order thermal rate constant k_{uni} is then

$$k_{\text{uni}} = \frac{\text{Rate}}{N} = \int \ldots \int g(E) \, L \, d\epsilon_1 \ldots d\epsilon_n$$

$$= \omega \int \ldots \int \frac{L}{\omega + L} \exp[-E/(kT)] \prod_{i=1}^{n} \frac{d\epsilon_i}{kT} \qquad (2.8)$$

where integration extends over the domain satisfying Eq. (2.2).

In the limit of high pressure, $\omega \gg L$ and Eq. (2.8) becomes

$$k_{\text{uni}} \to k_\infty = \int \ldots \int L \exp[-E/(kT)] \prod_{i=1}^{n} \frac{d\epsilon_i}{kT} \qquad (2.9)$$

The lower limit of the integral is zero since L vanishes when the condition of Eq. (2.2) is not satisfied. It can be shown (Problem 2.3) that

$$k_\infty = \left(\frac{\sum_i \alpha_{1i}^2 \nu_i^2}{\alpha^2} \right) \exp[-E_0/(kT)] \qquad (2.10)$$

where $\alpha^2 = \sum_i \alpha_{1i}^2$ and $E_0 = q_0^2/\alpha^2$. The pre-exponential factor is thus given by an average of all frequencies.

2.2.2 The low-pressure limit

In the limit of low pressure such that $\omega \ll L$, Eq. (2.8) reduces to

$$k_{\text{uni}} \to k_0 = \omega \int \cdots \int \exp[-E/(\Bbbk T)] \prod_{i=1}^{n} \frac{\mathrm{d}\epsilon_i}{\Bbbk T} \tag{2.11}$$

The domain of integration in this integral *is* subject to Eq. (2.2), i.e. $\Sigma_i \alpha_{1i} \sqrt{\epsilon_i} \geq (E_0 \Sigma_i \alpha_{1i}^2)^{1/2}$. The result, after fairly complicated manipulation [5], is

$$k_0 = \omega (4\pi b)^{(n-1)/2} \exp(-b) \left[1 - \left(\frac{n-1}{2} \right)(2b)^{-1} + \cdots \right] \prod_{i=1}^{n} \mu_i \tag{2.12}$$

where $b = E_0/(\Bbbk T)$ and $\mu_i = |\alpha_{1i}|/(\Sigma_i \alpha_{1i}^2)^{1/2}$. The condition of Eq. (2.2) is much too stringent (as Slater himself realized [6]) because, compared with experiment, k_0 calculated from Eq. (2.12) is too small. This was verified for a number of thermal reactions: the decompositions cyclobutane $\to 2CH_2 = CH_2$ and $N_2O_5 \to NO_2 + NO_3$, as well as the isomerization cyclobutene $\to 1, 3$-butadiene [7], and also for the decompositions $H_2O_2 \to OH + OH$, $N_2O \to N_2 + O$, and $C_2H_6 \to CH_3 + CH_3$ [8].

A more reasonable condition would be to integrate Eq. (2.11) over energies such that $\Sigma_i \epsilon_i \geq E$, with the result (Problem 2.4)

$$k_0 = \omega \frac{b^{n-1}}{(n-1)!} \exp(-b) \left(1 + \frac{n-1}{b} + \cdots \right) \tag{2.13}$$

This is in fact the RRK prescription (see Problem 1.5), which represents a more realistic approach. By comparison, the Slater result in Eq. (2.12) yields a smaller k_0 principally because some of the factors μ_i may be very small on account of symmetry restrictions. On a more fundamental level, the essential difference is that, in the Slater formulation, each ϵ_i is a constant of motion, whereas in the RRK formulation it is only the sum $\Sigma_i \epsilon_i = E$ (there is more on this in Section 3.1.6).

The reason for the eventual demise of Slater's theory is the unrealistic stringency of his condition for dissociation (Eq. (2.2)), a consequence of treating vibrations as strictly harmonic. Experiment has shown repeatedly that, except in unusual circumstances, the internal energy of a molecule very rapidly becomes scrambled due to anharmonic interactions so that the notion of normal modes disappears. Slater's theory cannot be easily modified to take into account anharmonic and quantum effects and at the same time maintain a detailed picture of molecular dynamics and with it the mathematically appealing analysis. Solc [9–12] has investigated consequences of a modified version of Slater's model in which molecular oscillators are separated into sets, with exchange of energy allowed within a set, but not among sets; another modification is the possibility of non-equilibrium distribution of energy among modes with different effective temperatures [13].

2.2.3 Non-random dissociation

The main legacy of Slater's theory is his approach to the analysis of the process of dissociation which influenced thinking about circumstances when the redistribution of internal energy in the reacting molecule is not random.

In particular, Slater showed that the treatment that leads to Eq. (2.8) is the result of a random incidence of dissociation, as is implied by the presence of the factor $L/(\omega + L)$ which is the analogue of Eq. (1.19) in Chapter 1.

If molecular states are specified in great detail, the occurrence of the dissociation configuration in Slater's theory ceases to be random, and the time from $t = 0$ to the first upward zero of $F(t)$ becomes determinate [14], i.e. a function of the initial phases Ψ_i. Let the time to the first upward zero of $F(t)$ be t_1. A molecule will dissociate only if it suffers no collision in the time interval $(0, t_1)$, given the assumption that every collision deactivates. The probability of *no* collision in $(0, t_1)$ is $\exp(-\omega t_1)$. Since t_1 depends on the phases,

$$\phi(\omega) = \omega \int_0^1 \cdots \int_0^1 \exp(-\omega t_1)\, d\Psi_1, \ldots, d\Psi_n \tag{2.14}$$

represents the phase average of $\exp(-\omega t_1)$, and the rate constant can be written

$$k_{\text{uni}} = \int \cdots \int \phi(\omega) \exp[-E/(kT)] \prod_{i=1}^n \frac{d\epsilon_i}{kT} \tag{2.15}$$

where integration with respect to the ϵ_i's extends over the domain satisfying Eq. (2.2).

If we let t_r be the rth gap between the zeros of $F(t)$, the asymptotic average over the gaps is

$$\lim_{N \to \infty} \frac{1}{N} \sum_r^N \exp(-\omega t_r) = \langle \exp(-\omega t) \rangle \tag{2.16}$$

It can then be shown that the long-time average of $\phi(\omega)$ is ([1], p. 197)

$$\phi(\omega) = L[1 - \langle \exp(-\omega t) \rangle] \tag{2.17}$$

and Eq. (2.15) for the "exact" rate constant becomes

$$k_{uni} = \int \ldots \int L[1 - \langle \exp(-\omega t) \rangle] \exp[-E/(\hbar T)] \prod_{i=1}^{n} \frac{d\epsilon_i}{\hbar T} \qquad (2.18)$$

where integration is subject to Eq. (2.2).

2.2.4 The distribution of gaps

For the actual evaluation of this expression the distribution of gaps must be known. However, note first that, if the pressure is very high, $\langle \exp(-\omega t) \rangle \to 0$, then, from Eq. (2.17), we have $\phi(\omega) \to L$ and Eq. (2.15) reduces to Eq. (2.9). At very low pressure it follows from Eq. (2.14) that $\phi(\omega) \to \omega$, so that Eq. (2.15) reduces to Eq. (2.11). Therefore the distribution of gaps affects only the "fall-off," i.e. $\phi(\omega)$ appears only in the expression for k_{uni} at intermediate pressures, and the distribution of gaps is without effect on the high- and low-pressure limits.

Consider now the actual distribution of gaps $h(t)$, assumed for convenience to be continuous. Thus

$$\langle \exp(-\omega t) \rangle = \int_0^{\infty} \exp(-\omega t)\, h(t)\, dt \qquad (2.19)$$

Slater [1] suggested for $h(t)$ a gamma-type form with parameter u

$$h(t) = \frac{(uL)^u t^{u-1} \exp(-uLt)}{\Gamma(u)}, \qquad 1 \le u < \infty \qquad (2.20)$$

normalized in $0 \le t \le \infty$, from which Eq. (2.19) yields

$$\langle \exp(-\omega t) \rangle = \left(1 + \frac{\omega}{uL}\right)^{-u} \qquad (2.21)$$

It is instructive to consider the effect of particular values of u on the nature of $h(t)$. For $u = 1$, we have

$$h(t)_{u=1} = L \exp(-Lt) \qquad (2.22)$$

Since L is Slater's analog of what was called rather non-committally the rate constant k in Chapter 1, $h(t)_{u=1}$ is the analog of $P'(t)$ in Eq. (1.12) (Problem 2.5). For $u = 1$ it follows from Eq. (2.21) that $\langle \exp(-\omega t) \rangle = L/(\omega + L)$, which is the probability of random dissociation in the presence of collisions. Using this result in k_{uni} of Eq. (2.18), we can write

$$L[1 - \langle \exp(-\omega t) \rangle] = \omega \int_0^\infty \exp(-\omega t)\, L \exp(-Lt)\, dt$$

$$= -\omega \int_0^\infty \exp(-\omega t)\, \frac{dP(t)}{dt}\, dt \qquad (2.23)$$

Equation (2.23) is the analog of Eq. (1.20), using $P'(t) = L \exp(-Lt)$ as Slater's version of the survival probability.

At the other extreme, $u \to \infty$ transforms $h(t)$ into a δ-function centered at $Lt = 1$ (Fig. 2.2); then $\langle \exp(-\omega t) \rangle = \exp(-\omega/L)$, which means that gaps are "regular," i.e. all of length $t = 1/L$. If ω is small, $\exp(-\omega/L) \approx 1 - \omega/L$, and Eq. (2.18) reduces to Eq. (2.11) in the low-pressure limit, as it should. A gamma-type distribution has been found for the time interval between two successive passages of a trajectory through the critical surface in the decomposition of N_2O [15].

The publication of Slater's concept of gap distribution was soon followed by a number of numerical checks. Model calculations have shown [16] that, in classical theories of unimolecular reactions, the gap distribution is adequately represented by the gamma distribution, Eq. (2.20), which approaches the random limit $u \to 1$ as energy approaches the critical energy E_0. Trajectory calculations [17] on triatomics revealed harmonic models to be unrealistic, since most molecules failed to dissociate; at the same time, the random-gap assumption could be confirmed for anharmonic models only. Thus, while these calculations tended to cast doubt on Slater's theory as a representation of the behavior of real molecules, his approach was nevertheless fruitful in providing the impetus for a critical examination of the fundamental assumptions in rate theory.

In particular, it focused interest on the general problem of the distribution of lifetimes $P'(t)$. It may be noted that $h(t)$ of Eq. (2.20) represents the first term in the expansion of the incomplete gamma function $\Gamma(u, uLt)$ [18, p. 263], multiplied by $uL/\Gamma(u)$. Using the full expansion, we can define a more general $P'(t)$ (cf. [17]):

$$P'(t) = \frac{uk}{(u-1)!} \Gamma(u, ukt) = \frac{uk}{(u-1)!} \int_{ukt}^\infty x^{u-1} \exp(-x)\, dx \qquad (2.24)$$

where k represents the specific-energy unimolecular rate constant for molecules of energy E, i.e. the analog of Slater's L, and u, as before, is the parameter $1 \leq u < \infty$. The two distributions $P'(t)$ and $h(t)$ in dimensionless form are compared in Fig. 2.2. They are identical for $u = 1$, the random case, but quite different for all other values

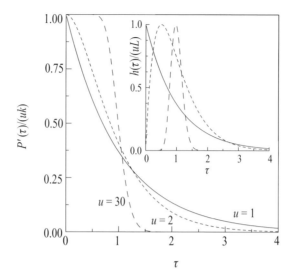

Fig. 2.2. A plot of the dimensionless lifetime distribution $P'(\tau)/(uk)$ (Eq. (2.24)) versus dimensionless time $\tau = kt$, for $u = 1$ (solid line), $u = 2$ (short-dashed line) and $u = 30$ (long-dashed line). Inset: Slater's dimensionless gap distribution $h(\tau)/(uL)$ (Eq. (2.20)) versus the dimensionless gap length $\tau = Lt$ for $u = 1, 2$, and 30 (with the same identification of curves). Here the maximum of each distribution is arbitrarily set to unity. For $u = 1$ (random case, solid lines) the two distributions $P'(\tau)/(uk)$ and $h(\tau)/(uL)$ are identical. For $u \to \infty$, $P'(\tau)/(uk)$ approaches the rectangular distribution $P'(\tau)/(uk) = 1$ for $0 \le \tau \le 1$, and zero otherwise, whereas $h(\tau)/(uL)$ approaches a δ-function at $\tau = 1$.

of u. The random case is characterized by the $P'(t)$ that has the largest negative slope at $t = 0$.

References

[1] N. B. Slater, *Theory of Unimolecular Reactions* (Cornell University Press, Ithaca, 1959). For an elementary exposition see K. J. Laidler, *Theories of Chemical Reaction Rates* (McGraw-Hill, New York, 1969), pp. 129ff.
[2] M. Kac, *Amer. J. Math.* **65**, 609 (1943).
[3] N. B. Slater, *Proc. Leeds Phil. Soc.* **6**, 268 (1955).
[4] N. B. Slater, *Proc. Roy. Soc. Edinburgh* **64A**, 161 (1955).
[5] N. B. Slater, *Phil. Trans. Roy. Soc.* **A246**, 57 (1953).
[6] N. B. Slater, in *The Transition State*, Special Publication No. 16 (The Chemical Society, London, 1962).
[7] E. Thiele and D. J. Wilson, *Can. J. Chem.* **37**, 1035 (1959).
[8] E. K. Gill and K. J. Laidler, *Proc. Roy. Soc.* **A250**, 121 (1959). Equation (3) is missing the factor $\exp[-\epsilon^*/(kT)]$.
[9] M. Solc, *Chem. Phys. Lett.* **1**, 160 (1967).
[10] M. Solc, *Mol. Phys.* **11**, 579 (1966).
[11] M. Solc, *Mol. Phys.* **12**, 101 (1967); cf. N. B. Slater, *Mol. Phys.* **12**, 107 (1967).
[12] M. Solc, *Coll. Czech. Chem. Commun.* **50**, 2635 (1985).

[13] A. Metsala, *Chem. Phys.* **189**, 637 (1994).

[14] N. B. Slater, *J. Chem. Phys.* **24**, 1256 (1956).

[15] M. Ya. Gol'denberg, N. M. Kuznetsov and B. V. Pavlov, *Doklady A. N. SSSR*, **252**, 38 (1980).

[16] E. Thiele, *J. Chem. Phys.* **36**, 1472 (1962); *J. Chem. Phys.* **38**, 1959 (1963).

[17] D. Bunker, *J. Chem. Phys.* **40**, 1946 (1964).

[18] M. Abramowitz and I. A. Stegun, *Handbook of Mathematical Functions* (NBS Applied Mathematics Series 55, Washington, 1972).

Problems

Problem 2.1. The determination of the critical energy E_0 is a problem of a minimum with constraint: finding a set of ϵ_i's such that $\Sigma_i \epsilon_i = E$ is a minimum subject to $\Sigma_i \alpha_{1i} \sqrt{\epsilon_i} = q_0$. Use the method of Lagrange multipliers to show that $\epsilon_{i0} = \alpha_{1i}^2 q_0^2 / (\Sigma_i \alpha_{1i}^2)^2$.

Problem 2.2. Show that, in a canonical ensemble of a collection of independent harmonic oscillators, the fraction of those having energy between ϵ and $\epsilon + d\epsilon$ at temperature T is $\exp[-\epsilon/(kT)]\,[d\epsilon/(kT)]$. Generalize this result to obtain Eq. (2.5).

Problem 2.3. The Kac formula [2] for L is

$$L = \frac{1}{4\pi^2} \int\limits_{-\infty}^{+\infty}\!\!\int y^{-2} \cos(xq_0) \left(\prod_{i=1}^{n} J_0(a_i x) - \prod_{i=1}^{n} J_0\left(a_i \sqrt{x^2 + 4\pi^2 y^2 v_i^2} \right) \right) dx\,dy \quad (2.25)$$

where $a_i = \alpha_i \sqrt{\epsilon_i}$ and $J_0(z)$ is the Bessel function of order zero. Show that this L substituted into Eq. (2.9) leads to Eq. (2.10).

Problem 2.4. Prove Eq. (2.13) and compare it with the result of Problem 1.5. (Note that n in Eq. (2.13) represents the number of oscillators, whereas in Chapter 1 the same symbol represents the number of quanta.)

Problem 2.5. Show that the average lifetime $\langle t \rangle$ obtained using $h(t)$ of Eq. (2.20) is $\langle t \rangle = 1/L$, i.e. independent of the value of the parameter u.

3

The RRKM rate constant $k(E)$

This chapter develops the general argument for obtaining the statistical, or RRKM, specific-energy unimolecular rate constant $k(E)$. This rate constant, in its various forms and disguises, is the principal subject of this textbook. It is the fundamental information that forms the basis for further developments in later chapters.

3.1 Classical rate constant

Over the years various derivations of $k(E)$ have appeared in the literature [1–7]. The present derivation (cf. [4]) is in somewhat greater detail than the original derivation by Rice and Ramsperger [8], which occupies a mere two pages in the original publication.

3.1.1 The model

Consider a general type of unimolecular reaction given schematically as M $\rightarrow$ products, where the reactant species M is composed of n *internal* degrees of freedom that contain internal energy E. If non-linear molecule M is constituted of N atoms, $n = 3N - 6$, there being six *external* degrees of freedom, three for translation of the center of mass (which is of no interest in the present context), and three for external rotations, meaning rotations of the molecule as a whole. For linear M we have $n = 3N - 5$, with only two degrees of freedom of external rotation. For the purpose of the present demonstration we shall consider only the internal degrees of freedom, the consideration of external rotations being deferred to Section 3.4.

3.1.2 The potential-energy surface

If the molecular potential energy were to be plotted against the n internal degrees of freedom of the molecule in question, the result would be an $(n + 1)$-dimensional

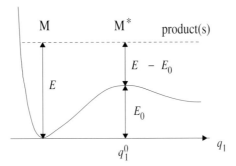

Fig. 3.1. The classical representation of a cut through the potential-energy surface for the process M $\rightarrow$ product(s) along the direction of the reaction coordinate q_1. The potential is assumed to be vibrational. M* is the transition state, located at $q_1 = q_1^0$. E_0 is the height of the potential-energy barrier separating reactant from product(s). If E is the internal energy of M, the internal energy of the transition state is $E - E_0$. In a semi-classical representation the potential, and hence also the energies, would be understood to be in excess of zero-point energies.

(hyper-)surface, called the potential-energy surface [9], which in principle gives information about all possible reaction channels of a given reactant. Obviously for $n > 2$ such a surface cannot be visualized, but slices of it in a given direction can. In this way the general topography of the potential-energy surface can be explored piecewise, in particular as regards the path of least energy that leads from reactant to specified products, which is commonly referred to as the reaction coordinate. In general, the potential-energy profile along the reaction coordinate has the form shown in Fig. 3.1, the most important feature of which is the potential-energy maximum (which may be very small in particular cases, *vide infra*) that separates reactant from products. This maximum is a two-dimensional representation of what in higher dimensions is a saddle point, that is a maximum in one direction (the reaction coordinate), and a minimum in all other directions.

3.1.3 Phase space

The potential-energy surface gives a static picture of a given reaction. For the purpose of dealing with the dynamics of the reaction it is convenient to construct a $2n$-dimensional (hyper-)space ϕ_{2n} (the "phase space") for the n internal degrees of freedom of the molecule M, one dimension for each generalized coordinate $\mathbf{q} = (q_1, \ldots, q_n)$ and one for each conjugate momentum $\mathbf{p} = (p_1, \ldots, p_n)$. The interest of this representation is that the instantaneous state (the "phase") of molecule M is represented by a single point in ϕ_{2n} (the "phase point"), which during the temporal evolution of M into products can be thought of as describing a trajectory in phase space. Note that phase space is not ordinary space, and a generalized coordinate is not necessarily a Cartesian coordinate (e.g. it would be the angle of twist in *cis–trans* isomerization). Objects of lower dimensionality than ϕ_{2n} are likewise referred to as (hyper-)"surfaces."

We assume that the system evolves in a time-independent potential (assumed implicitly to be vibrational), so that the Hamiltonian $H(\mathbf{q}, \mathbf{p})$ of the system is simply the total energy E, which is a constant of motion:

$$H(\mathbf{q}, \mathbf{p}) = E = \text{constant} \tag{3.1}$$

At constant E phase points in ϕ_{2n} will therefore evolve on a $(2n - 1)$-dimensional surface S_{2n-1} of constant energy.

3.1.4 The transition state

The process of decomposition of a molecule M containing sufficient energy E will be considered to proceed through an intermediate structure M* called in the literature interchangeably the "transition state," "critical configuration," or "activated complex":

$$\mathrm{M} \xrightarrow{k(E)} \mathrm{M}^* \to \text{products} \tag{3.2}$$

This transition state or critical configuration (nomenclature to be used interchangeably henceforth) is located in principle on top of the potential-energy barrier of height E_0 separating the reactant from the products (Fig. 3.1). Thus "sufficient energy" for decomposition means $E \geq E_0$.

In practical terms, the transition state is thus conceived as an entity in which the reactant is no longer fully bound, and the product(s) not yet fully formed.

The critical configuration may be thought of as a $(2n - 1)$-dimensional dividing surface C_{2n-1} that separates reactant and product regions in phase space, defined for convenience by

$$C_{2n-1}(\mathbf{q}, \mathbf{p}) = 0 \tag{3.3}$$

so that the dissociation can be described as a flux of phase points crossing the dividing surface in the direction of decomposition (the "forward" direction). The assumption is that, once a trajectory has crossed the dividing surface in the direction of products, it will never return to the reactant region of phase space.

Since the threshold to reaction E_0 corresponds to the height of a barrier (from a classical point of view) that the system must overcome in order to react, the surface C_{2n-1} can be crossed only by phase points evolving on surfaces S_{2n-1} of energy $E \geq E_0$. The intersection between $C_{2n-1} = 0$ and one of the S_{2n-1} surfaces defines a $(2n - 2)$-dimensional surface X_{2n-2} specified by

$$H(\mathbf{q}, \mathbf{p}) - E_0 \geq 0 \qquad (3.4)$$

Therefore, in view of Eq. (3.1), the phase point that has reached X_{2n-2} can be described by the Hamiltonian

$$H'(\mathbf{q}, \mathbf{p}) = E - E_0 \qquad (3.5)$$

3.1.5 The rate constant

The unimolecular process is thus the passage of the phase point over the potential-energy barrier, which is assumed here to be given by the change in vibrational potential as reactant evolves into products; for rotational effects see Section 3.4 and Chapter 8.

It is useful now to define $\Phi(\mathbf{q}, \mathbf{p})$ (dimension s^{-1}), the number of times a phase point passes through X_{2n-2} per unit time in the forward direction. The rate constant $k(E)$ for phase points evolving on S_{2n-1} is then the *ensemble average* of $\Phi(\mathbf{q}, \mathbf{p})$:

$$k(E) = \langle \Phi \rangle = \frac{\int \ldots \int\limits_{X_{2n-2}} d\mathbf{q} \, d\mathbf{p} \, \rho(\mathbf{q}, \mathbf{p}) \, \Phi(\mathbf{q}, \mathbf{p})}{\int \ldots \int\limits_{S_{2n-1}} d\mathbf{q} \, d\mathbf{p} \, \rho(\mathbf{q}, \mathbf{p})} \qquad (3.6)$$

where $\rho(\mathbf{q}, \mathbf{p})$ is the distribution of phase points in S_{2n-1}. The domain of integration in the numerator is over the portion of phase space X_{2n-2} containing those phase points that cross the critical surface in the forward direction. The bottom integral normalizes $\rho(\mathbf{q}, \mathbf{p})$ with respect to phase points contained in S_{2n-1}. Since $\Phi(\mathbf{q}, \mathbf{p})$ has the dimension s^{-1}, it is sometimes referred to as "frequency," although no movement back and forth is involved.

3.1.6 The microcanonical ensemble

The initial excitation of the molecule M is assumed to be such that energy $E \geq E_0$ is randomized among internal degrees of freedom by an unspecified process of intramolecular redistribution supposed to operate on a time scale much shorter than the lifetime of the activated molecule M. In terms of classical statistical mechanics, such a distribution corresponds to a microcanonical ensemble of phase points in the bound-states part of phase space representing the states of the activated molecule. The distribution of phase points in a microcanonical ensemble ρ is by definition uniform, i.e.

$$\rho(\mathbf{q}, \mathbf{p}) = \text{constant} \qquad (3.7)$$

meaning that equal "volumes" of phase space contain equal numbers of phase points.

One can nowadays use classical mechanics to simulate the path (trajectory) of a collection of molecules in phase space [9, 10]. In terms of such phase-space trajectories the assumption of a microcanonical ensemble means that a RRKM trajectory is free to roam all over the constant-energy surface, in contrast with a "Slater trajectory" of Chapter 2, which is confined to a subspace of phase space where normal-mode energies remain at their initial values.

Thus $k(E)$ of Eq. (3.6), which may now be called more appropriately the *microcanonical* rate constant, reduces to

$$k(E) = \frac{\int \ldots \int\limits_{X_{2n-2}} d\mathbf{q}\, d\mathbf{p}\, \Phi(\mathbf{q}, \mathbf{p})}{\int \cdots \int\limits_{S_{2n-1}} d\mathbf{q}\, d\mathbf{p}} \tag{3.8}$$

In classical terms integration in the numerator is over all $(\mathbf{q}, \mathbf{p})$ such that Eq. (3.5) is satisfied, and integration in the denominator is over all $(\mathbf{q}, \mathbf{p})$ such that

$$H(\mathbf{q}, \mathbf{p}) = E \tag{3.9}$$

3.1.7 The frequency of passage Φ

It is always possible to take a linear combination of the set of generalized coordinates $\mathbf{q}$ such that the coordinate q_1 and its conjugate momentum p_1 refer to barrier crossing in the forward direction; thus q_1 becomes the reaction coordinate. Suppose that the transition state is located at $q_1 = q_1^0$; we then assume that the degree of freedom involving q_1 and p_1 separates from the total Hamiltonian, at least within a short distance dq_1 in the vicinity of q_1^0.

Thus the Hamiltonian of the critical configuration (Eq. (3.5)) separates into two parts, a term that refers to the $n-1$ degrees of freedom not directly involved in the reaction process (H^*), and a term that refers to the reaction coordinate (H_1):

$$H'(\mathbf{q}, \mathbf{p}) = H^*(\mathbf{q}^*; \mathbf{p}^*) + H_1\left(q_1^0; p_1\right) \tag{3.10}$$

Here $H^*(\mathbf{q}^*; \mathbf{p}^*)$, with $\mathbf{q}^* = q_2, \ldots, q_n$ and $\mathbf{p}^* = p_2, \ldots, p_n$ represents what we have called the transition state (henceforth any property of the transition state will be denoted by an asterisk), and $H_1(q_1^0; p_1)$ represents one-dimensional motion in the reaction coordinate. If there is no curvature along the reaction path in the vicinity of the transition state, $H_1(q_1^0; p_1)$ reduces to just the translational energy ϵ_t in the reaction coordinate; thus

$$H_1\left(q_1^0; p_1\right) \Rightarrow H_1(p_1) = \epsilon_t = \frac{p_1^2}{2\mu^*} \tag{3.11}$$

where μ^* is the reduced mass of M^*. As a result, $\Phi(\mathbf{q}, \mathbf{p})$ is a function of p_1 only: $\Phi(\mathbf{q}, \mathbf{p}) \Rightarrow \Phi(p_1)$. Now the velocity of the phase point in the vicinity of q_1^0 is p_1/μ^*, so that the number of passages per unit time is

$$\Phi(p_1) = \frac{p_1}{\mu^* \, dq_1} \tag{3.12}$$

In this way the dynamics of the process is reduced to a one-dimensional passage of the phase point over the potential hill. Figure 3.1 thus represents a slice of the potential-energy surface in the direction of q_1.

3.1.8 The classical rate constant

The net result of these assumptions is that the total energy of the transition state $E - E_0$ is partitioned between its internal degrees of freedom and the translational energy ϵ_t in the reaction coordinate. In the absence of any information on how to apportion this energy, assume that in principle ϵ_t can take any value between 0 and $E - E_0$, the total available energy. If we define a detailed rate constant $k(E, \epsilon_t) \, d\epsilon_t$ for a particle with total energy E and translational energy in the reaction coordinate within a narrow range $d\epsilon_t$ about ϵ_t, the unimolecular rate constant $k(E)$ then follows by merely integrating over all energetically allowed values of ϵ_t:

$$k(E) = \int_0^{E-E_0} k(E, \epsilon_t) \, d\epsilon_t \tag{3.13}$$

Consequently the numerator of $k(E)$ as given by Eq. (3.8) becomes

$$\int \cdots \int_{H' = E - E_0} d\mathbf{q} \, d\mathbf{p} \, \Phi(\mathbf{q}, \mathbf{p}) = \int_0^{E-E_0} \left(\int \cdots \int_{H^* = E - E_0 - \epsilon_t} d\mathbf{q}^* \, d\mathbf{p}^* \int_{H_1 = \epsilon_t}^{H_1 = \epsilon_t + d\epsilon_t} \Phi(p_1) \, dq_1 \, dp_1 \right) \tag{3.14}$$

The integral with respect to $\Phi(p_1)$ over positive momenta (the "forward" direction) is merely $d\epsilon_t$, so that the full classical expression for the rate constant from Eq. (3.8) becomes

$$k(E) = \frac{\int_0^{E-E_0} d\epsilon_t \left(\int \cdots \int_{H^* = E - E_0 - \epsilon_t} d\mathbf{q}^* \, d\mathbf{p}^* \right)}{\int \cdots \int_{H = E} d\mathbf{q} \, d\mathbf{p}} = \frac{\int \cdots \int_{0 \leq H^* \leq E - E_0} d\mathbf{q}^* \, d\mathbf{p}^*}{\int \cdots \int_{H = E} d\mathbf{q} \, d\mathbf{p}} \tag{3.15}$$

Here the denominator represents an n-dimensional surface integral for the reactant, while the numerator represents an $(n-1)$-dimensional volume integral for the transition state (recall the definition above of $d\mathbf{q}^* \, d\mathbf{p}^*$), so that the transition state is "missing" one degree of freedom, the one that became the one-dimensional translation along the reaction coordinate.

Some authors prefer to write integrals of the type shown on the RHS of Eq. (3.15) in the more compact form $\int \ldots \int d\mathbf{q} \, d\mathbf{p} \, \theta(E - H)$ (numerator) and $\int \ldots \int d\mathbf{q} \, d\mathbf{p} \, \delta(E - H)$ (denominator), where $\theta(E - H)$ is the Heaviside step function $\theta(E - H) = 1$ for $0 \le H \le E$, and zero otherwise (cf. Appendix 2, Section A2.2), and δ is the Dirac delta function $\delta(E - H) = 1$ for $E = H$, and zero otherwise.

It should be noticed that this classical expression for the rate constant has the dimension of energy, whereas $k(E)$ should have the dimension s^{-1}; thus an important element is lacking.

3.2 Semi-classical rate constant

It is now necessary to introduce into the classical expression for $k(E)$ a modicum of quantum-mechanical concepts, in the expectation that such "corrected" $k(E)$ will have the proper dimension. In this respect the Kassel form of $k(E)$ (Chapter 1), although it uses a quantum-oscillator model for the reactant, does not address the quantum-mechanical aspects of the fragmentation process.

3.2.1 Quantum considerations

The products $d\mathbf{q}^* \, d\mathbf{p}^*$ and $d\mathbf{q} \, d\mathbf{p}$ in Eq. (3.15) represent elements of volume in phase space. One consequence of Heisenberg's uncertainty principle [11], according to which the uncertainties in position and momentum are related by $\Delta q \, \Delta p \approx h$ (h is Planck's constant), is that phase space has a cellular strucure, each cell of volume $dp \, dq/h$ containing one quantum state (note that $dp \, dq/h$ is dimensionless). Thus classical integration in phase space, as in Eq. (3.15), is replaced by the counting of quantum states. In particular, if the reactant molecule has n degrees of freedom, we have for the transition state of $n - 1$ degrees of freedom

$$\int \cdots \int_{0 \le H^* \le E - E_0} d\mathbf{q}^* \, d\mathbf{p}^* = h^{n-1} G^*(E - E_0) \tag{3.16}$$

where $G^*(E - E_0)$ is the total number of quantum states of the transition state in the interval 0, $E - E_0$.

A further consequence of the position–momentum uncertainty is the appearance of zero-point energy, so that all energies involved are to be understood in excess of zero-point energy. (In this respect Fig. 3.1 has to be interpreted as representing energies after subtraction of zero-point energies.)

Another consequence of Heisenberg's uncertainty principle is that energy and time are connected in a similarly restrictive relation $\Delta E \, \Delta t \approx h$, which means that, in a non-stationary ensemble, energy can be specified only to within an uncertainty dE. This is of course the case of $k(E)$, which refers to an unstable molecule in the process of decomposition. Thus, while the denominator in Eq. (3.15) is classically an integral over the range where the Hamiltonian H has the value E *exactly*, the quantum version of this integral (after division by h^n) is over the less precise range $E \leq H \leq E + \delta E$, i.e. over an energy "shell" of "thickness" δE. We can write formally

$$\int \cdots \int_{E \leq H \leq E + \delta E} \frac{d\mathbf{q}\, d\mathbf{p}}{h^n} = \delta G(E) \tag{3.17}$$

where $\delta G(E) = G(E + \delta E) - G(E)$ is the difference between the total number of quantum states of the reactant at $E + \delta E$ and that at E. Taking the limit $\delta E \to 0$, we have

$$\lim_{\delta E \to 0} \frac{\delta G(E)}{\delta E} = N(E) \tag{3.18}$$

where $N(E)$ is the density of states (dimension energy^{-1}) of the reactant molecule (n degrees of freedom) at energy E, i.e. the number of quantum states per unit energy (for further discussion see Chapter 4). Hence the denominator in Eq. (3.15) becomes

$$\int \cdots \int_{H = E} d\mathbf{q}\, d\mathbf{p} = h^n N(E) \tag{3.19}$$

By the same token we can write the numerator on the LHS of Eq. (3.15):

$$\int_0^{E - E_0} d\epsilon_t \left(\int \cdots \int_{H^* = E - E_0 - \epsilon_t} d\mathbf{q}^*\, d\mathbf{p}^* \right) = h^{n-1} \int_0^{E - E_0} d\epsilon_t\, N^*(E - E_0 - \epsilon_t) \tag{3.20}$$

which is in fact the same as Eq. (3.16) but shows more explicitly how the translational energy ϵ_t is apportioned. We shall have occasion to return to Eq. (3.20) in Section 3.5.3, and later in Chapter 6.

The "corrected" version of the rate constant of Eq. (3.15), using Eqs. (3.16) and (3.19), with the zero of energy at the ground state of the reactant, is thus

$$k(E) = \frac{G^*(E - E_0)}{h\, N(E)} \tag{3.21}$$

which now has the expected dimension s^{-1} owing to the presence of Planck's constant h. In this form $k(E)$ is known as the RRKM microcanonical or specific-energy rate constant. It has numerous not-so-obvious ramifications that are explored below. Problem 3.1 illustrates the dimensionality problem using the simple case of the classical harmonic oscillator.

3.2.2 Classical aspects

It should be clear from the derivation given here that, despite the presence of Planck's constant in Eq. (3.21), RRKM is nevertheless a *classical* theory, into which quantization is introduced only *a posteriori*, a fact that is sometimes lost sight of but should be emphasized [2, 12]. Therefore the final result as given by Eq. (3.21) should more accurately be termed semi-classical.

The fundamentally classical aspect is not unreasonable since after all a reaction involves principally the movement of heavy nuclei. The implicit assumption is that the energy levels of the decomposing molecule are broadened just sufficiently to overlap, so that there is a continuum of states, a typical characteristic of a classical system. Therefore the sum of states $G^*(E - E_0)$ and also the density $N(E)$ are to be interpreted as essentially continuous or smoothed quantities, with consequences for their actual numerical evaluation which are discussed in more detail in Chapter 4.

3.2.3 The average rate

Consider the hypothetical case of a reaction in which it has been determined that, at a given energy, there are a initial (reactant) quantum states that are coupled (i.e. give rise) to b final (product) quantum states. In an ideal fully resolved situation we would know which initial state decays to which final state(s) with which rate constant. However, if all we know is only the number of initial and final states, as shown schematically in the diagram below,

$$\langle k \rangle \Rightarrow \left.\begin{array}{c} \underline{} \\ \underline{} \\ \underline{} \\ \underline{} \\ \underline{} \end{array}\right\} \xrightarrow{k_{\text{elem}}} \left\{\begin{array}{c} \underline{} \\ \underline{} \\ \underline{} \\ \underline{} \\ \underline{} \\ \underline{} \\ \underline{} \\ \underline{} \\ \underline{} \\ \underline{} \end{array}\right. \qquad (3.22)$$

a initial b final
states   states

the best we can do is determine an average rate constant $\langle k \rangle$ in terms of an overall elementary flux k_{elem}. In such a situation we can only count up the states and say that *on average* each initial state decays with rate constant $(a/b)k_{elem}$:

$$\langle k \rangle = \frac{\text{total number of final states}}{\text{number of initial states}} \times k_{elem} \tag{3.23}$$

Now $k(E)$ of Eq. (3.21) has a similar structure since it can be decomposed into constituent elements as follows (TS = transition state):

$$k(E) = \frac{1}{N(E)\delta E} \times G^*(E - E_0) \times \frac{\delta E}{h}$$

$$= \begin{pmatrix} \text{number of} \\ \text{initial states} \\ \text{in } E, E + \delta E \end{pmatrix}^{-1} \times \begin{pmatrix} \text{total number of} \\ \text{states of TS} \\ \text{in } E - E_0 \end{pmatrix} \times \begin{pmatrix} \text{elementary} \\ \text{flux} \end{pmatrix} \tag{3.24}$$

Comparison with Eq. (3.23) shows that the transition state is a device by means of which the enumeration of the unspecified final states is replaced by the counting of states in the transition state, where by hypothesis *every* state is present in the interval $0, E - E_0$ with equal probability.

Originally (that is, in the 1930s when the concept of the transition state was first invented [13]), the transition state was conceived merely as a theoretical construct useful to rationalize the progress of a chemical transformation in time and space, somewhat in the sense of the above paragraph. With modern advances in laser technology it is now possible to observe the transition state in the instant of formation (see Section 3.5.4).

The purpose of the demonstration in Eqs. (3.23) and (3.24) is to show that $k(E)$ of Eq. (3.21) is in fact an *average* rate constant (see also Eq. (3.6)), as has been recognized early [2, 15–17]. (Note that this is true of Slater's $k(E)$ as well, cf. Chapter 2, Eq. (2.4).) In this respect it should be noted that an average makes sense only if the number of states to be counted is not too small, which means that the theory is intended to apply principally to polyatomic molecules. Another facet of the average is that we can expect that there will be fluctuations about the average. Information-theoretic analysis suggests that these fluctuations may be fairly large since, in the statistical limit, the variance is [16, 17]

$$\langle [(k - k(E)]^2 \rangle = 2[k(E)]^2 \tag{3.25}$$

assuming that k, the individual microcanonical rate constants for states within the manifold of states of energy E, can be resolved.

Here we may note a connection with the formulation of the unimolecular reaction as a scattering problem. The complete information about a scattering process is contained in the scattering matrix $\mathbb{S}$. An individual matrix element $|S_{\beta\alpha}(E)|^2$ gives the probability of a transition from internal state α of reactant to internal state β of products at total energy E. For many chemical reactions at energies of interest there may be hundreds or thousands of energetically accessible internal states, and one is then interested only in the averaged quantity, the *cumulative reaction probability* $\mathcal{N}(E)$ [18], defined by

$$\mathcal{N}(E) = \sum_{\beta\alpha} |S_{\beta\alpha}(E)|^2 \qquad (3.26)$$

which is the analog of the sum of states $G^*(E - E_0)$ in Eq. (3.21).

With $\delta E / h$ as the elementary flux, $G^*(E - E_0)/h$ may be viewed as the forward flux through the critical surface. By the application of microscopic reversibility the same (forward) rate constant $k(E)$ can be obtained, under some conditions, from the reverse flux by using the properties of the actual product states. This is discussed in some detail in Chapter 8.

3.2.4 Special cases

The elementary flux or rate constant is seen to be $\delta E / h$, the hypothetical (average) rate at which one initial state would decay to one "final" state. This would be the case of dissociation of a diatomic molecule: if the diatomic molecule were represented by a harmonic oscillator of frequency v, $\delta E = hv$ would be the spacing of energy levels, so that the hypothetical microcanonical rate constant for the dissociation of a harmonic oscillator would be $k(E) = v$. This can be viewed as the maximum possible rate. A simple illustration of the above concepts is the subject of Problem 3.2.

Consider now the case of a polyatomic molecule. If the available energy were low enough that only one state is accessible in the transition state ($G^*(E - E_0) = 1$, i.e. $E = E_0$), the rate constant would be the threshold rate, given by

$$k(E_0) = \frac{1}{h\, N(E_0)} \qquad (3.27)$$

As energy increases above E_0, then with each additional accessible state in $G^*(E - E_0)$ the rate constant $k(E)$ increases proportionately. This shows that degrees of freedom, other than the reaction coordinate, function essentially as reservoirs of energy. See Section 3.5.5 for application of Eq. (3.27) to the dissociation of NO_2.

3.2.5 Reaction-path degeneracy

In some instances the potential-energy surface may possess a symmetry such that there may be several equivalent reaction paths, as, for example, in the reaction $CH_4 \rightarrow CH_3 + H$ in which the hydrogen atom in the products may be any one of the four equivalent hydrogens of methane. There are thus four equivalent reaction paths, with the result that the probability of dissociation, i.e. the rate constant $k(E)$, is increased fourfold. In order to allow for the symmetry properties of the potential-energy surface, the rate constant of Eq. (3.21) will therefore henceforth involve the additional factor α, usually called the statistical factor or reaction-path degeneracy:

$$k(E) = \frac{\alpha G^*(E - E_0)}{hN(E)} \tag{3.28}$$

The degeneracy α is related to rotational symmetries of reactant M (σ) and its transition state (σ^*) by $\alpha = \sigma/\sigma^*$. In the cited case of the reaction $CH_4 \rightarrow CH_3 + H$, $\sigma = 12$ (CH_4 as a spherical top) and $\sigma^* = 3$ ($CH_3 \cdots H$ as a prolate top) (cf. Appendix 1, Section A1.4.3). In most cases α can be determined by a direct count, but this can get complicated if there are geometrical and/or optical isomers, as discussed at some length in [5].

3.3 Implementation of the RRKM rate constant

The actual implementation of the RRKM rate constant requires consideration of several details specific to the type of reaction to be dealt with.

3.3.1 Reaction types

The potential-energy profile along the reaction coordinate determines the principal characteristics of a given unimolecular reaction, so that it is useful for classification of reaction types. According to this criterion any unimolecular reaction belongs perforce to one of three types.

Type 1. Reaction involves simple bond-breaking in a neutral species, which yields two radicals, for example $C_2H_6 \rightarrow CH_3 + CH_3$, or bond-breaking involving an ion, for example $C_2H_5Br^+ \rightarrow C_2H_5^+ + Br$. The principal feature is that, while there is no formal potential hill for the reverse association of fragments insofar as internal energy is concerned, external rotations cause the appearance of a centrifugal barrier (Section 3.4). Since Type 1 reactions are of considerable practical as well as theoretical interest, they are discussed further in Chapters 7 and 8.

Type 2. Reaction involves either simple bond-breaking with internal re-organization of one of the fragments, e.g. $C_2H_5 \rightarrow H + CH_2 = CH_2$, or complex bond-breaking, e.g. $C_2H_5Cl \rightarrow HCl + CH_2 = CH_2$ or $C_4H_8^+ \rightarrow CH_2CCH_2^+ + CH_4$. In all cases there is a potential hill for the back-association of fragments.

Type 3. Reaction is an isomerization, either *cis* $\rightarrow$ *trans*, e.g. *cis*-butene-2 $\rightarrow$ *trans*-butene-2, for which the reaction coordinate is the angle of torsion about the the C—C bond, or structural as in cyclopropane $\rightarrow$ propylene, which proceeds through a diradical intermediate. Again, in every case there is a potential hill to be overcome for the reverse reaction.

Most reactions of Type 2 and 3 neutral species have been studied in thermal systems, which are the subject of Chapter 5.

3.3.2 Pertinent degrees of freedom

The "final" form of the unimolecular rate constant in Eq. (3.28) involves a sum and density of states for what was referred to non-committally as "internal" degrees of freedom of the transition state and the reactant. These are the degrees of freedom among which the excitation energy is assumed to be randomized; such degrees of freedom we may call "pertinent." In the absence of any detailed information concerning which among the internal degrees of freedom are to be considered pertinent, we have to assume that in principle *all* vibrational degrees of freedom (less the reaction coordinate in the case of the transition state), are pertinent, including internal rotations, hindered or free, if such are present.

A particular case arises when there is vibration–rotation interaction due to Coriolis coupling, which has its origin in the interaction of vibrational and rotational angular momenta [21]. This effect is particularly noticeable for a symmetric top due to the coupling between the angular momentum of a doubly degenerate vibration and the angular momentum due to the component of overall rotation about the axis of symmetry of the top [22]. This is the prolate K-rotor of Appendix 1 (Section A1.4) with energy given by

$$E_r(K) = (A - B)K^2 \text{ (prolate top, } A > B) \qquad (3.29)$$

where A and B are rotational constants and K is the quantum number for the component of angular momentum along the axis of symmetry. The consequence of the vibration–rotation coupling is that K need not be a "good" quantum number, so that, as an approximation, it can be assumed that the rotational energy $E_r(K)$ becomes part of the pertinent (i.e. randomizable) degrees of freedom. The case of such an "active" K-rotor is discussed in more detail in Sections 4.5.6 and 4.5.7.

The vibration–rotation coupling concerns in particular transition states of Types 1 and 2, which are tops that become increasingly prolate as the interfragment distance increases. The effect on the rate constant of active and inactive K-rotors in reactant and transition state is analyzed in [23]. More or less direct evidence for active K-rotors comes mostly from observations on small molecules (e.g. H_2CO, NCNO [24] and NO_2 (Section 3.5.5)), but, for large symmetric-top molecules (e.g. C_2H_6), because of their complex spectra, the evidence is only indirect, in the sense that calculations assuming the existence of an active K-rotor usually give better agreement with experiment.

3.3.3 Sums and densities of states

The numerous assumptions that had to be made in order to arrive at the result of Eq. (3.21) may seem to make the rate constant $k(E)$ poorly adapted for the description of an actual unimolecular reaction of a real molecule.

This is not really the case, as has been witnessed over the years by successful applications of the theory to a large number of reactions. We have already mentioned that, since a reaction involves the movement of heavy nuclei, the fundamentally classical nature of the theory is not unreasonable. Thus the onus of a realistic description of the fragmentation process is to a large extent on how far a decomposing molecule may be modeled in realistic fashion.

The success of Eq. (3.28) as a true representation of the microcanonical rate constant $k(E)$ therefore depends to a large extent on how realistic the calculation of the density of states $N(E)$ and the sum of states $G^*(E - E_0)$ is. Classical densities and sums of states are often off by one or more orders of magnitude (see Table 4.1), so that a quantum-oscillator model of M and its transition state is necessary. Given that molecular parameters of the reactant are not usually known in sufficient detail, the calculation is usually done assuming that we have separable harmonic oscillators, but, if data are available, the calculation may include effects of anharmonicity, free or hindered internal rotors, and eventually coupling among the oscillators. These different possibilities are discussed in some detail in Section 4.5.

Thus the realism with which the calculation of the density of states $N(E)$ and of the sum of states $G^*(E - E_0)$ may be accomplished is limited only by the availability of relevant molecular data. While the quantum properties are fundamentally discrete, the underlying semi-classical aspect of the theory behind the rate constant $k(E)$ requires that $N(E)$ and $G^*(E - E_0)$ be used in their smoothed or quasi-continuous form.

3.3.4 Non-classical effects

The theory behind the $k(E)$ of Eq. (3.28) can also accomodate quantum effects that affect the rate in other ways, such as tunneling through the potential-energy barrier, and non-adiabatic transitions. The latter lifts the implicit assumption made in the derivation of Eq. (3.28) that the fragmentation of M proceeds entirely on a potential-energy surface of the same electronic multiplicity. Both tunneling and non-adiabatic transitions are discussed in Chapter 6.

One basic premise of the theory that cannot be easily removed is the assumption that we have a microcanonical ensemble since, in the absence of microcanonical equilibrium, the rate constant $k(E)$ cannot be evaluated by the simple counting of states.

3.3.5 Recapitulation

While the final result for the rate constant $k(E)$ given by Eq. (3.28) is deceptively simple, it has been arrived at by using numerous simplifying assumptions that merit attention.

(1) Decomposition takes place on the same potential-energy surface under a vibrational, or more generally, effective potential.
(2) Treating the transition state as a configuration of "no return" implies that phase points that reach the critical configuration from the reactant side never recross the dividing surface in the opposite direction. Since this cannot be guaranteed *a priori*, $k(E)$ at best yields only an *upper limit* to the actual rate constant. This aspect is addressed in Chapter 7.
(3) The distribution of internal states of reactant M is assumed to correspond to a microcanonical ensemble, which means that *every* state corresponding to energy E is represented *with equal probability*. As a result, it is merely sufficient to count the states (the totality of states for $G^*(E - E_0)$, or states per unit energy for $N(E)$). If the initial excitation does not produce a microcanonical ensemble, it is assumed that such an ensemble will be created prior to decomposition by rapid intramolecular vibrational redistribution (IVR) on a time scale that is short relative to the lifetime $(=1/k(E))$ of the decomposing molecule.

Consequently the initial distribution of excitation energy need not be specified and as a result $k(E)$ is a function of the total energy E only; thus all dynamical considerations are ignored. While the actual mechanism responsible for IVR is left unspecified, it is implicitly assumed that IVR is due to anharmonic effects, which, however, are not supposed to be strong enough to contribute significantly to the density of states (see also Section 4.5.5).

(4) The critical surface separating reactant and products is assumed to be a function of coordinates only. The process of dissociation is thus viewed as a one-dimensional translation along the "reaction coordinate" q_1, which is taken to be a bond stretch in simple bond fission, or the torsional angle in *cis–trans* isomerization, for example. This reaction coordinate, assumed to be a separable degree of freedom, is the only degree of freedom that is directly involved in the actual decomposition process in that it evolves into the relative translation of products as the reaction progresses. It corresponds to a vibrational degree of freedom of the transition state that is subject to a potential *maximum*, so that its vibrational frequency is imaginary; hence it is a vibration that is "missing" in the transition state. The translational energy ϵ_t in the reaction coordinate is treated classically and all other vibrational degrees of freedom of the transition state are treated quantum mechanically.

The theory, as presented up to this point, is not specific enough as to the nature of the transition state, beyond stating that it corresponds to the critical configuration at the top of the potential barrier. Calculation of the detailed properties of the transition state is in principle the province of quantum mechanics, but such calculations are not easy for a large molecule. Additionally, while quantum-mechanical methods are well adapted for the calculation of stable configurations (reactant and products), they are less so for the intermediate transition-state region, in particular the critical energy E_0.

In the absence of such calculation, one may use the expedient of estimating the properties of the transition state by analogy with a similar stable molecule, with allowance for the deformation produced by the incipient chemical transformation, and the loss of the degree of freedom corresponding to the reaction coordinate. Some guidance from experiment and the nature of products is then necessary, usually by adjusting vibrational frequencies likely to be affected by the reaction in order to obtain agreement with experimental results.

What might be called a zeroth-order approximation would consist of merely assuming the transition state to have the same vibrational frequencies as the reactant, with the frequency corresponding to the reaction coordinate deleted. This should be good enough for an order-of-magnitude estimate of the rate constant.

A better approximation is obtained by paying attention to how a reactant evolves into fragments. Thus, for example, in the dissociation $C_2H_6 \rightarrow CH_3 + CH_3$, the reaction coordinate is the C–C stretching vibration, and, while C–H vibrations are clearly unaffected in the fragmentation, deformation and rocking frequencies must be "softened" (i.e. lowered) since ultimately they evolve into fragment rotations or considerably lower CH_3 frequencies, as shown schematically in Table 3.1. This sort of transition-state specification is mostly good enough only for thermal reactions, as discussed in Chapter 5.

However, there are methods for assigning transition-state properties that are more solidly grounded in theory, which are the subject of Chapters 7 and 8.

3.4 Rotational effects

Equation (3.28) for the rate constant $k(E)$ has been obtained by neglecting the effect of overall or "external" rotational energy, that is, rotational energy that is *not* part of randomizable degrees of freedom. In this sense "external" rotations would exclude the active K-rotor discussed above.

3.4.1 The role of "external" rotations

Consider the process M $\rightarrow$ products, where reactant M has "external" rotational energy due to angular momentum J. The principal interest here is the entrance channel, that is, events that lead to the formation of the transition state at the potential extremum, while events past the maximum, i.e. the actual formation of the products (the exit channel), are of no direct concern in the present context. The

Table 3.1. *Simple transition-state vibrational assignment in*
$C_2H_6 \rightarrow 2CH_3$ *(vibrational frequencies are in* cm^{-1}*, with*
degeneracies in parentheses)

Reactant C_2H_6	Products $2CH_3$	Decision
C—H stretch	C—H stretch	
2954, 2896, 2985 (2), 2969 (2)	3044 (2), 3162 (4)	Leave as is
CH_3 deformation	CH_3 deformation	
1472 (2), 1469 (2)	1396 (4)	Leave as is
1388, 1379	580 (2)	Reduce
CH_3 rocking mode, C—C torsion		
821 (2), 1206 (2), 289	Rotations	Reduce
C—C stretch	Absent	
995	(reaction coordinate)	Omit

The first two columns represent the juxtaposition of reactant and product frequencies of a given kind as a guide to their likely evolution in
the transition state. The transition state is considered to be a modified
reactant: thus "Decision" means how a given *reactant* vibrational-mode
frequency, in view of its value in the products, is expected to change in
the transition state. At this level of approximation the actual decreases
in frequency can be determined only by comparing the calculated rate
constant with experiment.

problem is considered further in Chapters 8 and 9, but from the point of view of
events past the potential maximum.

Assuming central potential (see Chapter 9 for details), "external" rotations can be
taken into account by invoking the so-called effective potential, which is obtained
by simply adding vibrational and rotational potentials:

$$V_{\text{eff}}(r) = V(r) + J(J+1)B_e(r_e/r)^2 \qquad (3.30)$$

where r is the interfragment separation considered to be the reaction coordinate,
$V(r)$ is the vibrational potential along r, $J(J+1)B_e = E_r$ is the rotational energy
of the reactant with angular-momentum quantum number J, and $B_e = \hbar^2/(2\mu r_e^2)$
is the reactant equilibrium rotational constant ($\hbar = h/(2\pi)$, μ is the reduced mass,
and r_e is the equilibrium value of r, i.e. r at the minimum of $V(r)$). Implicitly we
consider that Eq. (3.30) applies to $r_e < r < r_{\text{max}}$ as defined below in connection
with Eq. (3.31). The factor $(r_e/r)^2$ allows for the change of rotational energy with
increasing r.

The rotational-energy expression used in Eq. (3.30) is exact for a linear reactant,
but, if the reactant is a symmetric top, it implies that K is not a "good" quantum
number, i.e. that the K-rotor is a randomizable degree of freedom as defined in the
previous section. If this is not the case, the K-rotor is part of "external" rotations

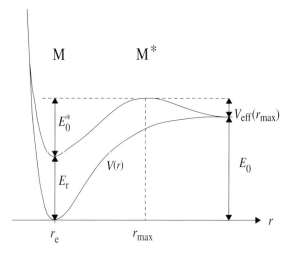

Fig. 3.2. A schematic representation of the rotational effect in a Type-1 reaction. The purely vibrational potential barrier $V(r)$, of height E_0 (lower curve), increases monotonically with interfragment distance r without a definable maximum. When the reactant has angular momentum, i.e. rotational energy E_r, the effective potential (upper curve) exhibits a maximum such that the potential barrier separating reactant and products is reduced from E_0 to E_0^*. As angular momentum increases, the location of the maximum r_{max}, and therefore also the transition state M*, shifts closer to r_e. The drawing neglects zero-point energies.

and the effective potential should be written in terms of both J and K [25, 26]. One possible simplification in such a case, which will allow the effective potential still to be written in the form of Eq. (3.30), is to average the rotational energy E_r over all K, as shown in Problem 3.3.

Since the effect of "external" rotations is different for different types of reaction, these will be discussed separately.

3.4.2 Type-1 reactions

The effective potential of Eq. (3.30) for a Type-1 reaction is shown schematically in Fig. 3.2 (subtraction of zero-point energies is assumed). The principal feature to be noted is that, for a reactant with rotational energy, i.e. with angular momentum, the effective height of the potential barrier separating the reactant from its products, i.e. the critical energy E_0^*, is lowered with respect to E_0, the threshold for a purely vibrational ($J = 0$) potential: $E_0^* < E_0$. (Note that both E_0^* and E_0 refer to the *internal* energy of the reactant.) In other words, angular momentum helps the reactant to "get over the hump."

3.4.3 The rotational barrier

The information of interest here is the relation between E_r and E_0^*, which will be referred to as "the barrier." Its general form will be written $E_0^* = E_0 - \mathcal{F}(E_r)$, where the explicit form of the function $\mathcal{F}(E_r)$ will depend on the form of the potential $V(r)$.

As an example for which the barrier can be obtained analytically, take $V(r)$ as the attractive part of the Lennard-Jones 6–12 potential (Appendix 4) in the form $V(r) = -2E_0(r_e/r)^6$; the effective potential is then

$$V_{\text{eff}}(r) = -2E_0(r_e/r)^6 + E_r(r_e/r)^2 \tag{3.31}$$

By differentiation we find that this potential has a maximum at $(r_e/r_{\text{max}})^4 = E_r/(6E_0)$. On substituting back into Eq. (3.31), we get $V_{\text{eff}}(r_{\text{max}})$, and, since $E_r + E_0^* = E_0 + V_{\text{eff}}(r_{\text{max}})$ (Fig. 3.2), the barrier is

$$E_0^* = E_0 - \left(E_r - \sqrt{\frac{2}{27E_0}}E_r^{3/2} \right) \tag{3.32}$$

where $\mathcal{F}(E_r)$ is the expression in parentheses. E_0^* for the general $1/r^n$ potential is the subject of Problem 3.4.

More realistic are the Morse and extended-Rydberg potentials (Appendix 4), for which it is necessary to obtain the barrier by determining E_0^* numerically as the difference between the maximum and minimum of the effective potential at a given E_r.

The various barriers, which differ significantly only at very low E_0^*/E_0, are shown in dimensionless form in Fig. 3.3. It can be seen that the rotational energy necessary to reduce the critical energy to zero is always larger than E_0, so that rotational energy is less effective than vibrational energy to overcome the potential

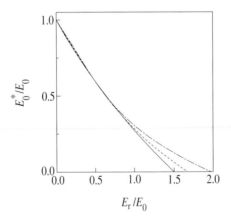

Fig. 3.3. Rotational barriers in dimensionless form for a Type-1 reaction. The solid line is for the Lennard-Jones potential (Eq. (3.32)), the dashed line is for the Morse potential, and the dot–dashed line is for the extended Rydberg potential. In every instance rotational energy is less effective than vibrational energy for overcoming the potential barrier separating reactant and products.

hill separating reactant and products. For example for $E_0^* = 0$ the solution of Eq. (3.32) is $E_r = 1.5E_0$.

3.4.4 Type-2 and -3 reactions

By definition, these are reactions with a potential barrier for the reverse reaction even in the absence of angular momentum. This barrier is a *vibrational* barrier, which is always considerably more pronounced than the purely centrifugal barrier in Type-1 reactions, so rotational effects in Type-2 and -3 reactions modify the position and height of the existing vibrational barrier only in a very minor way.

It is therefore sufficient to examine the effect of rotations only in the vicinity of the vibrational-barrier maximum. A useful approximation is to model this part of the barrier by a parabolic potential, for convenience referred to potential maximum as zero:

$$V(r) = -\frac{E_0(r - r_0)^2}{(r_e - r_0)^2} \tag{3.33}$$

where (Fig. 3.4) E_0 is the height of the vibrational barrier, with maximum located at internuclear distance r_0, and r_e is the equilibrium value of r. For a given J the

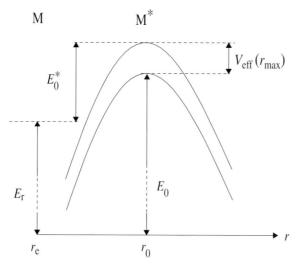

Fig. 3.4. A schematic representation of the rotational effect in the vicinity of the top of a parabolic potential barrier in a Type-2 or -3 reaction. E_0 is the height of the purely vibrational potential barrier when the reactant M has no rotational energy ($E_r = 0$). If the reactant has rotational energy ($E_r \neq 0$), the height of the barrier (only a portion of which is shown) increases by $V_{eff}(r_{max})$ but the reactant energy increases even more, by E_r, so that the height of the effective potential barrier (E_0^*) is reduced. The bottom of the vibrational potential well of the reactant is at r_e, while the maximum of the barrier remains at r_0, independently of E_r; this is also the location of the transition state M*.

effective potential is still given by Eq. (3.30), but with $V(r)$ of Eq. (3.33):

$$V_{\text{eff}}(r) = -\frac{E_0(r - r_0)^2}{(r_e - r_0)^2} + J(J + 1)B_e \left(\frac{r_e}{r}\right)^2 \tag{3.34}$$

The maximum of this effective potential will be at r_{max} given by the solution of $\partial V_{\text{eff}}(r)/\partial r = 0$, which leads to

$$r^3(r_0 - r) = J(J + 1)B_e r_e^2(r_e - r_0)^2/E_0 \tag{3.35}$$

For any reasonable E_0 ($E_0 \geq 10^4$ cm^{-1}) the RHS will be several orders of magnitude smaller than the LHS, regardless of the value of J, so that $r_{\text{max}} \cong r_0$. Thus, in contrast with type-1 reactions, the maximum of the effective potential in Type-2 and -3 reactions does not shift significantly with J, so that

$$V_{\text{eff}}(r_{\text{max}}) \cong J(J + 1)B_e \left(\frac{r_e}{r_0}\right)^2 \tag{3.36}$$

Consequently, from Fig. 3.4, $E_0^* = E_0 + J(J + 1)B_e(r_e/r_0)^2 - E_r$, and hence

$$E_0^* \cong E_0 - E_r[1 - (r_e/r_0)^2] \tag{3.37}$$

E_0^* now decreases linearly with E_r since r_e/r_0 in a given reaction is a constant.

3.4.5 The J-shifting approximation

Consideration of rotational effects has the result that the microcanonical rate constant $k(E)$ becomes a function both of internal (E) and of rotational (E_r) energy; thus $k(E) \Rightarrow k(E, E_r)$. Equivalently, $k(E)$ can be written as a function of E and angular momentum J as $k(E, J)$. The conclusion to be drawn from the above results is that overall or "external" rotations have the effect of decreasing E_0, the critical energy for reaction, in all types of reaction.

A simple device of some antiquity [2, Chapter 7], but which has only recently received a renewal of attention [27, 28], is to introduce E_r- or J-dependence into the microcanonical rate constant by simply shifting the energy zero in the transition state:

$$k(E, J) \quad \text{or} \quad k(E, E_r) = \frac{\alpha G^*(E - E_0^*)}{hN(E)} \tag{3.38}$$

In this representation the energy zero of E is the minimum (in principle corrected for zero-point energy) of the *effective* potential at the specified E_r or J (see Fig. 7.1). The total energy of the reactant is $E + E_r$, of which only E, the internal (essentially vibrational) energy, is randomizable. E_0^* is the E_r- or J-dependent minimum E necessary to surmount the potential barrier separating reactant and products, so that the energy available for randomization in the transition state is $E - E_0^*$, which is therefore E_r- or J-dependent. In this way the E_r- or J-dependence of the rate constant arises uniquely via the rotational decrease of E_0.

There is a somewhat subtle point connected with rotational degeneracy that affects the form of Eq. (3.38) in the rather special case in which the reactant is a spherical top. Discussion of this point is deferred until Section 4.5.7.

Since E_0^* is less than E_0, "external" rotations cause the microcanonical rate constant $k(E, J)$ at a given energy to increase in general with J (or $k(E, E_r)$ to increase with E_r), but this depends on the energy scale chosen.

Accordingly the plot of $\log k(E, J)$ (as defined in Eq. (3.38)) versus $E - E_0^*$ produces a collection of curves with a common origin at $E - E_0^* = 0$, monotonically increasing with J at each E. The plot of $\log k(E, J)$ versus $E + E_r - E_0$ ($E + E_r$ is the *total* reactant energy) is similar, but as a rule the rate constant $k(E, J)$ is found to actually *decrease* with increasing J over a range of energies above the threshold (see Fig. 8.11 in Chapter 8), as the result of the interplay among E, E_r, and E_0^*: at fixed total reactant energy $E + E_r$, increasing J, i.e. increasing E_r, leaves less energy for E. This effect has been confirmed by experiment in the case of the dissociation $NO_2 \rightarrow NO + O$ [20].

3.5 Experimental evidence

The basic theory of the microcanonical rate constant $k(E)$, as presented in the previous sections, was developed in its general outline long before technology had advanced to the point where $k(E)$, the microcanonical rate constant, could actually be determined experimentally as such, or at least retrieved from some average. It is a testimony to the perspicacity of RRK that they devised a successful theory for $k(E)$ for which there was very little experimental foundation at the time. This section presents experimental evidence in favor of (and some against) the RRKM version of $k(E)$.

3.5.1 Fragmentation of ions

Early probings of the theory on the more fundamental level concerned the study of ionic reactions in the mass spectrometer [29]. Ions behave just like neutral molecules, they can be moved about and mass- and energy-analyzed by electric and magnetic fields regardless of their size, which allows limited access to $k(E)$ that is not too highly averaged, under the assumption that any electronic excitation that may have been produced in the ionization step is rapidly degraded into internal energy of the ground-state ion. Early results, discussed in [2, Chapter 10], seemed to confirm in general the statistical nature of ionic fragmentation, that is, randomization of internal energy prior to decomposition, and fragmentation according to the RRKM prescription for $k(E)$.

3.5.2 PEPICO

A more recent technique that can produce ions in well-defined states is photoionization [7, Chapter 5], especially in its incarnation as TPEPICO (threshold photoelectron–photoion coincidence). When a neutral molecule is ionized (by vacuum-UV radiation), an electron is ejected with a distribution of kinetic energies, which is reflected in the internal energy distribution of the ion left behind. If ions are measured in coincidence with threshold electrons, i.e. electrons of zero kinetic energy, the ionic internal energy is precisely known. The fragmentation of the thus energy-selected ions is then analyzed in a time-of-flight mass spectrometer.

A reaction that has been the subject of much scrutiny is the dissociation of the 1-butene ion $C_4H_8^+$ prepared by this technique from 1-butene [30] (actually any of the butenes will do, since, before dissociation, there is rearrangement to a common structure). The three lowest-threshold fragmentation channels are

$$C_4H_8^+ \rightarrow C_3H_4^+ + CH_4 \qquad E_0 = 15\,060\,\text{cm}^{-1} \qquad \text{(a)}$$

$$C_4H_8^+ \rightarrow C_4H_7^+ + H \qquad E_0 = 16\,050\,\text{cm}^{-1} \qquad \text{(b)}$$

$$C_4H_8^+ \rightarrow C_3H_5^+ + CH_3 \qquad E_0 = 18\,370\,\text{cm}^{-1} \qquad \text{(c)}$$

For rotationally cold ions produced from a cold-seeded molecular beam of 1-butene, there is excellent agreement between RRKM theory (Eq. (3.28)) and experiment for reactions (b) and (c), using transition-state frequencies obtained from *ab initio* calculations. For reaction (a) *ab initio* calculation was unable to locate the transition-state, so that, in order to obtain agreement with experiment, the transition-state frequencies had to be adjusted.

A room-temperature molecular beam of 1-butene produced ions with a 298-K rotational distribution, so that $k(E, J)$ (Eq. (3.38)) could be determined, which improved agreement between theory and experiment for room-temperature data. The data were insensitive to whether or not the K-rotor was active (Section 3.4).

3.5.3 Kinetic energy release

Ions have the additional advantage that the presence of an electric charge makes it possible to determine the kinetic energy of fragments, which is very difficult with neutral molecules. This adds another measurement that can be used to check on the statistics of unimolecular decay. From Eq. (3.13) the probability that fragments are formed with relative kinetic energy ϵ_t when the total energy is E can be written formally

$$P(E, \epsilon_t) = \frac{k(E, \epsilon_t)\, d\epsilon_t}{\int\limits_0^{E-E_0} k(E, \epsilon_t)\, d\epsilon_t} \qquad (3.39)$$

which assumes that the translational energy in the transition state is preserved in the fragments. If $k(E, \epsilon_t)$ is expressed with the help of Eqs. (3.19) and (3.20), the probability simplifies to

$$P(E, \epsilon_t) = \frac{N^*(E - E_0 - \epsilon_t)\, d\epsilon_t}{\int\limits_0^{E-E_0} N^*(E - E_0 - \epsilon_t)\, d\epsilon_t} \qquad (3.40)$$

which therefore depends only on the properties of the transition state. The average energy release is then $\langle \epsilon_t \rangle = \int \epsilon_t P(\epsilon_t, E)\, d\epsilon_t$ (see Problem 3.5 for a simple explicit expression for $\langle \epsilon_t \rangle$). This $\langle \epsilon_t \rangle$ was used (after further averaging over E) in the early days (before the advent of PEPICO) when only $\langle\langle \epsilon_t \rangle\rangle$, the average of $\langle \epsilon_t \rangle$ over a distribution of energies E, could be determined. Since this distribution was not very well known, claims that $\langle\langle \epsilon_t \rangle\rangle$ in several cases was statistical (i.e. based on Eq. (3.40)), were suspect. Later PEPICO results confirmed that $\langle \epsilon_t \rangle$ is indeed often statistical, e.g. in the decomposition of the isomeric acetaldehyde and ethylene oxide ions $C_2H_4O^+$ into $CH_3^+ + HCO$ and $HCO^+ + CH_3$ [31].

The probability of Eq. (3.40) is but a crude version of a more general probability distribution given in Eq. (8.101) (Section 8.10). Since the actual probability is more informative than $\langle \epsilon_t \rangle$, further discussion is deferred to Chapter 8.

3.5.4 Laser experiments

Techniques that permit more detailed probing of molecular states were developed after the invention of the tunable laser. The principle is "pump and probe": one or more lasers are used to prepare the reactant in a well-defined energy state above threshold; this is the "pump" part. After a variable delay, another laser, tuned to a particular product frequency, is used to detect the product and the state it is in; this is the "probe" part. The process must take place at very low pressure to avoid collisions, and at low temperature (achieved by nozzle expansion) to eliminate interference from rotational states. Both scalar and vector (spatial-distribution) properties of fragments can be determined in principle.

This technique is most useful for probing the fragmentation of small molecules since the spectra obtained from large molecules are too difficult to interpret. The fragmentations of small molecules so examined include $NCNO \rightarrow NC + NO$, $CH_2CO \rightarrow CH_2 + CO$, and $CF_3NO \rightarrow CF_3 + NO$ [32].

3.5.5 Reaction of NO_2

The workhorse system that has been the subject of most scrutiny is the dissociation of the Type-1 reaction $NO_2 \rightarrow NO + O$ [33, 34]. Although there are complications due to electronic states, these laser experiments reveal much detail of the fragmentation dynamics. For a non-rotating molecule, the experimental microcanonical rate constant is found to be consistent with Eq. (3.28), and several transition-state parameters can be deduced from experimental observation of room-temperature samples at $<1000\,cm^{-1}$ above threshold: the O–NO distance in the transition state ($\cong 1.7\,nm$) is about 40% larger than that in ground-state NO_2 (1.19 nm), and the O–NO bending frequency (759 cm^{-1} in NO_2) is decreased to $\cong 100\,cm^{-1}$. With increasing excitation energy the transition state is found to move to shorter O–NO distance [35], but this can be accounted for theoretically only by a variational treatment, which will be discussed later in Chapter 7.

The effect of rotation of NO_2 reactant can also be seen in room-temperature experiments, in which it is found that rotations do indeed lower the threshold energy, as would be expected for a Type-1 reaction (Section 3.4), and that the K-rotor (NO_2 is roughly a symmetric top) is approximately active due to Coriolis coupling [34] (see the discussion in Section 3.3).

A threshold rate has also been determined experimentally in the dissociation of NO_2. By actual counting of energy levels in the absorption spectrum of NO_2 just below threshold ($E_0 = 25\,129\,cm^{-1}$), the density of states at threshold is $N(E_0) \approx 0.2\,(cm^{-1})^{-1}$ [33]; it then follows from Eq. (3.27) that $k(E_0) \approx 1.5 \times 10^{11}\,s^{-1}$. Direct experimental determination [20, 34] gives $k(E_0) \cong 2 \times 10^{11}\,s^{-1}$, in satisfactory agreement with theory.

3.5.6 Femtosecond chemistry

This is the name given to a recent advance in laser technology that pushes the time resolution down to the femtosecond (10^{-15} s) range, where the ephemeral transition state is fleetingly available for direct observation, thus proving its factual existence. One of the first transition states so observed was $[I \cdots CN]^*$ in the reaction $ICN \rightarrow [I \cdots CN]^* \rightarrow I + CN$ [36]. A particularly interesting case is the dissociation $HgI_2 \rightarrow [I \cdots Hg \cdots I]^* \rightarrow I + HgI$, for which it was possible to observe how the symmetric and anti-symmetric stretch modes of $[I \cdots Hg \cdots I]^*$ evolve, respectively, into the single vibrational degree of freedom of HgI and relative translation [37].

3.5.7 Failure of RRKM

So far we have mentioned only examples for which predictions of RRKM theory are verified by experiment. While this is true in the overwhelming majority of cases, there are nevertheless instances in which RRKM theory fails because the internal energy of the decomposing molecule fails to randomize over all internal degrees of freedom. Thus the fundamental assumption of the theory regarding intramolecular vibrational redistribution (IVR) is not fulfilled. The result is that the decomposing molecule acts as if it were smaller (i.e. had fewer internal degrees of freedom) than it actually is, so that the calculated RRKM rate constant is too large compared with experiment. (Compare this with Fig. 1.1, which shows $k(E)$ increasing with *decreasing* number of degrees of freedom s.)

Such failure of randomization occurs sometimes in chemical activation (see Section 5.7) if there is a bottleneck between the site where energy is deposited and the rest of the molecule. As an example, consider the case of tetraallyl tin, with four identical ligands disposed symmetrically around the central tin atom, activated by reaction with a fluorine atom [38]:

$$^{18}F + Sn(CH_2CH{=}CH_2)_4 \rightarrow (CH_2CH{=}CH_2)_3SnCH_2CHFCH_2^* \rightarrow CH_2{=}CHF + R$$

It is found that the excited radical decomposes into $CH_2{=}CHF + R$ about three orders of magnitude faster than calculated by RRKM theory. This appears to be due to excitation energy being redistributed only in the F-containing side-chain, where energy is initially deposited by the activating reaction, because the central tin atom effectively blocks the flow of energy into the other three ligands.

A somewhat different case is the dissociation of the molecule HeI_2:

$$HeI_2 \rightarrow He + I_2$$

for which RRKM *overestimates* the rate constant by over three orders of magnitude. Choosing this reaction as an example of failure of RRKM is somewhat unfair because HeI_2 is not a molecule in the usual sense, but rather a weakly bound van der Waals complex produced in supersonic free-jet expansion.

The model generally adopted is that of a T-shaped complex in which He is held by a very weak van der Waals bond [39] (Table 3.2). The reaction is initiated by depositing energy in the I–I bond and monitored by following the appearance of I_2. At $1280\,cm^{-1}$ of energy in the I–I bond, simple RRKM (Eq. (3.28)) gives for the rate constant $\cong 7.5 \times 10^{11}\,s^{-1}$, whereas the experimental value is $\cong 2.2 \times 10^8\,s^{-1}$ [40]; similar results are found at higher energies.

The discrepancy of three orders of magnitude between the calculated RRKM rate constant and the experimental rate constant is due to the factor of six mismatch between the I–I and He–I frequencies (Table 3.2). In this case there is therefore a considerable intramolecular bottleneck in the flow of energy from the site where energy is initially deposited to the bond to be broken, i.e. the reaction coordinate. An analogous discrepancy between RRKM rate and experiment is found for other van der Waals complexes of the type XY_2, where X is Ar or Ne, and Y_2 is I_2 or Cl_2.

Table 3.2. *Dissociation energy (D) and vibrational frequency (ω) of T-shaped* HeI_2

	I	$D(I–I) = 4911\,cm^{-1}$;	$\omega(I–I) = 128\,cm^{-1}$
He $\cdots\cdot$|		$D(He–I) = 18\,cm^{-1}$;	$\omega(He–I) = 26\,cm^{-1}$
	I		

In order to account for the intramolecular bottleneck in these complexes, Gray *et al.* [41] modified RRKM theory by replacing the concept of the transition state with a "separatrix," which represents essentially the phase-space diagram for the last bound state of the system.

References

[1] P. J. Robinson and K. A. Holbrook, *Unimolecular Reactions* (Wiley, New York, 1972).

[2] W. Forst, *Theory of Unimolecular Reactions* (Academic Press, New York 1973).

[3] J. Troe, in *Physical Chemistry. An Advanced Treatise*, Vol. VI B. W. Jost, Ed. (Academic Press, New York, 1975); J. Troe, in *Int. Rev. Sci. Phys. Chem.*, Ser. II, Vol. 9. (Butterworths, London, 1976); W. L. Hase, in *Dynamics of Molecular Collisions*, Part B, Ch. 3, W. H. Miller, Ed. (Plenum Press, New York, 1980); D. G. Truhlar, W. L. Hase and J. T. Hynes, *J. Phys. Chem.* **87**, 2664 (1983).

[4] J. C. Rayez and W. Forst, *J. Chem. Educ.* **66**, 311 (1989).

[5] R. G. Gilbert and S. C. Smith, *Theory of Unimolecular and Recombination Reactions* (Blackwell Scientific Publications, Oxford, 1990); A. J. Karas, R. G. Gilbert and M. A. Collins, *Chem. Phys. Lett.* **193**, 181 (1992).

[6] K. A. Holbrook, M. J. Pilling and S. H. Robertson, *Unimolecular Reactions*, 2nd Ed. (Wiley, Chichester, 1996).

[7] T. Baer and W. L. Hase, *Unimolecular Reaction Dynamics* (Oxford University Press, Oxford, 1996).

[8] O. K. Rice and H. C. Ramsperger, *J. Am. Chem. Soc.* **49**, 1617 (1927).

[9] See, for example, *Potential Energy Surfaces and Dynamics Calculations*, D. G. Truhlar, Ed. (Plenum Press, New York, 1981).

[10] T. D. Sewell and D. L. Thompson, *Int. J. Mod. Phys.* B **11**, 1067 (1997).

[11] See, for example, R. C. Tolman, *Principles of Statistical Mechanics* (Oxford University Press, Oxford, 1962), pp. 180ff.

[12] S. Nordholm, *Chem. Phys.* **129**, 371 (1989); L. E. B. Borjesson, S. Nordholm and L. L. Andersson, *Chem. Phys. Lett.* **186**, 65 (1991).

[13] H. Eyring, *J. Chem. Phys.* **3**, 107 (1935); *Chem. Rev.* **17**, 65 (1935). For a perspective see G. A. Petersson, *Theor. Chem. Acc.* **103**, 190 (2000).

[14] W. Forst and Z. Prasil, *J. Chem. Phys.* **51**, 3006 (1969).

[15] M. Quack and J. Troe, *Ber. Bunsenges.* **79**, 469 (1975).

[16] W. H. Miller, *Chem. Rev.* **1987**, 19 (1987).

[17] R. D. Levine, *Ber. Bunsenges. Phys. Chem.* **92**, 222 (1988).

[18] W. H. Miller, *Acc. Chem. Res.* **26**, 174 (1993).

[19] I. Bezel, P. Ionov and C. Wittig, *J. Chem. Phys.* **111**, 9267 (1999).

[20] B. Abel, B. Kirmse, J. Troe and D. Schwarzer, *J. Chem. Phys.* **115**, 6522 (2001).

[21] G. Herzberg, *Molecular Spectra and Molecular Structure. II. Infrared and Raman Spectra of Polyatomic Molecules* (D. van Nostrand, Princeton, N.J. 1960), p. 402.

[22] D. K. Sahm and T. Uzer, *Chem. Phys. Lett.* **163**, 5 (1989).

[23] L. Zhu and W. L. Hase, *Chem. Phys. Lett.* **175**, 117 (1990); L. Zhu, W. Chen, W. L. Hase and E. W. Kaiser, *J. Phys. Chem.* **97**, 311 (1993).

[24] C. G. Schlier, *Mol. Phys.* **62**, 1009 (1987); A. B. McCoy, D. C. Burleigh and E. L. Sibert, *J. Chem. Phys.* **95**, 7449 (1991); W. S. McGivern and S. W. North, *J. Chem. Phys.* **116**, 7027 (2002).

[25] A. F. Wagner and J. M. Bowman, *J. Phys. Chem.* **91**, 5314 (1987).

[26] D. L. Shen and H. O. Pritchard, *J. Chem. Soc. Faraday Trans.* **86**, 3171 (1990); *J. Chem. Soc. Faraday Trans.* **87**, 3595 (1991).

[27] Q. Sun, J. M. Bowman, G. C. Schatz, J. R. Sharp and J. N. L. Connor, *J. Chem. Phys.* **92**, 1677 (1990).

[28] S. L. Mielke, G. C. Lynch, D. G. Truhlar and D. W. Schwenke, *Chem. Phys. Lett.* **216**, 441 (1993).

[29] For an elementary account see T. Baer, *J. Am. Soc. Mass Spectrom.* **8**, 103 (1997). For a perceptive review of future directions see J. C. Lorquet, *Mass Spectrom. Rev.* **13**, 233 (1994).

[30] J. A. Booze, M. Schweinsberg and T. Baer, *J. Chem. Phys.* **99**, 4441 (1993).

[31] I. Powis, *Acc. Chem. Res.* **20**, 179 (1987).

[32] H. Reisler and C. Wittig, in *Advances in Chemical Kinetics and Dynamics*, Vol. 1. J. R. Barker, Ed. (JAI Press, Greenwich, CT, 1992), p. 139.

[33] S. I. Ionov, G. A. Brucker, C. Jaques, Y. Chen and C. Wittig, *J. Chem. Phys.* **99**, 3420 (1993).

[34] I. Bezel, P. Ionov and C. Wittig, *J. Chem. Phys.* **111**, 9267 (1999).

[35] S. A. Reid, A. Sanov and H. Reisler, *Faraday Disc.* **102**, 129 (1995).

[36] L. R. Khundkar and A. H. Zewail, *Ann. Rev. Phys. Chem.* **41**, 15 (1990); A. H. Zewail, *Femtochemistry: Ultrafast Dynamics of the Chemical Bond* (World Scientific, Singapore, 1994). A more elementary account is given by J. Baggott, *New Scientist*, June 1989, p. 17.

[37] M. Dantus, R. M. Bowman, M. Gruebele and A. H. Zewail, *J. Chem. Phys.* **91**, 7437 (1989).

[38] P. Rogers, D. C. Montague, J. P. Frank, S. C. Tyler and F. S. Rowland, *Chem. Phys. Lett.* **89**, 9 (1982).

[39] S. K. Gray, S. A. Rice and D. W. Noid, *J. Chem. Phys.* **84**, 3745 (1986).

[40] W. Sharfin, K. E. Johnson, L. Wharton and D. H. Levy, *J. Chem. Phys.* **71**, 1283 (1979).

[41] S. K. Gray, S. A. Rice and M. J. Davis, *J. Phys. Chem.* **90**, 3470 (1986). See also S. A. Rice and M. Zhao, *Int. J. Quant. Chem.* **58**, 593 (1996).

Problems

Problem 3.1. The Hamiltonian of a classical harmonic oscillator of mass m and force constant k is

$$\frac{p^2}{2m} + \frac{kq^2}{2} = E$$

which at constant E represents an ellipse in the p–q phase space (Fig. 3.5). Show that the phase-space "volume" integral $\iint \mathrm{d}p \, \mathrm{d}q \, \theta(E - H)$, in this case the actual area of the

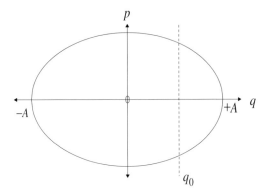

Fig. 3.5. The phase space of a classical harmonic oscillator at total energy E. The ellipse represents the equation $p^2/(2m) + kq^2/2 = E$, where m is mass and k is a force constant. The limits on the coordinate q at E are $\pm(2E/k)^{1/2}$, designated as $+A$ and $-A$. At energy $E_0 = E$, the corresponding coordinate value is $q_0 = (2E_0/k)^{1/2}$, which is shown as a dashed line.

ellipse, is classically $2\pi E \sqrt{m/k}$, which has the dimension energy × time. Show that similarly the "surface" integral is $\iint \mathrm{d}p\mathrm{d}q \, \delta(E - H) = 2\pi \sqrt{m/k}$, the dimension of which is time.

   In order to get the analogous semi-classical results for the sum and density of states, the dimensions of which are (pure) number and energy^{-1}, respectively, it is therefore necessary to divide the classical results by $h(=$ energy × time$)$.

Problem 3.2. Consider a classical oscillator with cut-off (see Fig. 2.1) at energy $E > E_0$. The phase space of this oscillator is shown in Fig. 3.5, with bound states to the left of the dashed line and dissociated states to the right. Show that the lifetime to dissociation t (i.e. the inverse of the rate constant) of this oscillator is

$$t = \frac{1}{2\nu} + \frac{1}{\pi \nu} \, \cos^{-1} \left[(1 - E_0/E)^{1/2}\right]$$

if the activation is such that the entire phase space is visited (i.e. the equivalent of the microcanonical ensemble), that is, along the portion of the circumference of the ellipse to the left of the dashed line, passing over both positive and negative momenta. If only part of the phase space is visited, the lifetime will be shorter. At threshold $E = E_0$ and $\cos^{-1}(0) = \pi/2$, so that $t = 1/\nu$. The same results were deduced above by a different argument.

Problem 3.3. The rotational energy $E_r(J, K)$ of a prolate symmetric top with rotational constants $A > B$ is given by Eq. (A1.18) in Appendix 1. Show that the average of $E_r(J, K)$ over all K is

$$\langle E_r(J, K)\rangle_K = J(J + 1)\frac{1}{3}(A + 2B)$$

which amounts to J-rotor energy with a re-defined rotational constant.

Problem 3.4. Show that, for the potential $V(r) = -C_n(r_e/r)^n$, the equivalent of Eq. (3.32) is

$$E_0^* = E_0 + E_r \left[\frac{n-2}{n} \left(\frac{2E_r}{nC_n} \right)^{2/(n-2)} - 1 \right]$$

Problem 3.5. Starting with Eq. (3.40), show that

$$\langle \epsilon_t \rangle = \frac{\int_0^{E-E_0} G^*(x)\,dx}{G^*(E-E_0)}$$

Using the semi-classical expression for the sum of states (Eq. (4.23)), show that, for a transition state having v vibrational degrees of freedom and zero-point energy E_z, this expression leads to

$$\langle \epsilon_t \rangle = \frac{(E - E_0 - E_z)}{v+1}$$

This means that (statistical) $\langle \epsilon_t \rangle$ should increase linearly with excess energy, as is found experimentally in several cases [31].

4

Calculation of energy-level densities

In a statistical theory, such as the one that is the subject of the present textbook, it is implicitly admitted that no information exists as to the probability of any particular quantum state. Under such circumstances, the only workable hypothesis is that all states are equally probable, which is the essence of the microcanonical ensemble posited in the previous chapter; the consequence is that it is sufficient just to count the number of states under specified constraints. In the present chapter we shall consider only the constraint that the total energy shall have a specified value. The additional constraint of total angular momentum is the subject of Chapter 8.

4.1 General considerations

4.1.1 Sources of data

The degree of detail with which the counting of states can be performed depends on the molecular parameters available for a given molecule. As a minimum, vibrational frequencies and rotational constants (or moments of inertia) are necessary in order to calculate rotational and vibrational sum of states, in the harmonic and free-rotor approximations (e.g. Eq. (4.97) and most of Section 4.5). Less frequently, we may have information about the anharmonicity of vibrations (Section 4.5.4), the potential that hinders rotations (Section 4.3), and electronic states.

These parameters are generally obtained from the analysis of infrared and Raman spectra, as described in the classic book by Herzberg [1]. Other sources of data include various NIST databases (NIST is the National Institute of Standards and Technology, Gaithersburg, MD, USA), notably NIST JANAF Thermochemical Tables, Beilstein and Gmelin Databases (for organic and inorganic molecules, respectively), and an older compilation by Shimanouchi [2]. New data are published from time to time in the *Journal of Physical and Chemical Reference Data*. At present the methods of quantum mechanics are sufficiently reliable for the calculation of molecular parameters of stable molecules for which no experimental determination exists.

At energies of chemical interest the number of states in polyatomic molecules, as seen in infrared or Raman spectra, is often very large, or difficult to identify, so that there is no simple way to determine the number of states experimentally even in a molecule as small as NO_2. The solution is to use data derived from experiment (e.g. harmonic frequencies) to perform an exact count, which then serves as a benchmark in lieu of an actual experimental count, as illustrated in Fig. 4.1 and Table 4.1.

4.1.2 Definitions and notation

The symbol E will represent energy (in units of wavenumbers (cm^{-1})), understood throughout to be in excess of the zero-point energy, unless specifically mentioned otherwise. The lowest possible energy of a system is therefore $E = 0$. The following notation is used: $W(E)$ (dimensionless) represents the number of quantum states of the given system *at* energy E. If the ground state of a system is non-degenerate, we have $W(0) = 1$. A quantity related to $W(E)$ is the quantum-mechanical partition function, which may be considered to be the weighted sum of $W(E)$'s:

$$Q = \sum_i W(E_i) \exp[-E_i/(kT)] \tag{4.1}$$

where E_i are the energy levels of the system, k is Boltzmann's constant, T is temperature, and the sum is over energies of all levels. The cumulant of $W(E)$, i.e. the *total* number of states between 0 and E, will be designated by $G(E)$ (likewise dimensionless):

$$G(E) = \sum_0^E W(E) \tag{4.2}$$

The continuum analog of the discrete partition function Q in Eq. (4.1) is defined semi-classically by

$$Q = \int_0^\infty N(E) \exp[-E/(kT)] \, dE \tag{4.3}$$

where $N(E)$ (dimension: energy^{-1}) is the density of states, i.e. the number of states per unit energy range at E. It is related to the number of states through δE, the spacing between two adjacent energy levels around E:

$$W(E) = N(E) \delta E \tag{4.4}$$

Here and in Eq. (4.3) both $N(E)$ and E are considered to be smooth continuous variables. A related quantity is the continuum analog of $G(E)$ in Eq. (4.2), the total number of states at E, defined as the integral of $N(E)$:

$$G(E) = \int_0^E N(E)\,\mathrm{d}E \tag{4.5}$$

Conversely, $N(E)$ can be regarded as the derivative of $G(E)$:

$$N(E) = \frac{\mathrm{d}G(E)}{\mathrm{d}E} \tag{4.6}$$

These relations are quite general and apply to any type of degree of freedom. In order to identify the nature of the partition function, or sum or density of states, subscripts "v," "r," and "t" will be used when necessary to distinguish among vibrational, rotational, or translational degrees of freedom, respectively. Similarly, the subscript "vr" will identify combined (but *not* necessarily coupled) vibrational and rotational degrees of freedom. Occasionally these subscripts may be given numerical values, e.g. $r = 2$ will signify two rotational degrees of freedom.

The notation $G(E)$ used here for the sum of states follows that of Tolman [3, pp. 492ff], but there is no agreement in the literature on uniform notation. Several authors use $N(E)$ or $W(E)$ for the *sum* of states (Eq. (4.2)) and $\rho(E)$ for the density (Eq. (4.6)).

4.1.3 Discrete energy spectra

If a system is highly quantized, as in the case of vibrational degrees of freedom, the energy spectrum will be highly discrete, as shown in Table 4.1 and illustrated in Fig. 4.1, which represents a typical case. The number of vibrational states $W_v(E)$ as a function of E is seen to be a "stick" spectrum, the height of each "stick" being equal to the number of states at that particular energy. The discontinuous line in Fig. 4.2 represents the "exact" $G_v(E)$ of Table 4.1 obtained by an exact count according to Eq. (4.2), for the same system as that in Fig. 4.1, which shows graphically that $G_v(E)$ is a staircase function, which, at low energies, is very much unlike the smooth function implied by Eq. (4.5).

In this context, the concept of an "exact" density of vibrational states requires a comment. For $G_v(E)$ as the staircase function in Fig. 4.2, the corresponding "exact" density $N_v(E)$ from Eq. (4.6) will be a string of weighted delta functions. Since this form is inconvenient for rate-constant calculations, an alternative formulation would be to consider a range ΔE of energies and define $N_v(E) \approx \Delta G_v(E)/\Delta E$, an example of which is given in Table 4.1. However, then the "exact" density becomes ill-defined [4, 5], depending as it does on the chosen step length or "graining" ΔE, i.e. the difference $\Delta G_v(E)$ between $G_v(E)$ at two different values of E (assuming that there are actually states at the chosen energies). In keeping with the

Table 4.1. *An exact harmonic state count in methane for $E < 5300\,cm^{-1}$ at "frequencies" ω_i in cm^{-1} (degeneracies) 1306 (3), 1534 (2), 2917, and 3020 (3)*

n_1	n_2	n_3	n_4	$E\,(\mathrm{cm}^{-1})$	$W_v(E)$	$G_v(E)$	$N_v(E)\,((\mathrm{cm}^{-1})^{-1})$
0	0	0	0	0	1	1	–
1	0	0	0	1306	3	4	0.002 297
0	1	0	0	1534	2	6	0.008 772
2	0	0	0	2612	6	12	0.005 566
1	1	0	0	2840	6	18	0.0263
0	0	1	0	2917	1	19	0.012 987
0	0	0	1	3020	3	22	0.029 126
0	2	0	0	3068	3	25	0.0625
3	0	0	0	3918	10	35	0.011 765
2	1	0	0	4146	12	47	0.052 632
1	0	1	0	4223	3	50	0.038 961
1	0	0	1	4326	9	59	0.087 379
1	2	0	0	4374	9	68	0.1875
0	1	1	0	4451	2	70	0.025 974
0	1	0	1	4554	6	76	0.058 252
0	3	0	0	4602	4	80	0.083 333
4	0	0	0	5224	15	95	0.024 116

n_i ($i = 1, 2, 3, 4$) is the number of quanta in oscillator i (oscillators i in order are given in the heading).
$N_v(E_{k+1})$ is defined as $[G_v(E_{k+1}) - G_v(E_k)]/(E_{k+1} - E_k)$.

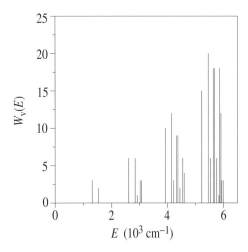

Fig. 4.1. The energy dependence of $W_v(E)$ for vibrational states in methane, showing exact harmonic counts for energies between 0 and 6000 cm^{-1}. The "frequencies" in cm^{-1} are 3020 (3), 2917, 1534 (2), and 1306 (3) (with degeneracies in parentheses).

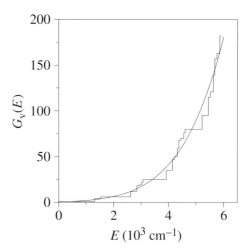

Fig. 4.2. The energy dependence of the total number of vibrational states $G_v(E)$ in methane between 0 and 6000 cm^{-1}. The discontinuous staircase line is the exact harmonic $G_v(E)$ obtained as the cumulant of $W_v(E)$ in Table 4.1; the continuous line is a smooth-function approximation calculated from the first-order approximation $I_1(E, k = 1)$ (Eq. (4.69)). The second-order approximation $I_2(E, k = 1)$ (Eq. (4.70)) would be indistinguishable on the scale of the graph.

semi-classical nature of the RRKM theory, which has been emphasized in Section 3.2, and to avoid the wild oscillations of such "exact" $N_v(E)$, it is necessary to consider $N(E)$ (for whatever degree of freedom) as basically a semi-classical quantity, defined as the derivative of a *smooth*-function version of $G(E)$.

Therefore the sum of states $G(E)$, regarded as the primary information, will henceforth be the center of interest, while the density of states $N(E)$ will be considered a secondary derived information, always obtainable as the derivative of $G(E)$ via Eq. (4.6).

4.1.4 Methods of calculation

There are in principle three approaches for the calculation of the sum of states.

(i) Exact enumeration of quantum states, i.e. as the cumulant of $W(E)$ (Eq. (4.2)). Two methods of exact enumeration are discussed in Section 4.5 in connection with the determination of the vibrational sum of states.

(ii) Evaluation of the volume in phase space. The sum of states for a system of N degrees of freedom at total energy E is defined as the volume in phase space, which is given semi-classically by

$$G(E) = \frac{1}{h^N} \int \cdots \int_{0 \leq H \leq E} \prod_{i=1}^{N} dq_i \, dp_i \qquad (4.7)$$

(cf. Eq. (3.16)). For the Hamiltonian $H = H(\mathbf{q}, \mathbf{p})$, the integration (or summation in the quantum analog) is over all p_i's and q_i's such that $0 \leq H \leq E$. Equation (4.7) is completely general; it looks quite straightforward, but in fact is difficult to evaluate for all but the simplest systems (Problem 4.1). As an illustration that has some pedagogical merit, Appendix 1 uses Eq. (4.7) to calculate $G(E)$ for a number of simple semi-classical cases of interest in the present context.

(iii) One of the approximate methods. The semi-classical nature of the microcanonical rate constant $k(E)$ implies – as shown in Chapter 3 – a quasi-continuous density and sum of states. It is therefore necessary and sufficient to obtain reasonably accurate smooth-function approximations rather than execute a computationally intensive exact count [6]. Exact counting of states (which requires additional smoothing) gives an illusion of superior accuracy, for this in no way corrects or improves the inherent semi-classical nature of the theory.

Several approximate methods, of varying accuracy and complexity, for the counting of energy levels have been proposed over the years, most of which are discussed in [7, Chapter 6] and [8], and will not be repeated here. Instead the focus here will be on the Laplace-transform method which is based on the inversion of the partition function. It is the most accurate and most general, and remarkably simple to use. The development given here assumes a rudimentary knowledge of the general properties of the Laplace transformation, as summarized in Appendix 2.

4.1.5 The Laplace transform

Operationally, Eq. (4.3) can be considered to be the Laplace transform $\mathcal{L}$ of the density $N(E)$ with $s = 1/(\mathcal{k}T)$ as the transform parameter; conversely $N(E)$ is then the inverse transform $\mathcal{L}^{-1}$ of the partition function $Q(s)$ considered as a function of s:

$$Q(s) = \mathcal{L}\{N(E)\}; \quad N(E) = \mathcal{L}^{-1}\{Q(s)\} \tag{4.8}$$

The integration theorem of Laplace transformation gives directly, in view of Eq. (4.5) (Appendix 2, Section A2.2),

$$G(E) = \mathcal{L}^{-1}\left\{\frac{Q(s)}{s}\right\} \tag{4.9}$$

so that Eqs. (4.8) and (4.9) can be combined by means of the parameter k (not to be confused with $\mathcal{k}$, which is Boltzmann's constant, or k, the rate constant, which is used only in Chapters 1 and 2) to give

$$\left.\begin{array}{l} N(E) \\ k=0 \\ G(E) \\ k=1 \end{array}\right\} = \mathcal{L}^{-1}\left\{\frac{Q(s)}{s^k}\right\} \tag{4.10}$$

Equation (4.9) implies that

$$\mathscr{L}\{G(E)\} = \int_0^\infty G(E) \exp[-E/(kT)] \, dE = kTQ \qquad (4.11)$$

a result that obviates the actual integration and will be useful later in connection with thermal reactions (Section 5.8). Equation (4.10) is a special case of a more general class of repeated integrals of $N(E)$ (Appendix 2, Section A2.2). In particular, the second integral of the density is

$$\mathscr{L}^{-1}\left\{\frac{Q(s)}{s^2}\right\} = \int_0^E G(E) \, dE \qquad (4.12)$$

which appears in the calculation of the average translational energy of fragment ions (cf. Problem 3.5) as well as of reactive cross-sections and rate constants of bimolecular reactions [9].

The interest of these manipulations is that classical rotational and quantum harmonic partition functions, which are generally adequate in most cases, are not difficult to compute and then to invert according to Eqs. (4.8) and (4.9). Anharmonicity and other complicating factors can be taken into account without much trouble as long as a partition function can be written down (cf. Section 4.5). Likewise the integrals of $G(E)$ are easier to obtain directly from Eqs. (4.11) and (4.12) than by numerical integration.

Although the actual calculation of the inverse Laplace transform for an explicit form of the partition function is deferred until Section 4.4, it is useful to consider first a number of important general results.

4.1.6 Convolution

The Laplace-transform technique allows a straightforward determination of the combined density or sum of states of two independent systems indexed 1 and 2. Suppose that the individual partition functions are Q_1 and Q_2, where $Q_1(s) = \mathscr{L}\{N_1(E)\}$ and $Q_2(s) = \mathscr{L}\{N_2(E)\}$; the partition function of the combined system is then simply the product $Q_{1,2}(s) = Q_1(s)Q_2(s)$ since the systems are independent. According to the convolution theorem (Appendix 2, Section A2.2)

$$N_{1,2}(E) = \mathscr{L}^{-1}\{Q_{1,2}(s)\}$$

$$= \mathscr{L}^{-1}\{Q_1(s)\, Q_2(s)\} = \int_0^E N_1(E-x)\, N_2(x) \, dx \qquad (4.13)$$

The combined density of states of two independent systems is thus obtained by the convolution of the *densities* of states of the two component systems. Since the LHS refers to the product of partition functions, the labeling of the systems is arbitrary, so that interchanging the indices under the integral sign leads to the same result.

For the determination of the sum of states, the convolution theorem gives

$$G_{1,2}(E) = \mathcal{L}^{-1}\left\{\frac{Q_{1,2}(s)}{s}\right\}$$

$$= \mathcal{L}^{-1}\left\{\frac{Q_1(s)Q_2(s)}{s}\right\} = \int_0^E G_1(E - x)\, N_2(x)\, dx \qquad (4.14)$$

so that the sum of states for the combined system is given by the convolution of the *density* of states of one part and the *sum* of states of the other part. Since it is immaterial whether we consider the inverse transform in Eq. (4.14) as that of $[Q_1(s)/s]Q_2(s)$ or of $Q_1(s)[Q_2(s)/s]$, permutation of the indices under the integral sign again has no effect on the result.

In view of Eq. (4.4), $N_2(x)\, dx$ in Eqs. (4.13) and (4.14) can be replaced by $W_2(x)$, the discrete *number* of states in the interval dx, so that the integrals become summations. We then have the discrete analog of $G_{1,2}(E)$ (Eq. (4.14)):

$$G_{1,2}(E) = \sum_{x=0}^{E} G_1(E - x)\, W_2(x) \qquad (4.15)$$

Similarly, Eq. (4.13) yields

$$N_{1,2}(E) = \sum_{x=0}^{E} N_1(E - x)\, W_2(x) \qquad (4.16)$$

If both sides of Eq. (4.16) are further multiplied by dx, we have the fully discrete analog of $N_{1,2}(E)$ (Eq. (4.13)):

$$W_{1,2}(E) = \sum_{x=0}^{E} W_1(E - x)\, W_2(x) \qquad (4.17)$$

Again, permutation of the indices in the summations has no effect on the result.

The relations (4.15)–(4.17) are useful when one or both component systems are highly quantized, whereas (4.13) and (4.14) apply when both component systems are semi-classical.

4.2 Semi-classical degrees of freedom

Quantization, i.e. the discreteness of the energy spectrum, is small enough in semi-classical degrees of freedom that energy becomes virtually a continuous variable, as in a purely classical system. In the present context this refers principally to rotations for which the spacing between energy levels is typically $\approx 5\,\text{cm}^{-1}$ or less (except in H_2), which is so small that rotations can be assumed to behave classically without significant error. This is even more true of translations, and is in contrast with vibrations, for which the spacing may be $\approx 1000\,\text{cm}^{-1}$ or more. Nevertheless, as a concession to quantum mechanics, there will appear, in what are basically classical formulas, Planck's constant h and (in rotations) the symmetry number σ; hence the prefix "*semi*" (cf. Section 3.2.2).

The absence of significant quantization in these semi-classical degrees of freedom has the advantage that summations can be replaced without much error by integrations that lead to closed-form analytical formulas. Several of these are derived below, using throughout rotational constants defined in energy units.

4.2.1 Independent free rotors

By means of a semi-classical version of Eq. (4.7), it is shown in Eq. (A1.13) of Appendix 1 that the sum of states for r free rotations can be put in the compact analytical form

$$G_r(E) = \frac{q_r E^{r/2}}{(r/2)!} \tag{4.18}$$

In this representation an r-dimensional rotor is counted as r rotations. Here q_r stands for the corresponding partition function (Q_r) from which the kT term has been dropped: $q_r = Q_r(kT)^{-r/2}$; hence q_r has the dimension energy$^{-r/2}$. The same result for $G_r(E)$ is obtained more directly from the inverse Laplace transform of $Q_r(s)/s$ ($s = 1/(kT)$) as $q_r \mathcal{L}^{-1}\{s^{-(1+r/2)}\}$ (Appendix 2, Section A2.2). The density of rotational states then follows by differentiation, or alternatively by taking the inverse transform of $Q_r(s)$. By means of the parameter k the two results can be put in the compact form

$$\left.\begin{array}{l} N_r(E) \\ k=0 \\ \\ G_r(E) \\ k=1 \end{array}\right\} = \frac{q_r E^{k-1+r/2}}{(k-1+r/2)!} \tag{4.19}$$

Note that, for two rotations ($r = 2$) the density ($k = 0$) of rotational states is a constant independent of energy: $N_r(E) = q_r$, while the sum of states ($k = 1$) is proportional to energy: $G_r(E) = q_r E$.

In order to appreciate the error of the analytical formula of Eq. (4.19), the sum $G_r(E)$ can be evaluated exactly. For a single two-dimensional rotor ($r = 2$) with rotational constant B and symmetry number σ, we have $q_r = 1/(\sigma B)$ (Appendix 1). Let J be the rotational quantum number, and let J^* be the nearest integer solution of $E = J(J + 1)B$, i.e. $J^* = -\frac{1}{2} + \frac{1}{2}\sqrt{1 + 4E/B}$. For each J there are $2J + 1$ states, so that (including the symmetry number)

$$G(E)_{\text{exact}} = \sum_{J=1}^{J^*}(2J + 1) = (1 + J^*/\sigma)(1 + J^*)$$

$$= \frac{E}{\sigma B} + \frac{1}{2}\left(1 + \sqrt{1 + 4E/B}\right) \tag{4.20}$$

For $r = 2$, Eq. (4.19) yields $G(E) = E/(\sigma B)$, and $G(0) = 0$, in both cases a small undercount with respect to the above exact result. The difference is an additive factor that increases only as the square root of energy.

For a single one-dimensional rotor ($r = 1$) with rotational constant A, $q_r = (1/\sigma)(\pi/A)^{1/2}$ (Appendix 1). Let m^* be the nearest integer solution to $E = m^2 A$, i.e $m^* = \pm(E/A)^{1/2}$. Every energy level except the first ($m^* = E = 0$) is therefore doubly degenerate; in addition, for reasons of symmetry, only every σth m^* will appear, so that (using a degeneracy of 2 for each $m^* > 0$, the symbol $\Sigma 1$ meaning the sum is equal to the number of its terms),

$$G(E)_{\text{exact}} = 1 + \frac{2}{\sigma}\sum_1^{m^*}1 = 1 + 2m^*/\sigma = 1 + \frac{2}{\sigma}\left(\frac{E}{A}\right)^{1/2} \tag{4.21}$$

which differs from Eq. (4.19) for $r = 1$ only by the additive constant of unity at all energies. The result $G(0) = 0$ at $E = 0$ (instead of the correct quantum result $G(0) = 1$) is common to all semi-classical approximations, including the two-dimensional rotor in Eq. (4.20).

Observe that the density of states for a one-dimensional rotor, obtained as the derivative of Eq. (4.21) with respect to E, is $N(E) = (1/\sigma)(EA)^{-1/2}$, which goes off to infinity at $E = 0$. Fortunately this is of little consequence: while the density of states at zero energy is not required in rate calculations, the smooth-function approximation remains reasonable at energies as low as the lowest rotational level, as shown in Section 4.6.

4.2.2 Coupled rotors

In equations for the rotational sum (or density) of states given so far it has been assumed that rotations are independent, that is, sums or densities of states for rotors of different dimensionalities combine according to the rules of convolution, i.e.

subject only to the total available energy (Problem 4.2). This is not necessarily the case, an example being the three-dimensional rotor known as a symmetric top, which is discussed in some detail in Appendix 1, Section A1.4.

Briefly, the rotational energy of such a top, with rotational constants A and B, is characterized by quantum numbers J and K:

$$E_r(J, K) = J(J + 1)B + (A - B)K^2 \qquad (4.22)$$

which applies both to a prolate top ($A > B$) and to an oblate top ($B > A$). The quantum number J refers to a two-dimensional rotor with what may be taken as a doubly degenerate rotational constant B. The quantum number K refers to the component of angular momentum along the molecule-fixed axis, and refers to a one-dimensional rotor with rotational constant $A - B$. K is constrained to the interval $-J \leq K \leq +J$ so that, in the symmetric top, the one-dimensional rotor (the "K-rotor") is *coupled* (not combined) with a two-dimensional rotor (the "J-rotor"). By coupling, in the more general sense, is meant that the energy of a degree of freedom depends on the energy (or quantum number) of another.

For simplicity, one frequently makes the approximation that the K-rotor is separable, i.e. that the J- and K-dependent parts of $E_r(J, K)$ can be dealt with independently (Problem 4.3), as discussed in Section 3.3.2 (Eq. (3.29)). The question of how to apportion the overall symmetry number of the separable symmetric top then arises. In such a case, the symmetry number may be thought of for convenience as the product of two parts, the "symmetry number" σ_J of the J-rotor, and the "symmetry number" σ_K of the K-rotor, such that $\sigma = \sigma_J \sigma_K$. For example, in rigid ethane with $\sigma = 6$, the symmetry number may be considered as composed of $\sigma_J = 2$ and $\sigma_K = 3$, the latter justified by noting that the K-rotor refers to the top symmetry axis which has a threefold symmetry for rotation about the C–C bond.

Of practical importance in rate calculations is the effect of the J–K coupling on the number of vibrational–rotational states, which is deferred until Section 4.5.

4.2.3 Independent harmonic oscillators

For v harmonic oscillators, where ν_i is the frequency of the ith oscillator, Eq. (4.7) yields $G_v(E)$ given by Eq. (A1.2) in Appendix 1, known as the "classical" formula. It has the advantage of providing analytical results for combination with other degrees of freedom (Problem 4.4). For a polyatomic molecule the classical formula is a particularly poor approximation at all energies (Table 4.2). The problem is principally the neglect of the zero-point energy $E_z = \frac{1}{2}\Sigma_i h\nu_i$ of the v oscillators, so that a better approximation (known as the "semi-classical"

Table 4.2. *Values for the vibrational sum of states $G_v(E)$ obtained by using various approximations*

				Steepest-descents method	
$E(\text{cm}^{-1})$	Exact	Classical	Semi-classical	First order	Second order
		Small molecule: NO_2			
750	2	0.043	1.81	1.71	1.50
1500	4	0.35	3.87	3.71	3.45
3000	12	2.77	11.7	11.6	11.1
4018	21	6.66	20.8	20.7	20.0
4902	32	12.1	31.7	31.8	30.8
		Large molecule: C_2H_6			
578	3	4.8×10^{-24}	717.2	2.97	2.58
995	7	8.4×10^{-20}	1125.5	7.62	6.93
2509	122	1.4×10^{-12}	5287.9	125.9	120.5
4999	3769	3.5×10^{-7}	52047	3850.9	3750.4

Energies (in excess of the zero-point energy) at which there are actually states by direct counting are chosen.
Classical approximation: Eq. (A1.2) in Appendix 1; semi-classical: Eq. (4.23); and steepest-descents: Eqs. (4.69) and (4.87).
Frequencies (in cm^{-1}) are 1325, 1634, and 750 for NO_2; those for C_2H_6 from Table 3.1.

approximation) is [11]

$$\left.\begin{array}{l} N_v(E) \\ k=0 \\ G_v(E) \\ k=1 \end{array}\right\} = \frac{(E + E_z)^{v-1+k}}{(v-1+k)! \prod_{i=1}^{v} h\nu_i} \tag{4.23}$$

A still better approximation is obtained by replacing E_z with aE_z, where a is an empirical energy-dependent factor [12]. Using the convolution procedure as in Problem 4.4, Eq. (4.23) can be generalized for the semi-classical density or sum of states of a combined system of v oscillators and r rotors :

$$\left.\begin{array}{l} N_{vr}(E) \\ k=0 \\ G_{vr}(E) \\ k=1 \end{array}\right\} = \frac{q_r(E + aE_z)^{v-1+k+r/2}}{(v-1+k+r/2)! \prod_{i=1}^{v} h\nu_i} \tag{4.24}$$

Note that, with $q_r = 1$, the *density* of states ($k = 0$) for a system of oscillators and two rotors ($r = 2$) is formally equivalent to the *sum* of vibrational states ($k = 1$, $r = 0$) for the same system. This is a quite general result.

4.2.4 Independent anharmonic oscillators

A good model of an anharmonic oscillator is the Morse classical oscillator [13] with energy levels given by

$$\epsilon = \left(n + \frac{1}{2}\right)\omega - \left(n + \frac{1}{2}\right)^2 \omega\chi \tag{4.25}$$

where n is the quantum number, $\omega = h\nu$ is the vibrational quantum (in the following called for short "frequency" in energy units, while the symbol ν is reserved for the actual frequency in s^{-1}) and χ is the (dimensionless) anharmonicity coefficient which is related to D, the classical dissociation energy, by $D = \omega/(4\chi)$. The energies ϵ (including D) are referred to potential minimum as the zero of energy, so that $\epsilon = E + E_z$, where $E_z = \frac{1}{2}\omega(1 - \frac{1}{2}\chi)$.

On solving Eq. (4.25) for n we have

$$n = \frac{2D}{\omega}\left(1 - \sqrt{1 - \frac{\epsilon}{D}}\right) - \frac{1}{2} \tag{4.26}$$

The sum of states at $\epsilon < D$ (but assigned to E) is then $G(E) = n + \frac{1}{2}$ from which the density of states of a single oscillator is directly

$$N(E) = \frac{dn}{d\epsilon} = \frac{1}{\omega\sqrt{1 - \epsilon/D}} \tag{4.27}$$

The combined density of states of two Morse oscillators, indexed 1 and 2, has to be obtained by convolution (Eq. (4.13)), with the result

$$N(E) = \frac{(D_1 D_2)^{1/2}}{\omega_1 \omega_2}\left[\sin^{-1}\left(\frac{\epsilon - D_1 + D_2}{a}\right) - \sin^{-1}\left(\frac{D_2 - D_1 - \epsilon}{a}\right)\right] \tag{4.28}$$

where $a = D_1 + D_2 - \epsilon$, and $\epsilon < D_1, D_2$. For the combined density of states of three Morse oscillators this result would have to be convoluted further, with a sufficiently messy result to discourage further calculations.

It is therefore obvious that a better and more general approach is needed, one that would consider explicitly the discreteness of the energy spectrum, and at the same time provide smoothing, as illustrated in Fig. 4.2. It is not difficult to appreciate that the evaluation of the corresponding phase-space volume by use of the quantum equivalent of Eq. (4.7) would be impractical, so Sections 4.4 and 4.5 are devoted to other methods for obtaining reasonably accurate results for the sum and density of vibrational states.

4.3 Hindered rotors

In some molecules rotations are hindered by the presence of a periodic potential hill that opposes free rotation. The result is of course that the formulas derived above for free rotations no longer apply, in particular for energies below the potential hill. The calculation is more involved but, in the simpler cases, it is nevertheless possible to obtain an analytical solution by inversion of the corresponding partition function.

A common example of a one-dimensional hindered rotor is the internal rotation about the C—C axis in molecules like C_2H_6 and C_2H_4 in which there is rotation of one CH_3 or CH_2 group relative to the other. This rotation is opposed by a potential hill, which, as a function of the torsional angle, has three identical minima in the case of ethane, and two identical minima in the case of ethylene. The potential hill in ethylene is considerably larger (≈ 9000 cm^{-1} [1, p. 227]) than that in ethane (≈ 1000 cm^{-1} [14]).

Since the case of the two-dimensional hindered rotor is somewhat simpler to deal with, it will be discussed first. Subscripts "*nf*" and "*nh*" ($n = 1$ or 2) will serve to distinguish properties of free ("f") and hindered ("h") one- and two-dimensional rotors, respectively.

4.3.1 A two-dimensional hindered rotor

In its most general form, the semi-classical partition function in polar coordinates for a two-dimensional hindered rotor at temperature T is

$$Q_{2h} = \frac{1}{h^2} \int \cdots \int \exp[-H/(kT)] \, dp_\phi \, dp_\theta \, d\phi \, d\theta \qquad (4.29)$$

where h is Planck's constant and H is the Hamiltonian

$$H = \frac{p_\theta^2}{2I} + \frac{p_\phi^2}{2I \sin^2 \theta} + V(\theta, \phi) \qquad (4.30)$$

in which I is the moment of inertia of the hindered rotor and $V(\theta, \phi)$ is the hindering potential.

If the hindering potential is assumed to be independent of angle ϕ, integration with respect to ϕ ($0 \le \phi \le 2\pi$) and the momenta p_θ and p_ϕ (both from $-\infty$ to $+\infty$) can be performed immediately, with the result

$$Q_{2h} = \frac{kT}{2\sigma B} \int_0^\pi \exp[-V(\theta)/(kT)] \sin \theta \, d\theta \qquad (4.31)$$

where $B = \hbar^2/(2I)$ is the rotational constant ($\hbar = h/(2\pi)$) and σ is the symmetry number. If $V(\theta) = 0$ we recover the two-dimensional *free*-rotor partition function $Q_{2f} = kT/(\sigma B)$ (see Appendix 1).

A hindering potential $V(\theta)$ with a maximum at π (e.g. that used to model the ion–dipole interaction in $CH_3^+ + HCN$ [16]) is

$$V(\theta) = \frac{1}{2}V_0(1 - \cos\theta) \tag{4.32}$$

where V_0 is a constant representing the barrier height. When it is substituted into Eq. (4.31) this yields for the partition function

$$Q_{2h} = \frac{(kT)^2}{\sigma B V_0}\{1 - \exp[-V_0/(kT)]\} \tag{4.33}$$

from which the sum of states can be obtained [17] in terms of the transform parameter $s = 1/(kT)$, by taking the inverse Laplace transform of $Q_{2h}(s)/s$. With $q_{2f} = Q_{2f}/(kT) = 1/(\sigma B)$, we have (Appendix 2, Section A2.2)

$$G(E) = \mathcal{L}^{-1}\left\{\frac{Q_{2h}(s)}{s}\right\} = \begin{cases} q_{2f}\dfrac{E^2}{2V_0} & \text{for } E < V_0 \\[2ex] q_{2f}\left(E - \frac{1}{2}V_0\right) & \text{for } E > V_0 \end{cases} \tag{4.34}$$

The density of states follows immediately by taking the derivative of $G(E)$ with respect to E: $N(E) = q_{2f}(E/V_0)$ for $E < V_0$, and $N(E) = q_{2f}$ for $E > V_0$.

The case of the CH_3 moiety in the process $CH_4 \rightarrow H + CH_3$ is an example of two-dimensional hindered rotation of particular interest. This fragmentation can be considered as a process during which the hindering potential decreases progressively from a very high value in the reactant ($\approx 37\,000\,cm^{-1}$ in CH_4 [15], essentially a potential for harmonic vibrations) to zero (i.e. free rotations) in the products $H + CH_3$. Modes that evolve from vibrations to rotations are called transitional modes, and are more fully discussed in Chapter 7.

Such two-dimensional hindered rotation can be modeled with the potential [15]

$$V(\theta) = V_0 \sin^2\theta \tag{4.35}$$

which has a maximum at $\pi/2$. With this form of $V(\theta)$ the integral in Eq. (4.31) cannot be obtained in closed form. Expanding the exponential and term-by-term integration yields the partition function for a two-dimensional hindered rotor as the infinite series [15]

$$Q_{2h} = Q_{2f}\sum_{n=0}^{\infty}\left(-\frac{2V_0}{kT}\right)^n\left(\prod_{i=0}^{n}(1+2i)\right)^{-1} \tag{4.36}$$

In principle, this series for Q_{2h}, as a function of $s = 1/(kT)$, could be Laplace-inverted term by term to obtain the density of states (see Section 4.5). A direct route to the sum of states is to note that from Eqs. (4.31) and (4.35) we have formally, again with $q_{2f} = Q_{2f}/(\ell T)$,

$$G(E) = \frac{1}{2}q_{2f} \, \mathcal{L}^{-1} \left\{ \frac{1}{s^2} \int_0^\pi [\exp(-s V_0 \sin^2 \theta)] \sin \theta \, d\theta \right\} \quad (4.37)$$

By the device of exchanging the order of integration and Laplace inversion [18], the sum of states can be obtained analytically, even though no closed-form analytical formula is available for the partition function. With $a = V_0 \sin^2 \theta$, the inverse tranform is (Appendix 2, Section A2.2)

$$\mathcal{L}^{-1} \left\{ \frac{\exp(-s V_0 \sin^2 \theta)}{s^2} \right\} = \begin{cases} 0 & \text{for } 0 < E < a \\ E - a & \text{for } E \geq a \end{cases} \quad (4.38)$$

so that, for $E \geq a$,

$$G(E) = \frac{1}{2}q_{2f} \int_0^\pi (E - V_0 \sin^2 \theta) \sin \theta \, d\theta \quad (4.39)$$

which yields directly

$$G(E) = q_{2f} \left(E - \frac{2}{3} V_0 \right), \quad E \geq V_0 \quad (4.40)$$

When $E < a$, the limiting angle θ_0 for which $E = V_0 \sin^2 \theta$ is given by $\sin \theta_0 = \sqrt{E/V_0}$ [18]. This means that, between θ_0 and $\pi - \theta_0$, we have $E < a$, and therefore this part of the integral in Eq. (4.37) is zero, while outside these limits we have $E \geq a$, where the integral is finite and is given again by Eq. (4.39), but with the integrand now in two parts, the first with integration limits from 0 to θ_0, and the second from $\pi - \theta_0$ to π. On account of the symmetry of the sine function between 0 and π, the integrals of the two parts are equal, with the result

$$G(E) = q_{2f} \left\{ E + (V_0 - E)\sqrt{1 - E/V_0} \right.$$
$$\left. - \frac{1}{3} V_0 [2 + (1 - E/V_0)^{3/2}] \right\} \quad \text{for } E < V_0 \quad (4.41)$$

At $E = V_0$, both Eq. (4.40) and Eq. (4.41) yield $G(E) = q_{2f} V_0/3$.

The density of states then follows directly by differentiation:

$$N(E) = \begin{cases} q_{2f} \left(1 - \sqrt{1 - E/V_0} \right) & \text{for } E < V_0 \\ q_{2f} & \text{for } E \geq V_0 \end{cases} \quad (4.42)$$

Note that, at energies above the barrier ($E > V_0$), the density (and at $E \gg V_0$ also the sum) of states is simply the density or sum of states for a free two-dimensional rotor.

Actual numerical application of these expressions to the fragmentation $CH_4 \rightarrow$ $CH_3 + H$ is discussed in Section 4.5.8 and shown further below in Fig. 4.8.

At very low energies, such that $E \ll V_0$, we have from Eqs. (4.41) and (4.42), after expansion of fractional powers to second order,

$$G(E) = q_{2f}\frac{E^2}{4V_0}, \qquad N(E) = q_{2f}\frac{E}{2V_0} \qquad (4.43)$$

which is equivalent to the sum and density of states, respectively, of a doubly degenerate harmonic oscillator with "frequency" $\omega = (2V_0/q_{2f})^{1/2}$.

The quantum equivalent of Q_{2h} (Eq. (4.36)) can be obtained only in the form of a rather complicated summation over individual energy levels [19].

4.3.2 A one-dimensional hindered rotor

Many years ago Pitzer [20] considered this problem as it applies to molecules in which symmetric tops are attached to a rigid frame, e.g. ethane and butene-2. The hindering potential function that has been found generally useful for these cases is (cf. Eq. (4.32))

$$V(\phi) = \frac{1}{2}V_0(1 - \cos(\sigma\phi)) \qquad (4.44)$$

where V_0 is again the height of the barrier. For one complete revolution ($0 \leq \phi \leq 2\pi$) of one moiety with respect to the other, this sinusoidal potential has σ equally spaced equivalent maxima and minima, with the minima midway between the members of each pair of maxima. The above form of $V(\phi)$ represents the lowest-order term in a Fourier expansion of the potential.

Before proceeding further it is instructive to obtain first the solution of the one-dimensional hindered rotor for the two extremes of low and high energies with respect to the height of the potential barrier V_0.

If we expand

$$\cos(\sigma\phi) = 1 - \frac{1}{2!}(\sigma\phi)^2 + \frac{1}{4!}(\sigma\phi)^4 - \cdots$$

for small ϕ and drop all terms beyond the quadratic, the potential simplifies to $V(\phi) \approx V_0\sigma^2\phi^2/4$, for which the solution of the corresponding Schrödinger equation is that for a harmonic oscillator with "frequency" (in cm^{-1}) $\omega = \sigma(AV_0)^{1/2}$, where A is the rotational constant of the one-dimensional rotor [21, p. 360]. This result for small ϕ is valid for oscillator energies near the bottom of the potential. Assuming this oscillator to behave classically, the sum of states is then (Appendix 1) $G(E) = E/\omega = E/[\sigma(AV_0)^{1/2}]$.

At energies above the barrier the solution of the Schrödinger equation yields energy levels of a one-dimensional free rotor but shifted downward by $\frac{1}{2}V_0$ [21, p. 359], sometimes referred to as the "Pitzer rotor" (cf. Fig. 4.3). Thus the sum of states at energy $E > V_0$ is proportional to $(E - \frac{1}{2}V_0)^{1/2}$ (cf. Eq. (4.19) with $r = 1$). The expectation is therefore that, with increasing energy, the nature of the hindered rotor will change from essentially that of a harmonic oscillator to that of a free rotor, with a change-over in the vicinity of $E \cong V_0$.

Discussed first will be the solution based on the classical partition function (Eq. (4.49) below), from which the analytical density and sum of states can be obtained by taking the inverse Laplace transform.

4.3.3 The classical solution

The classical partition function for this case is obtained from Eq. (4.29) by deleting the dimension with respect to angle θ; thus

$$Q_{1h} = \frac{1}{h} \int \int \exp[-H/(kT)]\, dp_\phi\, d\phi; \qquad H = \frac{p_\phi^2}{2I} + V(\phi) \qquad (4.45)$$

Integration with respect to the momentum p_ϕ in the expression for the partition function, Eq. (4.45), can be done immediately, so that, using Eq. (4.44) for the potential, we have

$$Q_{1h} = \frac{(2\pi I k T)^{1/2}}{h} \int_0^{2\pi/\sigma} \exp\{-V_0[1 - \cos(\sigma\phi)]/(2kT)\}\, d\phi \qquad (4.46)$$

As mentioned above, internal rotation involves in most cases the relative motion of two groups (with moments of inertia, say, I_1 and I_2) about their common bond. The moment of inertia I to be used in Eqs. (4.45) and (4.46) and in the rotational constant $A = \hbar^2/(2I)$ is the *reduced* moment of inertia given by $I = I_1 I_2/(I_1 + I_2)$. Using this definition of A, Eq. (4.46) becomes, in terms of new variables $b = V_0/(2kT)$ and $x = \sigma\phi$,

$$Q_{1h} = \left(\frac{kT}{\pi A}\right)^{1/2} \frac{\exp(-b)}{\sigma} \int_0^\pi \exp(b \cos x)\, dx \qquad (4.47)$$

Now

$$\int_0^\pi \exp(b \cos x)\, dx = \pi I_0(b) \qquad (4.48)$$

where $I_0(b)$ is the modified Bessel function of order zero [22, p. 376, item 9.6.16].

The final result is thus

$$Q_{1h} = Q_{1f}\exp(-b)\,I_0(b) \tag{4.49}$$

where $Q_{1f} = (1/\sigma)(\pi k T/A)^{1/2}$ is the partition function for the corresponding one-dimensional *free* rotor of symmetry number σ (see Appendix 1). Equation (4.49) represents the limiting classical case, i.e. the limit of high temperature or large moment of inertia, so that numerical values of thermodynamic functions calculated from Eq. (4.49) correspond in Pitzer's tables [20] to the column labeled "$1/Q_{1f} = 0$."

Since only the inverse transform of $\exp(-b)\,I_0(b)$ is available analytically, it is necessary to proceed in steps in the inversion of Q_{1h} of Eq. (4.49), taking advantage of the principle of convolution (Eq. (4.14)). This procedure was used in dealing with the three one-dimensional hindered rotors in the *tert*-butyl radical t-C_4H_9 [23]. With $a = V_0/2$ and $s = 1/(kT)$, we have [24, p. 297, item 12.3.1]

$$\mathcal{L}^{-1}\{\exp(-b)\,I_0(b)\} = \begin{cases} \dfrac{1}{\pi(V_0 E - E^2)^{1/2}}, & 0 < E < V_0 \\[2mm] 0, & E > V_0 \end{cases} \tag{4.50}$$

which yields $N_1(E)$, the density of states corresponding to $\exp(-b)\,I_0(b)$, considered as subsystem 1. Subsystem 2 is then the one-dimensional free rotor with partition function Q_{1f}, the sum of states of which is $G_2(E) = q_{1f}E^{1/2}/\frac{1}{2}!$. By the rules of convolution, the sum of states corresponding to Q_{1h} is the convolution of $N_1(E)$ with $G_2(E)$:

$$G(E) = \int_0^E N_1(x)\,G_2(E-x)\,\mathrm{d}x = \frac{q_{1f}}{\pi\frac{1}{2}!}\int_0^E \left(\frac{E-x}{x(V_0-x)}\right)^{1/2}\mathrm{d}x \tag{4.51}$$

The result of the integration is [25, p. 232, item 9, and p. 231, item 3]

$$G(E) = \begin{cases} \dfrac{4q_{1f}V_0^{1/2}}{\pi^{3/2}}\left[\mathscr{E}(E/V_0) - \left(1 - \dfrac{E}{V_0}\right)\mathscr{K}(E/V_0)\right], & E < V_0 \\[4mm] \dfrac{4q_{1f}E^{1/2}}{\pi^{3/2}}\mathscr{E}(V_0/E), & E > V_0 \end{cases} \tag{4.52}$$

where $\mathscr{K}(m)$ and $\mathscr{E}(m)$ are the complete elliptic integrals of the first and second kinds, respectively [22, p. 591, items 17.3.11 and 17.3.12], while $q_{1f} = (1/\sigma)(\pi/A)^{1/2}$ represents, as usual, $Q_{1f}/(kT)^{1/2}$. An example of the energy dependence of $G(E)$ calculated from Eq. (4.52) is shown in the upper panel of Fig. 4.3.

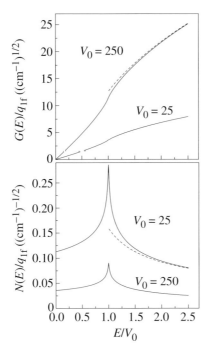

Fig. 4.3. Plots of $G(E)/q_{1f}$ (upper panel) and of $N(E)/q_{1f}$ (lower panel) versus E/V_0 for two values of the hindering potential barrier, $V_0 = 25\,\mathrm{cm}^{-1}$ and $V_0 = 250\,\mathrm{cm}^{-1}$ in a one-dimensional hindered rotor. The two panels have the same horizontal scale. Solid lines are calculated from Eqs. (4.52) and (4.53), respectively; dashed lines are calculated from the high-energy approximation $G(E)/q_{1f} = 2[(E - V_0/2)/\pi]^{1/2}$ and $N(E)/q_{1f} = 1/[\pi(E - V_0/2)]^{1/2}$ (the "Pitzer rotor"). Division by $q_{1f} = (1/\sigma)(\pi/A)^{1/2}$ renders $G(E)$ and $N(E)$ independent of the properties of the rotor. Note that, at $E/V_0 = 1$, the density $N(E)$ goes off to infinity, but the sum $G(E)$ remains finite with only a slight kink in the curve.

The density of states then follows by differentiation of $G(E)$ with respect to E:

$$
N(E) = \begin{cases} \dfrac{2q_{1f}}{\pi^{3/2}V_0^{1/2}}\,\mathcal{K}(E/V_0), & E < V_0 \\[2.5ex] \dfrac{2q_{1f}}{\pi^{3/2}E^{1/2}}\,\mathcal{K}(V_0/E), & E > V_0 \end{cases} \tag{4.53}
$$

using differentiation formulas of elliptic functions [25, p. 907, item 8.123]. Note that $\mathcal{K}(m)$ and $\mathcal{E}(m)$ of [25] are defined as power series in m^2, whereas [22] defines them as power series in m, which definition is used in Eqs. (4.52) and (4.53) above.

An example of the energy dependence of $N(E)$ calculated from Eq. (4.53) is shown in the lower panel of Fig. 4.3. For comparison, the dashed lines in Fig. 4.3 show results obtained from the "Pitzer rotor," for which $G(E) = 2q_{1f}[(E - \frac{1}{2}V_0)/\pi]^{1/2}$ and $N(E) = q_{1f}[\pi(E - \frac{1}{2}V_0)]^{-1/2}$. It can be seen that this rotor is an excellent approximation when the energy is about twice the hindering potential or higher.

At very high energies, i.e. when $E \gg V_0$, $V_0/E \rightarrow 0$ and then $\mathcal{E}(0) = \mathcal{K}(0) \rightarrow \pi/2$, so that the above $G(E)$ and $N(E)$ reduce to the sum and density of states, respectively, of a one-dimensional *free* rotor as given by Eq. (4.19) (this is true of the "Pitzer rotor" as well). At very low energies such that $E \ll V_0$, we have $\mathcal{K}(E/V_0) \approx (\pi/2)[1 + E/(4V_0)]$, $\mathcal{E}(E/V_0) \approx (\pi/2)[1 - E/(4V_0)]$ [22, *loc. cit.*], so that $G(E) = q_{1f}E/(\pi V_0)^{1/2}$ and $N(E) = q_{1f}/(\pi V_0)^{1/2}$, which are equivalent to the sum and density of states, respectively, of a classical harmonic oscillator with "frequency" $\omega = (\pi V_0)^{1/2}/q_{1f}$. Thus Eqs. (4.52) and (4.53) produce the expected *classical* result.

Note that, when $E \rightarrow V_0$, $G(E)$ is finite since $\mathcal{E}(1) = 1.35$, but the density $N(E)$ goes off to infinity since $\mathcal{K}(1) \rightarrow \infty$. Such discontinuity is undesirable, but it must be remembered that, in most actual applications, the density of states will be required for the combination of several degrees of freedom (vibrational and/or rotational) of the molecule under consideration, of which the one-dimensional hindered rotor will be just one. In such a case the effect of combining the hindered-rotor density with that of the other degrees of freedom is to smooth out the discontinuity at $E = V_0$.

For a simple demonstration consider the analytical solution for the combined density of states of a single one-dimensional hindered rotor and a single one-dimensional free rotor, as used in the case of the isopropyl radical and propylene [26]. The standard procedure is to invert the product of the corresponding partition functions, except that in the present case we take the hindered-rotor partition function in the form of the integral given by Eq. (4.47), and interchange integration and Laplace-transform inversion [26]. Apart from a constant (equal to the product of factors $q_{1f} = Q_{1f}/(\ell T)^{1/2}$ for each of the two rotors), the interchange amounts to

$$\frac{\text{constant}}{\pi} \mathcal{L}^{-1} \left\{ \int_0^\pi \frac{1}{s} \exp\left[-\frac{1}{2} s V_0 (1 - \cos x) \right] dx \right\}$$

$$= \frac{\text{constant}}{\pi} \int_0^\pi dx \, \mathcal{L}^{-1} \left\{ \frac{1}{s} \exp\left[-\frac{1}{2} s V_0 (1 - \cos x) \right] \right\} \qquad (4.54)$$

The result of the Laplace inversion is 0 or 1 (cf. Section A2.2 in Appendix 2), so that the density of states is

$$N(E) = \frac{\text{constant}}{\pi} \int_0^\pi dx \times \begin{cases} 0 & \text{for } E_0 < \frac{1}{2}V(1 - \cos x) \\ 1 & \text{for } E_0 > \frac{1}{2}V(1 - \cos x) \end{cases} \qquad (4.55)$$

Next we proceed as in the similar case of Eq. (4.38). Consider the result for $E < \frac{1}{2}V_0(1 - \cos x)$. The angle x_0 that defines $E = \frac{1}{2}V_0(1 - \cos x_0)$ is $x_0 = \cos^{-1}(1 - 2E/V_0)$; since the inverse cosine is defined only for $-1 \leq x \leq +1$, this means that $E \leq V_0$. Thus the integral $\int_0^\pi \mathrm{d}x$ is a constant (equal to π) for all $x > x_0$, but variable (equal to x_0) in the interval 0 to x_0. Hence

$$
N(E) = \text{constant} \times
\begin{cases}
\dfrac{1}{\pi} \cos^{-1}(1 - 2E/V_0), & E \leq V_0 \\[2ex]
1, & E > V_0
\end{cases}
\tag{4.56}
$$

The result is therefore that, at $E = V_0$, we no longer have a density that goes off to infinity but merely one that becomes constant, a result analogous to that encountered previously in connectiom with the two-dimensional hindered rotor. After convolution with two additional free rotors even the discontinuity at $E = V_0$ disappears (Problem 4.5).

4.3.4 The quantum solution

The partition function for the quantum one-dimensional hindered rotor can be obtained only by summation over the actual energy levels. However, since the Gibbs free energy F is related to the partition function Q by $F = -kT \ln Q$, the partition function is available in Pitzer's tables [20] disguised as $(F - F_f)/T$, the increase of free energy of the hindered rotor (F) relative to the free energy of the free rotor (F_f), divided by temperature. The tabulation is in terms of the dimensionless parameters $V_0/(kT)$ and $1/Q_{1f}$. Truhlar [27] has proposed a simple interpolation formula to recover the quantum partition function from these tables with fairly good accuracy. With energy zero at ground state, his hindered-rotor partition function in terms of $x = 1/(kT)$ is

$$
Q_{1h}(x) = \frac{\tanh\left[(\pi V_0 x)^{1/2}\right]}{1 - e^{-\omega x}} = Q_{1v}(x) \tanh\left[(\pi V_0 x)^{1/2}\right]
\tag{4.57}
$$

where $\omega = \sigma(AV_0)^{1/2}$ is the equivalent oscillator "frequency" mentioned above and thus $(1 - e^{-\omega x})^{-1} = Q_{1v}(x)$ is the corresponding vibrational partition function with energy zero at ground state. For small x, i.e. at high temperature, $\tanh x \to x$ and $\exp(-\omega x) \to 1 - \omega x$, so that $Q_{1h}(x) \to (\pi V_0 x)^{1/2}/(\omega x) = Q_{1f}$, which is the partition function of a one-dimensional classical free rotor with rotational constant $A = \omega^2/(\sigma^2 V_0)$, whereas $\tanh(\pi V_0 x)^{1/2} \to 1$ for large V_0, so that $Q_{1h}(x) \to Q_{1v}(x)$. These are the expected limiting values; at intermediate values Eq. (4.57) yields results that are slightly low but still of very acceptable accuracy.

Inversion of this partition function cannot be done analytically, but numerical inversion is fairly straightforward, as discussed below in Section 4.5.

4.4 Calculation of the inverse Laplace transform

Rotational partition functions are mostly of a form simple enough to permit analytical solution of the appropriate inverse Laplace transform (cf. Appendix 2, Section A2.2 and [24]). This is not the case with partition functions for other degrees of freedom, notably vibrations. It is therefore necessary to find means for numerical inversion of a partition function having an arbitrary form.

4.4.1 The inversion integral

With $s = 1/(\ell T)$ the transform parameter, let the total partition function of the system of interest be $Q(s)$ and let E be the energy at which the density or sum of states of the system is to be calculated. The explicit form of the Laplace inversion operator $\mathcal{L}^{-1}$ in Eq. (4.10) is given by the complex inversion integral

$$I(E, k) = \frac{1}{2\pi i} \int_{c-i\infty}^{c+i\infty} \frac{Q(s) \exp(s E) \, ds}{s^k} \tag{4.58}$$

where s is the complex variable $s = x + iy$ ($i = \sqrt{-1}$) and k is a parameter. If, as is invariably the case for every $Q(s)$ in the present context, the integrand has no singularities in the right half-plane, the path of integration is a straight line parallel to the imaginary axis with arbitrary abscissa c. By virtue of the integration theorem (Appendix 2, Section A2.2), $I(E, k = 0)$ represents $N(E)$, the density of states at E, and $I(E, k = 1)$ represents $G(E)$, the total number of states (or state count for short) at E (note again the distinction between k as a parameter in Eq. (4.58) (cf. also Eq. (4.19)) and ℓ, which is Boltzmann's constant).

While the inversion integral of Eq. (4.58) is discussed in most mathematics texts for functions of interest in electrical engineering, its evaluation in the context of state density was discussed first by Kubo [28, pp. 103 and 124] and later by others [29–31]. Equation (4.58) can be written as a real integral,

$$I(E, k) = \frac{1}{2\pi} \int_{-\infty}^{\infty} \frac{Q(x + iy) \exp[E(x + iy)] \, dy}{(x + iy)^k} \tag{4.59}$$

where the path of integration is the same as in Eq. (4.58). For $y = 0$ and in the direction of the real axis x, the (real) integrand will have a minimum at, say, $x = x^*$ (Fig. 4.4(a)), which suggests taking $c \equiv x^*$. If the complex integrand is now evaluated at $x = x^*$ in the imaginary direction as a function of y, the real part of the integrand is found to have a steep maximum while the imaginary part oscillates (Fig. 4.4(b)); there is thus a saddle point at x^*.

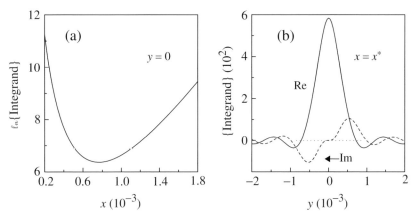

Fig. 4.4. The integrand in Eq. (4.59) for the density ($k = 0$) $N(E)$ of methane at $E = 5000 \, \text{cm}^{-1}$. (a) A semi-logarithmic plot of the integrand as a function of x (the real axis) at $y = 0$ (the origin of the imaginary axis). The integrand is therefore real. The minimum is at $x^* = 7.655 \times 10^{-4}$, and the value of the integrand at this point is 583.91. (b) A plot of the *real* (solid line marked Re) and *imaginary* (dashed line marked Im) parts of the same integrand at $x = x^*$ as a function of y (the imaginary axis). The height of the peak is 583.91, which corresponds to the minimum in the direction of the real x-axis at $y = 0$. There is thus a saddle point at x^*, i.e. the minimum in the direction of the real x-axis is at the same time the maximum of the real part in the direction of the imaginary y-axis. See also [34, Fig. 4.3.2, p. 58].

4.4.2 The method of steepest descents

Integrals with such sharply peaked integrands can be approximated by the method of steepest descents [28, *loc. cit.*, 30, 31], which is sometimes also called the saddle-point method. The name refers to the path of integration chosen so that the integrand falls off from the maximum at the saddle point at the greatest possible rate. In the present context it is convenient to work with the function $\phi(s)$ defined as the natural logarithm of the integrand in Eq. (4.58):

$$\phi(s) = \ln Q(s) - k \ln s + sE \tag{4.60}$$

so that the integral in Eq. (4.59) is simply

$$I(E, k) = \frac{1}{2\pi} \int_{-\infty}^{\infty} \exp[\phi(x^* + iy)] \, dy \tag{4.61}$$

The position of the saddle point (x^*) is given by the solution of $\partial \phi(x)/\partial x = 0$, i.e.

$$\phi'(x) = \frac{\partial \ln Q(x)}{\partial x} - \frac{k}{x} + E = 0 \tag{4.62}$$

On expanding $\phi(s)$ around x in a Taylor series we have

$$\phi(s) = \phi(x^*) + \sum_{n=2}^{\infty} \frac{1}{n!}(s - x^*)^n \frac{\partial^n \phi(x)}{\partial x^n}\bigg|_{x=x^*} \tag{4.63}$$

The first derivative ($n = 1$) of $\phi(x)$ is absent from this expansion because of Eq. (4.62). Since $s - x^* = iy$, the expansion, after substitution into Eq. (4.61), can be put in the form

$$I(E, k) = \frac{e^{\phi(x^*)}}{2\pi} \int_{-\infty}^{\infty} \exp(-b_2 y^2) \prod_{n=3}^{\infty} \exp[b_n(iy)^n]\, dy \tag{4.64}$$

where

$$b_n = \frac{1}{n!} \frac{\partial^n \phi(x)}{\partial x^n}\bigg|_{x=x^*} \tag{4.65}$$

The integral in Eq. (4.64) can be approximated [32] by first expanding each exponential in the product:

$$\exp[b_n(iy)^n] \approx 1 + b_n(iy)^n + \frac{1}{2!}b_n^2(iy)^{2n} + \frac{1}{3!}b_n^3(iy)^{3n} + \cdots \tag{4.66}$$

and then forming the product of the expansions for $n \geq 3$, the first few terms of which are

$$\prod_{n=3}^{\infty} \exp[b_n(iy)^n] \approx 1 + b_3(iy)^3 + b_4(iy)^4 + b_5(iy)^5 + \frac{1}{2}b_3^2(iy)^6 + \cdots \tag{4.67}$$

Integrals of odd powers of iy (i.e. of the oscillating imaginary part, cf. Fig. 4.4(b)) will vanish because of the symmetric integration limits in Eq. (4.64), and therefore $I(E, k)$ is real, as of course it should be.

Performing the integration in Eq. (4.64) of even powers from Eq. (4.67) yields

$$I(E, k) = \frac{e^{\phi(x^*)}}{2\pi} \sqrt{\frac{\pi}{b_2}} \left(1 + \frac{3}{4}\frac{b_4}{b_2^2} - \frac{15}{16}\frac{b_3}{b_2^3}\right) \tag{4.68}$$

On substituting back the actual derivatives from Eq. (4.65), keeping only the leading term in the large parentheses of Eq. (4.68), the final result known as the first-order steepest-descents approximation (Problem 4.6) is

$$\left.\begin{array}{l} N(E) \\ k=0 \\ G(E) \\ k=1 \end{array}\right\} = I_1(E, k) = \frac{\exp[\phi(x^*)]}{(2\pi\phi''(x^*))^{1/2}} = \frac{Q(x^*)\exp(x^* E)}{(x^*)^k (2\pi\phi''(x^*))^{1/2}} \tag{4.69}$$

where $\phi''(x^*)$ is the second derivative of $\phi(x)$ with respect to x at $x = x^*$. Including the other terms in Eq. (4.68) yields what is known as the second-order steepest-descents approximation [32]:

$$I_2(E, k) = I_1(E, k) \left(1 + \frac{1}{8} \frac{\phi^{(4)}(x^*)}{(\phi''(x^*))^2} - \frac{5}{24} \frac{(\phi^{(3)}(x^*))^2}{(\phi''(x^*))^3} \right) \tag{4.70}$$

where the higher derivatives for $n \geq 2$ are given by

$$(-1)^n \phi^{(n)}(x^*) = \left\{ \frac{\partial^n \ln Q(x)}{\partial x^n} + \frac{k(n-1)!}{x^n} \right\}_{x=x^*} \tag{4.71}$$

The inversion procedure for a given system with partition function $Q(x)$ ($x = 1/(kT)$) therefore consists of forming the logarithm of the integrand $\phi(x)$ in terms of the real variable x (i.e. the analog of Eq. (4.60)):

$$\phi(x) = \ln Q(x) - k \ln x + xE \tag{4.72}$$

Equation (4.62) is then solved for x^* and the various derivatives of $\phi(x)$ are calculated at $x = x^*$ (Eq. (4.71)). The sum and density of states to first or second order then follow from Eqs. (4.69) and (4.70), respectively.

Since $k = 0$ and $k = 1$ select the density and sum of states, respectively, we have approximately from Eq. (4.69) (Problem 4.7)

$$G(E) \approx N(E)/x^* \tag{4.73}$$

where x^* is understood to be the value obtained in the determination of $N(E)$ (i.e. from Eq. (4.62) with $k = 0$). In this way both $G(E)$ and $N(E)$ can be obtained by solving Eq. (4.62) just once. The $G(E)$ so determined is good to within a few percent of $G(E)$ obtained directly from Eq. (4.62) with $k = 1$. The converse, i.e. $N(E)$ calculated from $G(E)$ using x^* obtained in the determination of $G(E)$ (from Eq. (4.62) with $k = 1$), is given by (Problem 4.8)

$$N(E) \cong G(E) \left(x^* + \frac{\phi^{(3)}(x^*)}{2(\phi''(x^*))^2} \right) \tag{4.74}$$

In the approximate form $N(E) \approx G(E) x^*$ this relation performs poorly, but with both terms in the large parentheses it is generally within better than one percent of $N(E)$ obtained directly from Eq. (4.62) with $k = 0$. It is useful for complicated forms of $\phi(s)$ when Eq. (4.62) does not have a solution for $k = 0$ at small (or very large) E, but will almost always have one for $k \geq 1$. Thus problems with the direct calculation of the density $N(E)$ at very low energies can be avoided by using Eq. (4.74). A specific example is given in Section 4.6. Problems at large E with $k = 0$ occur when $\phi(s)$ is a polynomial rather than an infinite series.

4.4.3 The logarithmic formulation

For numerical reasons, it is sometimes advantageous to work with the variable $z = \exp(-x)$ [7, Chapter 6], [28, p. 124]; [33]) in which case the counterpart of $\phi(x)$ in Eq. (4.72) reads

$$\phi(z) = \ln Q(z) - k \ln(\ln z^{-1}) - E \ln z \tag{4.75}$$

and therefore

$$\phi'(z) = \frac{\partial \ln Q(z)}{\partial z} + \frac{k}{z^2 \ln z^{-1}} - \frac{E}{z} \tag{4.76}$$

The position of the saddle point is given by $\theta \equiv \exp(-x^*)$, which is the solution of $z\,\phi'(z) = 0$. The first-order formula becomes (Problem 4.9)

$$I_1(E, k) = \frac{Q(\theta)}{(\ln \theta^{-1})^k\, \theta^E\, (2\pi\theta^2\phi''(\theta))^{1/2}} \tag{4.77}$$

where

$$\theta^2\phi''(\theta) = \left\{ z\,\frac{\partial \ln Q}{\partial z} + z^2\,\frac{\partial^2 \ln Q}{\partial z^2} + \frac{k}{(\ln z^{-1})^2} \right\}_{z=\theta} \tag{4.78}$$

The logarithmic formulation yields results in all respects identical to the exponential formulation (Problem 4.10), occasionally with a slight reduction in machine time.

4.4.4 Thermodynamic considerations

For $k = 0$, Eq. (4.62) evaluated at $x = x^*$ becomes

$$-\left.\frac{\partial \ln Q(x)}{\partial x}\right|_{x=x^*} = E \tag{4.79}$$

From the point of view of ensemble theory, the energy E in the development presented so far is the energy at which the density/sum of states is calculated in a microcanonical ensemble. It follows from standard statistical mechanics [34] that the relation (4.79) also identifies the average energy of a system, represented by the partition function Q, in a constant-temperature bath of temperature $T^* = 1/(k\,x^*)$, i.e. in a *canonical* ensemble. In this sense the above E is a temperature-dependent constant, which, for the present purpose, may be designated $\langle E \rangle$.

Similarly, the second derivative $\phi''(x)$ evaluated at $x = x^*$ and $k = 0$ (Eq. (4.71)) will be recognized as related to $\partial\langle E \rangle/\partial T$, i.e. $C_v(T)$, the constant-volume specific heat of the canonical system at T^* (note that $\partial/\partial x = -k\,T^2\,\partial/\partial T$):

$$\left.\frac{\partial^2 \ln Q(x)}{\partial x^2}\right|_{x=x^*} = \left.\frac{-\partial\langle E \rangle}{\partial x}\right|_{x=x^*} = k\,T^2 C_v(T)\big|_{T=T^*} \tag{4.80}$$

Thus, for instance, Eq. (4.69) can be written equivalently as

$$N(E) = \frac{Q(T^*) \exp[E/(\mathit{k} T^*)]}{(2\pi \mathit{k} T^{*2} C_v(T^*))^{1/2}} \tag{4.81}$$

At the same time, quite generally,

$$\mathit{k} T^2 C_v(T) = \langle E^2 \rangle - \langle E \rangle^2 = \sigma_E^2 \tag{4.82}$$

which is the variance representing the fluctuation of energy in a canonical ensemble at temperature T.

It can be demonstrated [31], approximately and rather simply in the case of the density of states $N(E)$, that in a sense the steepest-descents method exploits a connection between a microcanonical ensemble of specified energy E and a canonical ensemble having the same *average* energy $\langle E \rangle$ [34]. The probability density for energies E in a canonical ensemble at $T^* = 1/(\mathit{k} x^*)$ is

$$P(E) = \frac{N(E)}{Q(x^*)} \exp(-x^* E) \tag{4.83}$$

The average energy in this ensemble, $\langle E \rangle$, is given by Eq. (4.79), and the variance σ_E^2 by Eq. (4.80). Suppose that $P(E)$ is now replaced by a rectangular function $p(E)$ such that

$$p(E) \approx 1/\sigma_E \quad \text{for } (\langle E \rangle - \sigma_E/2) < E < (\langle E \rangle + \sigma_E/2)$$
$$= 0 \quad \text{otherwise} \tag{4.84}$$

The relation between $P(E)$ and $p(E)$ is shown in Fig. 4.5. If we now equate $P(E)$ with $p(E)$, the result is (at E given by Eq. (4.79))

$$N(E) \approx \frac{Q(x^*)}{\sigma_E} \exp(x^* E) \tag{4.85}$$

which, except for missing the factor $(2\pi)^{1/2}$, is identical with Eq. (4.69) for the first-order approximation $I_1(E, k = 0)$ for the density if an explicit expression from Eqs. (4.71) and (4.80) is used for $\phi''(x)$. (If we had "guessed" $p(E)$ to be of the form $1/[\sigma_E(2\pi)^{1/2}]$ within $\pm\sigma_E(2\pi)^{1/2}$ centered at $\langle E \rangle$, we would have recovered the first-order steepest-descents result exactly.)

The thermodynamic approach has been used extensively by Klots [35].

4.5 Quantized systems

This section concerns primarily the counting of vibrational states since vibrations are the quantized systems of major interest here. Unless specifically mentioned otherwise, it will be assumed that vibrational degrees of freedom are *separable,* i.e. *independent.*

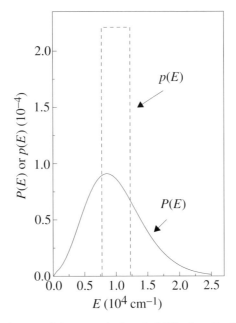

Fig. 4.5. Energy dependences of the canonical probability density $P(E)$ (Eq. (4.83)) and of the rectangular distribution $p(E)$ (Eq. (4.84)). Both are in units of $(\mathrm{cm}^{-1})^{-1}$ and are normalized to unity. The situation represented corresponds to the density of vibrational states in ethane (18 oscillators) at $10\,000\,\mathrm{cm}^{-1}$, at which $N(E) = 449.6\,(\mathrm{cm}^{-1})^{-1}$. The first-order S-D method yields $x^* = 8.4608 \times 10^{-4}$, i.e. $T^* = 1700.6\,\mathrm{K}$; $\sigma_E = 4523.9\,\mathrm{cm}^{-1}$; and $Q(x^*) = 1079$. Accordingly, $P(E)$ is a 1700-K probability density with average energy $\langle E \rangle = 10\,000\,\mathrm{cm}^{-1}$. The width of $p(E)$ is σ_E centered on $\langle E \rangle$ and its height is $1/\sigma_E$.

4.5.1 Exact enumeration

Comparison with exact results for an arbitrary collection of oscillators requires the actual enumeration of states. An exact formula for W_n the *number* of states of an s-fold degenerate quantum oscillator of frequency v (i.e. there are s identical oscillators, all of the same frequency v), containing n quanta of energy has already been given by Eq. (1.1) in Chapter 1 disguised as the number of ways to put n "balls" into s "boxes." In order to evaluate the corresponding *sum* of states G_n we have to find the *total* number of ways of placing $0, 1, 2, \ldots, n$ indistinguishable "balls" (quanta) into s indistinguishable "boxes" (oscillators); thus

$$G_n = \sum_{i=0}^{n} \frac{(i+s-1)!}{i!(s-1)!} = \frac{(n+s)!}{n!s!} \tag{4.86}$$

If we let $E = nhv$ and use the procedure used in Problem 1.4 of Chapter 1, W_n and G_n can be reduced, depending on approximations, to Eq. (A1.2) in Appendix 1, or Eq. (4.23) as in Problem 4.11. Neither approximation, although not assuming that the oscillators are necessarily identical, is accurate enough.

Since a real molecule does not consist entirely of identical oscillators, it is necessary to consider ways to enumerate vibrational states of a system consisting of oscillators having diverse vibrational quanta (i.e. frequencies). If the energy and the number of oscillators are not too high, the (exact) enumeration can be done by hand (cf. Table I in [7]), and in more complicated cases can be computed exactly by the remarkably efficient algorithm of Beyer and Swinehart ([36–38]; see also [39, pp. 183ff]). It calculates the sum of states essentially by progressive convolution of one degree of freedom after another, one at a time, each time updating the count. This algorithm yields the count at equal energy spacing (usually referred to as "graining") for *separable* degrees of freedom with known energy levels. It is not limited to harmonic oscillators and is good for rotors and hindered rotors.

Note that enumeration actually determines first the number of states $W(E)$ at each energy, and then the sum of states as the cumulant of $W(E)$ (Eq. (4.2)). The density of vibrational states is not amenable to exact counting, being by its very nature a string of delta functions, which creates problems (Section 4.1.3).

As stated in Section 4.1.4 (iii), what is required (and sufficient) in the context of unimolecular rate theory is a good smooth-function approximation. The approximation obtained from the inversion of the partition function fits the bill, because it is not only very good, but also very general. The specifics of this inversion will be the principal subject of the present section, using the method of steepest descents (henceforth abbreviated as "S-D" method), which was previously discussed in general terms in Section 4.4. This method provides in one fell swoop the smoothed density and any repeated integral of the density (e.g. the sum of states) of any classical or quantized system. The S-D method has been tested extensively on a great variety of systems and found to be generally reliable, except for a few isolated cases at very low energy.

4.5.2 Independent harmonic oscillators by the S-D method

The quantum partition function for a collection of v independent harmonic oscillators is

$$Q_v = \prod_{i=1}^{v} \{1 - \exp[-\omega_i/(k\,T)]\}^{-1} \tag{4.87}$$

where ω_i is the "frequency" of the ith oscillator. For this collection x^* is the solution of Eq. (4.62) with Q_v replacing Q. In general, and in particular with the above Q_v, Eq. (4.62) is a transcendental equation, the solution of which for x^* cannot be obtained analytically except for special forms of Q_v (Problems 4.12 and 4.15), but is readily found numerically; more on this in Section 4.6.

The higher derivatives of Q_v of Eq. (4.87) for substitution into Eq. (4.71) are obtainable analytically (Problem 4.13). In the logarithmic formulation (Eq. (4.75)) the derivatives are similar (Problem 4.14).

The vibrational partition function of Eq. (4.87) is defined with respect to energy zero at the ground vibrational state; this will be assumed throughout (Problem 4.16).

The performance of the S-D routine as a smooth-function approximation relative to exact counting at higher energies can be judged from Fig. 4.2, which compares the first-order approximation to the sum of states $G_v(E)$ in methane (continuous line, calculated from Eqs. (4.69) and (4.87)) with the exact count (staircase line). Numerical comparison of the S-D routine for $G_v(E)$ with the exact count is shown in Table 4.2 for a "small" and a "large" molecule at several energies.

4.5.3 The S-D method for the partition function as a sum

A partition function is by definition the sum of weighted negative exponentials containing the energy levels of the system in question divided by ℓT. While it is convenient to obtain the sum analytically, as, for example in Eq. (4.87), it is by no means essential for the S-D method.

In order to consider the most general case, let us define the following sum for a quantized system of m energy levels:

$$S^{(n)} = \sum_{i=0}^{m} (-1)^n a_i \epsilon_i^n \exp(-\epsilon_i x) \qquad (4.88)$$

where ϵ_i and a_i are the energy and degeneracy, respectively, of level i, and $x = 1/(\ell T)$; n is a parameter. The counter i starts at zero, so that, for energies counted from the ground state, $\epsilon_0 = 0$. The actual partition function of the system is $Q \equiv S^{(0)}$ and $S^{(n)}$ is the nth derivative of Q with respect to x. For the purpose of obtaining simple expressions for the various *logarithmic* derivatives of the partition function, it is convenient to define $D^{(n)} = S^{(n)}/Q$; then

$$\frac{\partial \ln Q}{\partial x} = D^{(1)}; \qquad \frac{\partial^2 \ln Q}{\partial x^2} = D^{(2)} - \left(D^{(1)}\right)^2 \qquad (4.89)$$

and the first-order S-D routine follows as usual (Eq. (4.72)), with some increase in machine time if the various sums consist of large numbers of terms.

This is not necessarily the case if there are only a few low-lying electronic states to be considered in the evaluation of a vibrational sum or density of states. For example, in NO the vibrational partition function has to be multiplied for this reason by the electronic partition function, which consists at ambient temperature of only two terms, with $a_0 = a_1 = 2$, $\epsilon_0 = 0$, and $\epsilon_1 = 121.1\ \mathrm{cm}^{-1}$ [40].

For the second-order S-D method the next two higher derivatives are needed, which are left for Problem 4.17.

4.5.4 Independent anharmonic oscillators by the S-D method

Oscillators in a real molecule are of course neither separable nor harmonic. As a first step to greater realism, assume that oscillators are still separable but *anharmonic*, represented as a collection of independent Morse oscillators. This supposes that the necessary ancillary information, in particular anharmonicity coefficients, are available.

The energy of a Morse oscillator of "frequency" ω (in principle *normal* "frequency"), quantum number n, and anharmonicity coefficient χ (customarily positive by definition and not to be confused with the variable x in the transform parameter $s = x + iy$), was given in Eq. (4.25); however, for the present purpose we need energy defined in excess of zero-point energy, so that now

$$\epsilon = \omega(n - n\chi - n^2\chi) \tag{4.90}$$

The partition function for such an oscillator is

$$Q_{\rm v} = \sum_{n=0}^{n_{\max}} \exp[-\epsilon/(kT)] \tag{4.91}$$

where

$$n_{\max} = \frac{1}{2}\left(\frac{1}{\chi} - 1\right) \tag{4.92}$$

Thus, unlike in the case of the quantum harmonic oscillator, the summation in the expression for the partition function (Eq. (4.91)) no longer runs to infinity since now we have an oscillator that dissociates when quantum number n becomes sufficiently high that $\partial \epsilon/\partial n = 0$ in Eq. (4.90). This is a specific example of a partition function represented by $S^{(0)}$ of Eq. (4.88) in which all levels are non-degenerate, i.e. $a_i = 1$ for all i.

For a collection of v such oscillators (indexed $m = 1, \ldots, v$), assumed to be independent, the total vibrational partition function $Q_{\rm v}$ is

$$Q_{\rm v} = \prod_{m=1}^{v} Q_{{\rm v}m}; \qquad Q_{{\rm v}m} = \sum_{n_i=0}^{n_{i,\max}} \exp[-\epsilon_{ni}/(kT)] \tag{4.93}$$

The S-D routine then proceeds normally, except that the equations analogous to Eq. (4.89) are now sums over the individual oscillators:

$$\frac{\partial \ln Q_{\rm v}}{\partial x} = \sum_{m} D_m^{(1)}; \qquad \frac{\partial^2 \ln Q_{\rm v}}{\partial x^2} = \sum_{m} \left[D_m^{(2)} - \left(D_m^{(1)}\right)^2\right] \tag{4.94}$$

and similarly for the higher derivatives of $Q_{\rm v}$ (cf. Problem 4.17).

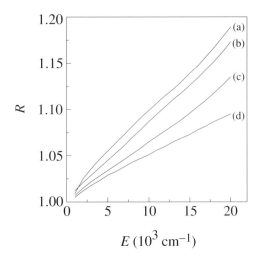

$$E \ (10^3 \ \mathrm{cm}^{-1})$$

Fig. 4.6. The energy dependence of the ratio of the Morse anharmonic and harmonic sums of states $R = G_v(E)_{\mathrm{anh}}/G_v(E)_{\mathrm{harm}}$ for four molecules, calculated by the S-D method: (a) water, (b) ozone, (c) NO_2, and (d) formaldehyde. The similar ratio of the densities $N_v(E)_{\mathrm{anh}}/N_v(E)_{\mathrm{harm}}$ in the same energy range is numerically very similar except for a slightly more pronounced curvature. Data are from [41].

The position of the saddle point x^* is then the solution of

$$\sum_m D_m^{(1)} - \frac{k}{x} + E = 0 \tag{4.95}$$

where E is the total energy of the v oscillators. The S-D results for the first- and second-order approximations $I_1(E, k)$ and $I_2(E, k)$ then follow from Eqs. (4.69)–(4.71), where the various derivatives of the function $\phi(x)$ involve summations over m of the individual derivatives of D_m. Comparison of this S-D Morse routine with state counting for fully coupled anharmonic oscillators shows that, in several instances, the routine is actually a very good approximation [41].

4.5.5 Anharmonic correction to the harmonic count

The Morse S-D routine can be used to obtain an appreciation of the error inherent in assuming that molecular vibrations are always harmonic. This is shown in Fig. 4.6 as the energy dependence of the ratio of the anharmonic and harmonic sum of states $R = G_v(E)_{\mathrm{anh}}/G_v(E)_{\mathrm{harm}}$ for four molecules, calculated, for the anharmonic part, from Eqs. (4.90)–(4.95).

The increase in the sum of states due to anharmonicity shown in Fig. 4.6 is not large, even at fairly high energies, which suggests that, in most instances, the effect of anharmonicity and non-separability of oscillators is lost since errors in the experimental rate constant ultimately used for comparing theory and experiment are likely to exceed the error resulting from the neglect of anharmonicity. This justifies the use of harmonic sums and densities of states in rate calculations, which was mentioned previously in Section 3.3.5.

4.5.6 Independent oscillators and free rotors by the S-D method

If rotations and vibrations are independent, the partition function for the combined rotation–vibration system is a simple product of the two partition functions: $Q_{vr} = Q_v Q_r$. Consequently, remembering that the partition function for r classical free rotors is $Q_r = q_r (\ell T)^{r/2}$ (cf. the text below Eq. (4.18)),

$$I_{vr}(E, k) = \mathcal{L}^{-1} \left\{ \frac{Q_r(s) Q_v(s)}{s^k} \right\} = q_r \mathcal{L}^{-1} \left\{ \frac{Q_v(s)}{s^{k+r/2}} \right\} \qquad (4.96)$$

Observe that this in no way modifies the inversion routines outlined above for harmonic or Morse oscillators, inasmuch as to obtain the density or sum of states of a rotation–vibration system requires (apart from q_r as a multiplicative constant) merely a re-definition of the parameter k as $k + r/2$ in Eq. (4.60) and the higher derivatives (Eq. (4.71)); in particular, the *density* $N_{vr}(E)$ of a combined vibrations-plus-two-rotations system with $q_r = 1$ is formally equivalent to the *sum* of states $G_v(E)$ for vibrations only. This has already been pointed out in connection with the semi-classical version (Eq. (4.24)).

On combining results for a collection of independent quantum oscillators (harmonic or Morse) and r classical free rotors, we have therefore for the first-order S-D approximation (cf. Eq. (4.69))

$$\left. \begin{array}{l} N_{vr}(E) \\ k = 0 \\ G_{vr}(E) \\ k = 1 \end{array} \right\} = \frac{q_r Q_v(x^*) \exp(x^* E)}{(x^*)^{k+r/2} (2\pi \phi''(x^*))^{1/2}} \qquad (4.97)$$

where x^* is the solution of Eq. (4.62) with k replaced by $k + r/2$. The second-order approximation follows from Eq. (4.70), where all the higher derivatives $\phi^{(n)}(x^*)$ are given by Eq. (4.71), again after the replacement $k \to k + r/2$. Thus the extension of the first- or second-order result for the vibrational-only sum/density to the vibrational–rotational sum/density is trivial.

If we go back to the explicit form of the transforms in Eq. (4.96), the above result for $k = 1$, for example, represents the integral

$$G_{vr}(E) = q_r \int_0^E N_v(E - y) \, y^{r/2} \, dy \qquad (4.98)$$

which will appear again in Section 8.7.1. This is a reminder that the S-D method yields the combined vibrational–rotational sum or density of states *directly*, without the need to calculate the convolution integrals in Eq. (4.13) or (4.14), a considerable computational advantage. This is also the case for other independent vibrational–rotational systems discussed below.

Ignored in all cases considered above is the centrifugal distortion that a rotor undergoes by virtue of the coupling between the rotations and vibrations of the molecule of which it is implicitly assumed to be a part. This simplification constitutes the so-called rigid rotor approximation, which is generaly a good one except in very exceptional circumstances.

4.5.7 A special case

A special case arises for the vibrational–rotational density or sum of states when the rotational quantum number of the molecule under consideration has a given value. Consider the case of a linear or symmetric-top molecule (rotational constant B) with *specified* rotational quantum number J, and therefore with specified rotational energy $E_r = J(J + 1)B$. Since the degeneracy of rotational levels is $2J + 1$ for each J, the partition function for such a system is $Q_v(2J + 1) \exp[-E_r/(kT)]$ (cf. the integrand of Eq. (A1.10) in Appendix 1), where Q_v is the partition function for internal (mostly vibrational) degrees of freedom of the molecule. It is now necessary to distinguish between the internal energy E_v of the molecule in question, and the total energy $E = E_v + E_r$ of the system. The inversion integral at specified J and E for this system is then, from Eq. (4.58),

$$I_{vr}(E, k, J) = \frac{2J + 1}{2\pi i} \int_{c-i\infty}^{c+i\infty} \frac{Q_v(s) \exp[s(E - E_r)] \, ds}{s^k} \qquad (4.99)$$

or, symbolically,

$$I_{vr}(E, k, J) = (2J + 1) \, \mathcal{L}^{-1} \left\{ \frac{Q_v(s)}{s^k} \exp(-s E_r) \right\} \qquad (4.100)$$

where the expression in braces is immediately recognized as simply the zero-shifted inverse transform of $Q_v(s)/s^k$ at E (cf. Appendix 2, Section A2.2). Thus, for $k = 0$, i.e. the density of states (provided that $E > E_r$), we have $I_{vr}(E, 0, J) = (2J + 1) N_v(E - E_r)$. The expression for the density of states of such a system with vibrational energy E_v and with specified J is therefore [42]

$$I_{vr}(E, 0, J) = (2J + 1) N_v(E_v) \qquad (4.101)$$

and simlarly for the sum of states, with $k = 1$ in Eq. (4.100),

$$I_{vr}(E, 1, J) = (2J + 1) G_v(E_v) \qquad (4.102)$$

These relations are quite general and in particular hold both for the first- and for the second-order S-D approximations to $N_v(E_v)$ and $G_v(E_v)$.

In the case of a spherical rotor with rotational constant B (e.g. CH_4), the rotational energy is again $E_r = J(J+1)B$ at specified J but the degeneracy of every rotational level is now $(2J+1)^2$ (cf. Appendix 1, Eq. (A1.7)), so that the analogous density of states is $I_{vr}(E, 0, J) = (2J+1)^2 N_v(E_v)$, and similarly the sum of states is $I_{vr}(E, 1, J) = (2J+1)^2 G_v(E_v)$.

The difference from the general results given in the previous section (when they are specialized for two rotational degrees of freedom) is that, in the present case, since J is specified, there are just $2J+1$ (or $(2J+1)^2$ for a spherical top) rotational states whatever the vibrational energy E_v of the system, whereas Eqs. (4.96) and (4.97) with $r = 2$ are the results of a true convolution in which *all* rotational states having energies between zero and the specified total energy E are combined with vibrational states in the same energy interval; consequently it was not necessary to distinguish between E_v and E_r.

These considerations have a bearing on the J-shifted rate constant of Chapter 3 (Eq. (3.38)) for Type-1 reactions. This will appear more clearly if the rate constant is written in terms of rotational quantum number J rather than rotational energy E_r. Since, as has been argued, the transition state is always a prolate top, at specified J the sum of states $G^*(E - E_0^*)$ will have $2J+1$ as factor. The same factor will appear in the denominator for the density of states of the reactant if it is a linear or symmetric-top molecule, except in the (rare) case that the reactant is a spherical top. Thus the equivalent of Eq. (3.38) becomes (with $E_0^*(J)$ written for E_0^* to emphasize the J-dependence)

$$k(E, J) = \frac{\alpha\,(2J+1)\,G^*[E - E_0^*(J)]}{h\,(2J+1)^m\,N(E)} \tag{4.103}$$

where $m = 1$ if the reactant is linear or a symmetric top, and $m = 2$ if the reactant is a spherical top. Since the case $m = 2$ is rare, we have for the great majority of reactant molecules that the factors $2J+1$ cancel out in Eq. (4.103) top and bottom, and Eq. (3.38) therefore applies as written.

The more general form of the rate constant given in Eq. (4.103) is used in Section 8.9 in a slightly different context.

4.5.8 *Oscillators and free and/or hindered rotors by the S-D method*

Assuming that oscillators and rotors are independent, the sum or density of states for a system consisting of a combination of oscillators and rotors, free and/or hindered, is obtained by standard S-D procedure, that is, inversion of the product of the corresponding partition functions $Q_v Q_r Q_{1h}$, where Q_v is given by Eq. (4.87), Q_{1h}

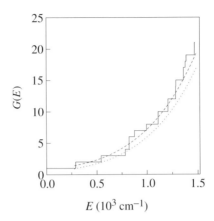

Fig. 4.7. The energy dependence of the sum of states $G(E)$ for separable oscillators + one hindered internal rotor in an ethane-like molecule ($V_0 = 1000\,\mathrm{cm}^{-1}$, $\sigma = 3$). Staircase line: exact count by use of the Beyer–Swinehart algorithm, using hindered-rotor eigenvalues from [44]; dashed line: S-D quantum approximation, based on Eq. (4.57); and dotted line, S-D classical approximation, based on Eq. (4.47).

by Eq. (4.49) for the classical and Eq. (4.57) for the quantum partition functions of the hindered one-dimensional rotors, and, as before, $Q_r = q_r(\ell\,T)^{r/2}$. The inversion makes use of the first and second derivatives of the hyperbolic tangent and modified Bessel functions. Explicit formulas have been published [43]; an example is shown in Fig. 4.7.

The case of the two-dimensional hindered rotor is somewhat different. One could proceed in principle by incorporating the series expansion for the hindered-rotor partition function (Eq. (4.36)) as a factor in the total partition function, and then proceed to invert by the S-D method. However, the series and its derivatives are tricky, with alternating positive and negative terms, such that, under some conditions, the equation $\phi'(x) = 0$ does not have a solution.

A better, and more robust, option is numerical convolution according to Eqs. (4.13) and (4.14), using for the hindered-rotor part the analytical results of Eqs. (4.40)–(4.42) (and S-D results for the vibrational + free-rotor part). This procedure was used in Fig. 4.8 for the sum of states for the transition state in the decomposition of methane, $CH_4 \rightarrow CH_3 + H$, assuming that the two oscillators at $1534\,\mathrm{cm}^{-1}$ (CH_3 deformations) can be treated as a hindered two-dimensional rotor with various values of V_0 in the hindering potential $V_0 \sin^2 \theta$. Figure 4.8 shows the expected increase of the sum of states with decreasing barrier height V_0.

4.5.9 *The combination of oscillators with a symmetric top K-rotor*

The three-dimensional rotor known as a symmetric top (Appendix 1, Section A1.4) presents a special case regarding the quantum number K and the corresponding K-rotor, as discussed in Sections 3.3 and 3.4.

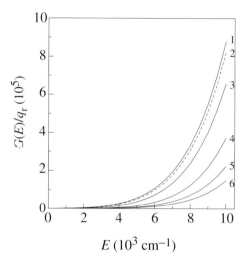

Fig. 4.8. The energy dependence of the sum of states $G(E)/q_r$ in units of cm^{-1} for the transition state in the decomposition of methane $CH_4 \rightarrow CH_3 + H$, modeled as CH_3, where two deformations at $1534\,cm^{-1}$ are replaced by a two-dimensional hindered rotor with hindering potential $V(\theta) = V_0 \sin^2 \theta$ (Eq. (4.35)). The curves are obtained by convolution according to Eq. (4.14), using for the hindered-rotor part the analytical results of Eq. (4.42) and S-D results for the vibrational part. The numbering is according to the value of the hindering potential V_0: 1, two free rotors ($V_0 = 0$); 2, $V_0 = 250\,cm^{-1}$; 3, $V_0 = 1000\,cm^{-1}$; 4, $V_0 = 3000\,cm^{-1}$; 5, $V_0 = 5000\,cm^{-1}$; and 6, $V_0 = 7000\,cm^{-1}$.

Assuming that the K-rotor is free and "separable," i.e. that the J- and K-dependent parts of $E_r(J, K)$ (Eq. (4.22)) can be dealt with separately [45–49], this means that the K-rotor is active, and its energy is considered to be part of the pool of randomizable energies (this is not the case for the J-rotor). If the angular momentum of the reactant is J, in such a case the restriction $-J \leq K \leq +J$ is ignored, so that the rotational energy of the active K-rotor

$$E_r(K) = (A - B)K^2 \tag{4.104}$$

is limited only by the total available energy rather than by the specified J [7, 39, 50, 51]. The corresponding vibrational–rotational sum/density of states can be obtained directly from Eq. (4.97) above with $r = 1$, which greatly simplifies the calculation and produces no serious error for high values of J, but causes an overcount at low values of J.

To remedy this overcount it is necessary to consider a properly J-restricted active K-rotor. The density or sum of states for such a system cannot be obtained from Eq. (4.97), which does not recognize the K-rotor restriction. In this context E now represents the total internal energy, which is the sum of vibrational energy (E_v, as defined above) plus the active K-rotor energy $E_r(K)$ (Eq. (4.104)):

$$E = E_v + E_r(K) \tag{4.105}$$

The "exact" sum/density of states at specified E, J within the active K-rotor approximation can be obtained by calculating the sum of a sequence of *vibrational* sums/densities of states at $E_v = E - E_r(K)$, for $K = 0, \pm 1, \pm 2, \ldots, \pm K_{max}$. (If E is defined as the *total* energy, internal + external rotational, then $E_v = E - J(J + 1)B_e - E_r(K)$.) Since the sum/density of states is defined only for positive E_v, the highest K is not necessarily $\pm J$ but is instead K_{max} defined as the largest integer for which $E - E_r(K) \approx 0$. In the case of a prolate top K_{max} is J or the integer nearest to $+\sqrt{E/(A - B)}$, whichever is *smaller*:

$$K_{max} = \min\left\{J, +\sqrt{E/(A - B)}\right\} \tag{4.106}$$

The sum of vibrational $+K$-rotor states $G_{vr}(E, J)$ at specified E, J is then (cf. Eq. (4.102))

$$G_{vr}(E, J) = \frac{1}{\sigma} \sum_{-K_{max}}^{+K_{max}} G_v[E - E_r(K)] \tag{4.107}$$

The RHS of this expression excludes the factor $2J + 1$ for reasons discussed in connection with Eq. (4.103). On account of symmetry considerations not all K-values need appear in the above summation, as discussed in Appendix 1. Equation (4.107) follows the usual practice, which is to execute calculations with all J-allowed values of K and then divide the result by a symmetry number σ [52].

Since, except for the sign, the same K-values appear on either side of $K = 0$, Eq. (4.107) can be written

$$G_{vr}(E, J) = \frac{1}{\sigma}\left(G_v(E) + 2\sum_{K=1}^{K_{max}} G_v[E - E_r(K)]\right) \tag{4.108}$$

An entirely analogous equation applies to the density of states $N_{vr}(E, J)$.

It is evident that the K-rotor restriction will be of consequence principally at low J and high E.

4.5.10 A general S-D routine for $G_{vr}(E, J)$

Equation (4.107) does not present any fundamental difficulty except that the sum/density-of-states S-D routine has to be run $K_{max} + 1$ times at each value of E. This is not necessary since, to obtain an excellent approximation to $G_{vr}(E, J)$ in a single S-D pass it suffices to invert the product of the vibrational partition function Q_v and the restricted K-rotor partition function $Q_{rK}(J)$ (Appendix 1, Section A1.4).

A more general partition function, of which $Q_{rK}(J)$ may be considered a special case, is one that, for the present purpose, we may call $Q_{xx}(J)$. The general form is

$$Q_{xx}(J) = \exp[-c/(kT)] \frac{b}{\sigma}(kT)^{r/2} \operatorname{Fn}\left\{\frac{a}{(kT)^{1/2}}\right\} \quad (4.109)$$

The inversion of $Q_v Q_{xx}(J)$ by the S-D method, as described below, can be specialized to particular cases merely by substituting into $Q_{xx}(J)$ appropriate values of the constants a, b, c, and r. One such particular case is the J-conserved partition function $Q_{xi}(J)$, the inversion of which is discussed later in Section 8.8.

In the application to $Q_{rK}(J)$, the constants to be used in Eq. (4.109) are $a = (J + \frac{1}{2})\sqrt{|A - B|}$, $r = 1$, and $c = 0$; σ is the same symmetry number that appears in Eq. (4.107). In addition, Fn stands for the error function erf in the case of the prolate top, in which case $b = \sqrt{\pi/(A - B)}$ (Appendix 1, Eq. (A1.20)), or for the modified Dawson function daw for the oblate top, in which case $b = 2/\sqrt{(B - A)}$ (Appendix 1, Eq. (A1.22)).

For the purpose of the inversion of $Q_v Q_{xx}(J)$ we define, as usual, the function $\phi(x)$ ($s = 1/(kT)$, x is the real part of s), which is the logarithm of the inversion integrand (Eq. (4.72)); thus

$$\phi(x) = \ln Q_v(x) + (E - c)x + \ln\left(\frac{b}{\sigma}\right) - (k + r/2)\ln x + \ln \operatorname{Fn}(a\sqrt{x}) \quad (4.110)$$

To keep the routine perfectly general, we have included the parameter k such that $k = 0$ selects the density, and $k = 1$ selects the sum of states. If we let $D(x) = \partial \ln \operatorname{Fn}(a\sqrt{x})/\partial x$, then

$$D(x) = \frac{a \exp(-a^2 x)}{\sqrt{\pi x}\, \operatorname{erf}(a\sqrt{x})} \qquad \text{if } \operatorname{Fn} = \operatorname{erf} \quad (4.111)$$

or

$$D(x) = \frac{a \exp(a^2 x)}{2\sqrt{x}\, \operatorname{daw}(a\sqrt{x})} \qquad \text{if } \operatorname{Fn} = \operatorname{daw} \quad (4.112)$$

The first derivative of $\phi(x)$ is

$$\frac{\partial \phi(x)}{\partial x} = \frac{\partial \ln Q_v(x)}{\partial x} + (E - c) - \frac{k + r/2}{x} + D(x) \quad (4.113)$$

and the second is

$$\phi''(x) = \frac{\partial^2 \ln Q_v(x)}{\partial x^2} + \frac{k + r/2}{x^2} - D(x)\left(D(x) \pm a^2 + \frac{1}{2x}\right) \quad (4.114)$$

The term $\pm a^2$ applies with the $+$ sign if $\operatorname{Fn} = \operatorname{erf}$, and with the $-$ sign if $\operatorname{Fn} = \operatorname{daw}$. If x^* is the solution of $\partial \phi(x)/\partial x = 0$ (Eq. (4.113)), the general result, to first order, for the density or sum of states of the combined system of oscillators and degrees of freedom represented by $Q_{xx}(J)$, is given by the standard first-order expression (Eq. (4.69)) which uses the above $\phi(x)$ and $\phi''(x)$ evaluated at $x = x^*$:

$$\left.\begin{array}{c} N_{vxx}(E, J) \\ k = 0 \\ \\ G_{vxx}(E, J) \\ k = 1 \end{array}\right\} = \frac{\exp[\phi(x^*)]}{(2\pi \phi''(x^*))^{1/2}} \tag{4.115}$$

If need be, the routine can be taken to second order (Eq. (4.70)) using third and fourth derivatives of $\phi(x)$ (Problem 4.18).

Given that the S-D method is an iterative routine, one has the choice of running Eqs. (4.110)–(4.115) for several energies E at a given J, or for several J at a given E. In the case when $Q_{xx}(J) \equiv Q_{rK}(J)$, this single-pass first-order S-D routine yields results that are virtually identical with those calculated from the "exact" Eq. (4.108).

4.6 An overview

It has been stated in Section 3.3.3 that, provided that the requisite basic information at the molecular level is available, a quite realistic unimolecular rate constant can be obtained if realistic densities (for the reactant) and sums of states (for the transition state) are calculated. The object of the preceding sections of this chapter was precisely to develop the tools for such calculations.

The advantage of the S-D technique developed here is that, by appropriate choice, a single routine is sufficient for calculations involving any combination of degrees of freedom, and this in a single pass at specified total energy.

4.6.1 General aspects

Inasmuch as this chapter is concerned primarily with the S-D approximation, it is of interest to consider some general aspects not hitherto mentioned. These are, first, practical considerations regarding the numerical implementation; and second, since various other approximation methods for the sum or density of states generally yield fairly acceptable results at high energies, but serious inaccuracies often appear at low energies, it is necessary to see how the S-D method performs under these conditions.

The S-D method provides a smooth-function approximation, so that, if the energy-level spectrum of the oscillator collection is sparse, vibrational-only sums and densities at low energy will not reflect the staircase character of the sum of states (cf. Fig. 4.1). However, this is not the case for the vibrational–rotational sum of states, owing to the much denser rotational level spectrum which has a smoothing effect (cf. Problem 4.5).

One not inconsiderable advantage of the S-D method with respect to the exact-count algorithm is that it can provide the sum or density of states at any single, or only a few, energies, whereas an exact count always has to be started at energy zero.

4.6.2 Practical aspects

The equation $\phi'(x) = 0$ (Eq. (4.62)) requires a numerical solution for x^* in all cases of interest. This is readily accomplished by iteration, e.g. by the method of Newton which has been found to be quite reliable: if x_1 is the initial estimate of x^*, the next better estimate is $x_2 = x_1 - \phi'(x_1)/\phi''(x_1)$; this cycle is repeated until convergence. Note that the second derivative is needed in any event for the final result. In general about five iterations are sufficient to calculate x^* to within 10^{-10}, if energies are in increasing or decreasing order, and if x^* obtained at one energy is used as the initial estimate for iteration at the next energy. This speeds up convergence since x^* is only weakly energy-dependent.

The first-order approximation $I_1(k)$ is very robust in the calculation both of $N(E)$ ($k = 0$) and of $G(E)$ ($k = 1$). For $G(E)$ it yields about a 1%–3% overcount relative to the exact result except at very low energies, for which the overcount is worse. The second-order approximation $I_2(k)$ eliminates most of the overcount for $G(E)$ and yields virtually the exact result, except at very low energies (1–10 cm^{-1}), for which it may produce a negative result for $N(E)$.

4.6.3 Vibrational S-D results at low energy

Consider the S-D routine for a system of independent oscillators (harmonic or anharmonic) at very low energies, say 1 or 10 cm^{-1}, i.e. at energies much less than the energy of the lowest-frequency oscillator. Under these conditions we have $\partial \ln Q(x)/\partial x \cong 0$, so that the solution of Eq. (4.62) is to a good accuracy $x^* = k/E$ and $\phi''(x^*) = k/(x^*)^2 = E^2/k$, and hence

$$I_1(E, k) = \frac{E^{k-1} \exp(k)}{k^{k-1/2}\sqrt{2\pi}} \tag{4.116}$$

For $k = 1$, i.e. for the purely vibrational sum of states $G_v(E)$ (equivalent to the vibrational–rotational density $N_{vr}(E)$ for two free rotors with $q_r = 1$), this yields $I_1(E, k = 1) = 1.084$ at all sufficiently low energies, i.e. a small overcount with respect to the expected result $G_v(E) = 1$, since obviously there can be only one state at all energies between $E = 0$ and the energy of the lowest-frequency os-cillator. For the second-order S-D correction we have $\phi^{(3)}(x^*) = -2k/(x^*)^3$ and

$\phi^{(4)}(x^*) = 6k/(x^*)^4$, under the same conditions, so that, from Eq. (4.70), the correction factor is $1 + (3/4k) - (5/6k) = 1 - 1/(12k)$, independently of energy. Consequently the second-order result is $I_2(E, k = 1) = 1.084 \times 11/12 = 0.994$. These results are obtainable down to $E = 1 \, \text{cm}^{-1}$ ($E = 0$ is excluded) and for a collection of oscillators with the lowest frequency as low as $20 \, \text{cm}^{-1}$.

At higher energies, say $100 \, \text{cm}^{-1}$, $\partial \ell_n Q(x)/\partial x$ is in general no longer negligible, so $x^* \cong k/E$ becomes a poor approximation. However, the first-order S-D result is still within a few percent of unity, while the second-order correction overcompensates somewhat, yielding $I_2(E, k = 1) \geq 1$; however, the second-order S-D result becomes better than the first-order result, and essentially exact, at or above the energy of the lowest-frequency oscillator.

Using the above low-energy ($E \approx 1\text{--}10 \, \text{cm}^{-1}$) results for $\phi^{(3)}(x^*)$, the vibrational *density* calculated from $G(E)$ (Eq. (4.74)) under the same conditions ($k = 1, x^* = 1/E$) is zero. This can be understood by noting that, by exact counting, $G_v(E)$ is a constant equal to unity for all energies below those of the lowest-frequency oscillator, so that the density, defined as the derivative $dG_v(E)/dE$ (Eq. (4.6)), is zero. By contrast, direct calculation of $N_v(E)$ by the S-D routine with $k = 0$ will yield acceptable results only at energies larger than about $500 \, \text{cm}^{-1}$, regardless of the lowest frequency. This is not a serious drawback in the usual rate-constant calculations since $N_v(E)$ refers to the density of the reactant molecule at $E \geq E_0$, where E_0 is typically at least $10^4 \, \text{cm}^{-1}$.

4.6.4 Vibrational–rotational S-D results at low energy

Let us now take $m = k + r/2$, and compute the vibrational–rotational sum of states ($k = 1$) for r free rotors (or the vibrational–rotational *density* for $r + 2$ free rotors), and assume that $q_r = 1$. At low energies such that no oscillator is excited we have $\partial \ell_n Q(x)/\partial x \cong 0$, so that the solution of Eq. (4.62) is $x^* \cong m/E$, but, now that $m > 1$, this is a good approximation to x^* at energies considerably higher than in the case $m = k = 1$ (vibrations only).

When the energy is low such that no oscillator is excited, the sum/density of states will be for rotations only. From Eq. (4.18) we know that the analytical result for the sum of states for r free rotors should be $G_{vr}(E) = E^{r/2}/(r/2)! = E^{m-1}/(m-1)!$ (we still assume that $m = 1 + r/2$). For example, at $E = 1 \, \text{cm}^{-1}, r = 5$ (i.e. $m = 3.5$), the exact result is $G_{vr}(E) = 1/2.5! = 0.3009$, while Eq. (4.116) yields for the first-order S-D result $G_{vr}(E) = 0.3081$, and application of the second-order correction ($1 - (12 \times 3.5)^{-1} = 0.976$) brings the S-D result to 0.3035, only a 1% overcount. The results improve with rising energy: for $E = 10 \, \text{cm}^{-1}$ and the same $m = 3.5$, Eq. (4.116) yields for the first-order result $G_{vr}(E) = 97.44$, compared with the exact result 95.153. Application of the second-order correction ($= 0.976$)

brings this to 95.12, which is essentially the exact result, and similarly for higher energies when eventually the vibrational part kicks in.

A somewhat special case is $m = \frac{1}{2}$, i.e. the density of states of a single free rotor, for which there is a problem when $E \to 0$ (cf. Section 4.2). Assuming that $q_r = 1$, and taking the lowest rotational level at, say, $5\,cm^{-1}$, the analytical result of Eq. (4.19) is 0.2523 at $5\,cm^{-1}$, and 0.1784 at $10\,cm^{-1}$, whereas Eq. (4.116) yields $I_1(E, \frac{1}{2}) = [e/(2\pi E)]^{1/2}$, which amounts to 0.2941 and 0.2080 at the two energies, respectively; this is still a useful result despite the low energy.

References

[1] G. Herzberg, *Molecular Spectra and Molecular Structure*. II. *Infrared and Raman Spectra of Polyatomic Molecules* (D. van Nostrand, Princeton, NJ, 1960).

[2] T. Shimanouchi, *Tables of Vibrational Frequencies*, Consolidated Volume I (NSRDS-NBS 39, Washington, 1972) and various updates.

[3] R. C. Tolman, *The Principles of Statistical Mechanics* (Oxford University Press, Oxford, 1962).

[4] M. Christianson, D. Price and R. Whitehead, *J. Phys. Chem.* **78**, 2326 (1974).

[5] B. J. Hayward and B. R. Henry, *Chem. Phys. Lett.* **38**, 158 (1976).

[6] S. Nordholm, *Chem. Phys.* **129**, 371 (1989).

[7] W. Forst, *Theory of Unimolecular Reactions* (Academic Press, New York, 1973).

[8] W. Forst, *Chem. Rev.* **71**, 339 (1971).

[9] S. H. Lin and H. Eyring, *Proc. Nat. Acad. Sci. USA* **68**, 402 (1971).

[10] P. J. Robinson and K. A. Holbrook, *Unimolecular Reactions* (Wiley, London, 1972).

[11] R. A. Marcus and O. K. Rice, *J. Phys. Coll. Chem.* **55**, 894 (1951).

[12] G. Z. Whitten and B. S. Rabinovitch, *J. Chem. Phys.* **38**, 2466 (1963); *J. Chem. Phys.* **41**, 1883 (1964); cf. [7], Ch. 6. See also G. Song, *Chem. Phys.* **145**, 111 (1990).

[13] e.g. S. Glasstone, *Theoretical Chemistry* (D. van Nostrand, New York, 1955), p. 181.

[14] N. Moazzen-Ahmadi, H. P. Gush, M. Halpern, H. Jagannath, A. Leung and I. Ozier, *J. Chem. Phys.* **88**, 563 (1988). See also S. H. Robertson, D. M. Wardlaw and D. M. Hirst, *J. Chem. Phys.* **99**, 7748 (1993).

[15] S. C. King, J. F. Leblanc and P. D. Pacey, *Chem. Phys.* **123**, 329 (1988).

[16] S. C. Smith, M. J. McEwan and R. G. Gilbert, *J. Chem. Phys.* **90**, 1630 (1989).

[17] M. J. T. Jordan, S. C. Smith and R. G. Gilbert, *J. Phys. Chem.* **95**, 8685 (1991).

[18] W. L. Hase and L. Zhu, *Int. J. Chem. Kinet.* **26**, 407 (1994).

[19] P. D. Pacey, *J. Chem. Phys.* **77**, 3540 (1982).

[20] K. S. Pitzer, *J. Chem. Phys.* **5**, 469 (1937); K. S. Pitzer and W. D. Gwinn, *J. Chem. Phys.* **10**, 428 (1942).

[21] H. Eyring, J. Walter and G. E. Kimball, *Quantum Chemistry* (Wiley, New York, 1963).

[22] M. Abramowitz and A. S. Stegun, *Handbook of Mathematical Functions* (NBS Applied Mathematics Series 55, Washington, 1970).

[23] V. D. Knyazev, I. A. Dubinsky, I. R. Slagle and D. Gutman, *J. Phys. Chem.* **98**, 5279 (1994).

[24] G. E. Roberts and H. Kaufman, *Table of Laplace Transforms* (W. B. Saunders Co., Philadelphia, 1966).

[25] I. S. Gradshteyn and I. M. Ryzhik, *Table of Integrals, Series and Products* (Academic Press, New York, 1965).

[26] H. S. Robertson and D. M. Wardlaw, *Chem. Phys. Lett.* **199**, 391 (1992). There is a misprint in Eq. (29).

[27] D. G. Truhlar, *J. Comp. Chem.* **12**, 266 (1991); Y. Y. Chuang and D. G. Truhlar, *J. Chem. Phys.* **112**, 1221 (2000). See also R. B. McClurg, R. C. Flagan and W. A. Goddard III, *J. Chem. Phys.* **106**, 6675 (1997); R. B. McClurg, *J. Chem. Phys.* **108**, 1748 (1998).

[28] R. Kubo, *Statistical Mechanics* (North-Holland, Amsterdam, 1965).

[29] W. Forst, Z. Prášil and P. St. Laurent, *J. Chem. Phys.* **46**, 3736 (1967).

[30] W. Forst and Z. Prášil, *J. Chem. Phys.* **51**, 3006 (1969).

[31] M. R. Hoare and T. W. Ruijgrok, *J. Chem. Phys.* **52**, 113 (1970).

[32] M. R. Hoare, *J. Chem. Phys.* **52**, 5695 (1970); *erratum J. Chem. Phys.* **55**, 3058 (1971).

[33] R. H. Fowler, *Statistical Mechanics* (Cambridge University Press, Cambridge, 1966), pp. 36ff. See also D. Rapp, *Statistical Mechanics* (Holt, Rinehart and Winston, New York, 1972), Ch. 2.

[34] See, for example, R. K. Pathria, *Statistical Mechanics* (Pergamon Press, Oxford, 1977).

[35] C. E. Klots, *Int. Rev. Phys. Chem.* **15**, 205 (1996).

[36] T. Beyer and D. F. Swinehart, *Commun. ACM* **16**, 379 (1973).

[37] S. E. Stein and B. S. Rabinovitch, *J. Chem. Phys.* **58**, 2438 (1973); *J. Chem. Phys.* **60**, 98 (1974).

[38] S. E. Stein and B. S. Rabinovitch, *Chem. Phys. Lett.* **49**, 183 (1977).

[39] T. Baer and W. L. Hase, *Unimolecular Reaction Dynamics* (Oxford University Press, New York, 1996).

[40] *JANAF Thermochemical Tables*, 2nd Ed. (NSRDS-NBS 37, Washington, 1971).

[41] W. Forst, *Chem. Phys. Lett.* **231**, 43 (1994).

[42] W. Forst and Z. Prášil, *J. Chem. Phys.* **53**, 3065 (1970).

[43] W. Forst. *J. Comput. Chem.* **17**, 954 (1996). See also W. Witschel and C. Hartwigsen, *Chem. Phys. Lett.* **273**, 304 (1997).

[44] E. B. Wilson Jr, *Chem. Rev* **27**, 17 (1940).

[45] J. H. Current and B. S. Rabinovitch, *J. Chem. Phys.* **38**, 783 (1963).

[46] W. Forst and P. St. Laurent, *Can. J. Chem.* **45**, 3169 (1967).

[47] W. H. Miller, *J. Am. Chem. Soc.* **101**, 6810 (1979).

[48] S. C. Smith, *J. Phys. Chem.* **97**, 7034 (1993).

[49] W. Forst, *J. Phys. Chem.* **95**, 3612 (1991).

[50] L. Zhu, W. Chen, W. L. Hase and E. W. Kaiser, *J. Phys. Chem.* **97**, 311 (1993).

[51] W. Forst, *Computers Chem.* **20**, 419 (1996).

[52] M. Quack and J. Troe, *Ber. Bunsenges. Phys. Chem.* **78**, 240 (1974).

Problems

Problem 4.1 Show that, if the Hamiltonian is $H = p^2/(2m)$ (a free particle of mass m in one dimension), Eq. (4.7) gives

$$G(E) = (2\ell/h)(2mE)^{1/2}$$

where ℓ is the linear dimension occupied by the particle.

Problem 4.2 Using Eq. (4.19) and the convolution of Eq. (4.14), show that the sum of states for a combined one- and two-dimensional rotor with rotational constants (symmetry numbers) $A(\sigma_A)$ and $B(\sigma_B)$, respectively, is

$$G_r(E) = \frac{4E^{3/2}}{3\sigma_B\sigma_A B\sqrt{A}}$$

Compare this result with $G_r(E) = 2E^{3/2}/(\sigma_B\sigma_A B\sqrt{A})$ obtained as the *product* of the sum of states for one- and two-dimensional rotors, and with the analytical formula in Appendix 1 (Eq. (A1.8)) if $\sigma = \sigma_A\sigma_B$.

Problem 4.3 Show that the partition function for a prolate symmetric top (Eq. (4.22)) that respects the constraint $-J \le K \le +J$, neglecting the symmetry numbers, is

$$Q_r = \left(\frac{\pi(kT)^3}{(A-B)B^2}\right)^{1/2} \int_0^\infty \exp(-y)\,\mathrm{erf}\left(\sqrt{y\frac{A-B}{B}}\right) dy$$

where erf(x) is the error function (Appendix 1, Eq. (A1.23)). Evaluation of the integral analytically ([25], p. 649) yields $\sqrt{1-B/A}$; thus $Q_r = [\pi(kT)^3/(AB^2)]^{1/2}$. (Ignoring the K-constraint amounts to taking the integral equal to unity.) Use this Q_r to obtain the corresponding sum of states $G_r(E)$ and compare it with the results of the preceding problem.

Problem 4.4 Obtain the following expression for the combined density of states of r rotors and v classical oscillators, using the convolution of Eq. (4.13) and appropriate derivatives of formulas in Appendix 1:

$$N_{vr}(E) = \frac{q_v q_r E^{v-1+r/2}}{\Gamma\left(\frac{r}{2}\right)\Gamma(v)} \mathcal{B}\left(\frac{r}{2}, v\right)$$

where $\mathcal{B}(r/2, v)$ is the beta function [22, p. 258, items 6.2.1 and 6.2.2]:

$$\mathcal{B}\left(\frac{r}{2}, v\right) = \int_0^1 t^{r/2}(1-t)^{v-1}\,dt = \frac{\Gamma\left(\frac{r}{2}\right)\Gamma(v)}{\Gamma\left(\frac{r}{2}+v\right)}$$

Problem 4.5 Show that the density for one free and one hindered rotor (Eq. (4.55)), convoluted with the density of two additional free rotors, yields

$$N_r(E) = \begin{cases} q_r \dfrac{V_0}{2\pi}[(1-y^2)^{1/2} - y\cos^{-1}y], & E < V_0 \\ q_r(E - V_0/2), & E \ge V_0 \end{cases}$$

where $y = 1 - 2E/V_0$ and q_r is the kT-independent part of the product of partition functions for the four rotors (including q_{1f} for the hindered rotor). This density is a smooth function of E throughout the range of energies.

Note. In the following, the first-order S-D method is adopted throughout; $\omega_i = hv_i$. Except when noted otherwise, energy is understood to be in excess of the ground-state energy.

Problem 4.6 (a) For a particle of mass m in unit three-dimensional volume, derive by the S-D method the following analytical expression for translational states:

$$I_1(E, k) = \frac{q_t E^{n+k-1}\exp(n+k)}{(n+k)^{n+k-1/2}\sqrt{2\pi}}$$

where $n = \frac{3}{2}$ and $q_t = (2\pi m)^{3/2}/h^3$.

(b) Obtain $I_1(E, k)$ directly by evaluating the inverse transform of $q_t \mathcal{L}^{-1}\{s^{-(n+k)}\}$ $(n = \frac{3}{2})$ and compare it with above result after replacing the factorial by Stirling's approximation $n! \approx (n/e)^n \sqrt{2\pi n}$.

Problem 4.7 As an example of the approximation in Eq. (4.73), consider the system of Problem 4.15, where $x^* = (v + k)/E$. Use $x^*(1) = (v + 1)/E$ to obtain $G(E)$ "exact" (as $I_1(E, k = 1)$), and $x^*(0) = v/E$ to obtain $N(E)$ "exact" (as $I_1(E, k = 0)$). Show that the ratio

$$R = \frac{G(E) \text{ "exact"}}{G(E) \text{ "approx"}} = e \left(\frac{v}{v+1} \right)^{v+1/2} \approx 1 \ (e = 2.718\ldots)$$

where $G(E)$ "approx" $= N(E)$ "exact"$/x^*(0)$. Actual evaluation gives $R = 0.96$ for $n = 1$, and $R = 0.999$ for $n = 10$.

Problem 4.8 Derive and verify Eq. (4.74) using data of Problem 4.15.

Problem 4.9 Show that the correspondences between the various derivatives of $\phi(x^*)$ in the exponential and logarithmic formulations are

$$\phi''(x^*) \Rightarrow \theta^2 \phi''(\theta) = D^{(1)} + D^{(2)} + kC^2$$
$$\phi^{(3)}(x^*) \Rightarrow -2\theta^2 \phi''(\theta) - \theta^3 \phi^{(3)}(\theta) = D^{(1)} + 3D^{(2)} + D^{(3)} + 2kC^3$$
$$\phi^{(4)}(x^*) \Rightarrow 4\theta^2 \phi''(\theta) + 5\theta^3 \phi^{(3)}(\theta) + \theta^4 \phi^{(4)}(\theta) = D^{(1)} + 7D^{(2)} + 6D^{(3)} + D^{(4)} + 6kC^4$$

where $D^{(n)} = \theta^n \, \partial^n \ln Q(\theta)/\partial\theta^n$, $C = (\ln \theta^{-1})^{-1}$, and $\theta = \exp(-x^*)$.

Problem 4.10 Show that the equivalent of Eq. (4.74) in the logarithmic formulation is

$$N(E) \cong G(E) \left\{ \ln z^{-1} - \frac{1}{z^2 \phi''(z)} - \frac{z^3 \phi^{(3)}(z)}{2(z^2 \phi''(z))^2} \right\}_{z=\theta}$$

Problem 4.11 The approximation used in Problem 1.4, adapted to Eq. (4.86) but using the notation of Chapter 1, is $(n + s)!/n! \approx n^s$. A considerably better approximation is

$$\frac{(n+s)!}{s!} \approx \left(n + \frac{s}{2} \right)^s$$

where it is sufficient that n (the number of quanta) be merely larger than s (not *very much* larger than s, which is the number of oscillators). Show that the energy analog of Eq. (4.86) becomes, for s degenerate oscillators of frequency v (cf. Eq. (4.23)),

$$G_v(E) \approx \frac{(E + E_z)^s}{s!(hv)^s}$$

where $E_z = shv/2$ is the zero-point energy of the s oscillators.

Problem 4.12 Show that, for an n-fold-degenerate quantum harmonic oscillator of "frequency" ω, the first-order S-D method gives for the density

$$N(E) = \frac{[1 - \exp(-x^*\omega)]^{1-n} \exp(x^*E)}{\omega \sqrt{2\pi n} \exp(-x^*\omega)}$$

where

$$x^* = \frac{1}{\omega} \ln \left(1 + \frac{n\omega}{E} \right)$$

By approximating the logarithm when $1 \gg n\omega/E$, i.e. at high energies, we find $x^* \cong n/E$.

Problem 4.13 If Q_V is the partition function for a collection of independent quantum harmonic oscillators (Eq. (4.87)), show that the successive derivatives of $\ln Q_V$ for substitution into Eq. (4.71) are

$$\frac{\partial \ln Q_V}{\partial x} = -\sum_i F_i$$

$$\frac{\partial^2 \ln Q_V}{\partial x^2} = \sum_i \left(\omega_i F_i + F_i^2 \right)$$

$$\frac{\partial^3 \ln Q_V}{\partial x^3} = -\sum_i \left(\omega_i^2 F_i + 3\omega_i F_i^2 + 2F_i^3 \right)$$

$$\frac{\partial^4 \ln Q_V}{\partial x^4} = \sum_i \left(\omega_i^3 F_i + 7\omega_i^2 F_i^2 + 12\omega_i F_i^3 + 6F_i^4 \right)$$

where $F_i = \omega_i \exp(-\omega_i x)[1 - \exp(-\omega_i x)]^{-1}$.

Problem 4.14 For a collection of independent quantum harmonic oscillators (Eq. (4.87)), show that the relations equivalent to $\phi^{(n)}(x)$ (Eq. (4.71)) in the logarithmic formulation are entirely analogous to those deduced in Problem 4.13, except for a different definition of D_i:

$$\phi''(x) \Rightarrow \sum_i \left(\omega_i D_i + D_i^2 \right) + kC^2$$

$$-\phi^{(3)}(x) \Rightarrow \sum_i \left(\omega_i^2 D_i + 3\omega_i D_i^2 + 2D_i^3 \right) + 2kC^3$$

$$\phi^{(4)}(x) \Rightarrow \sum_i \left(\omega_i^3 D_i + 7\omega_i^2 D_i^2 + 12\omega_i D_i^3 + 6D_i^4 \right) + 6kC^4$$

where $D_i = \omega_i z^{\omega_i}/(1 - z^{\omega_i})$, and $C = (\ln z^{-1})^{-1}$.

Problem 4.15 Consider a collection of v harmonic oscillators ("frequencies" ω_i) at high temperature $(x \to 0)$, i.e. in the classical limit. Show that $x^* = (v + k)/E$ and

$$I_1(E, k) = \frac{E^{v-1+k} e^{v+k}}{(v + k)^{v+k-1/2} \sqrt{2\pi} \prod_i \omega_i}$$

Problem 4.16 Shift the energy zero of a collection of v classical oscillators to aE_z, where $E_z = \frac{1}{2}\Sigma_i \omega_i$ is the zero-point energy of the oscillators and a is a positive constant. Obtain the corresponding partition function, from which, using the S-D method and Appendix 2, Section A2.2, show that the result for the sum or density of states is

$$I_1(E, k) = \frac{(E + aE_z)^{v+k-1}}{\Gamma(v + k) \prod_i \omega_i}$$

where E retains the usual definition as energy in excess of the ground-state energy. This is the semi-classical approximation to the sum or density of states with adjustable zero-point energy (cf. text after Eq. (4.23)) which forms the basis of the well-known Whitten–Rabinovitch empirical formula [12] in which a is made energy-dependent.

Problem 4.17 Show that the higher derivatives of the general partition function Q defined as $S^{(0)}$ of Eq. (4.88), using $D^{(n)} = S^{(n)}/Q$, are

$$\frac{\partial^3 \ln Q}{\partial x^3} = D^{(3)} - 3D^{(1)}D^{(2)} + 2(D^{(1)})^3$$

$$\frac{\partial^4 \ln Q}{\partial x^4} = D^{(4)} - 4D^{(1)}D^{(3)} + 12D^{(2)}(D^{(1)})^2 - 3(D^{(2)})^2 - 6(D^{(1)})^4$$

Problem 4.18 Show that the third and fourth derivatives of $\phi(x)$ in Eq. (4.110), using $D \equiv D(x)$ defined by Eq. (4.111), are

$$\frac{\partial^3 \phi(x)}{\partial x^3} = \frac{\partial^3 \ln Q_v(x)}{\partial x^3} - \frac{2(k+n)}{x^3} + D\left(a^4 + \frac{a^2}{x} + \frac{3}{4}x^2\right) + D^2\left(3a^2 + \frac{3}{2x}\right) + 2D^3$$

$$\frac{\partial^4 \phi(x)}{\partial x^4} = \frac{\partial^4 \ln Q_v(x)}{\partial x^4} + \frac{k+n}{x^4} - D\left(a^6 + \frac{3a^4}{2x} + \frac{9a^2}{4x^2} + \frac{15}{8x^3}\right)$$

$$- D^2\left(7a^4 + \frac{7a^2}{x} + \frac{15}{4x^2}\right) - D^3\left(12a^2 + \frac{6}{x}\right) - 6D^4$$

5

Unimolecular reactions in a thermal system

5.1 Introduction

In a thermal system reactant accumulates or loses energy by collisional energy transfer. As a result of the randomness of collisions, this produces, depending on temperature, a reactant with a more or less wide initial distribution of internal energies and angular momenta. The consequence is that microscopic information regarding subsequent events on the molecular level is largely obliterated and temperature is left as the principal macroscopic variable by which to characterize a thermal system. This is not as much of a disadvantage as it may seem, for thermal systems nevertheless provide much information about reaction mechanisms and thermodynamics; in fact, thermal systems were historically the first to be examined experimentally. Moreover, situations corresponding to a thermal system underlie many natural phenomena, as well as systems of practical importance (e.g. combustion), so that thermal systems are of considerably more than passing interest even if they are ill-suited for probing details of molecular mechanics on the microscopic level.

5.1.1 An overview of thermal systems

The term "thermal system" will be used here in the larger sense of any system in which collisions are the principal mechanism of accumulation or degradation of energy, and the state of the system is monitored by determining some average overall (i.e. bulk) property. Collisional energy transfer *per se* is a vast subject that will only be touched upon as the need arises (for details the reader is referred to a number of available reviews [1–3]). In the following our consideration will be limited to aspects of collisional transfer directly linked to thermal reactions.

One early example of a thermal reaction is the simple bulb experiment in which the reactant is introduced into a heated bulb and the contents are analyzed at suitable time intervals; this is often referred to as the static system. If the temperature is high enough, the reactant will accumulate by collisions sufficient energy to react; hence, provided that the reaction is not too rapid, information about the rate constant for decomposition and the nature of products can be obtained. A classic example is the study of the structural isomerization of methyl isocyanide $CH_3NC \rightarrow CH_3CN$ by Rabinovitch and co-workers [4], which has played an important part in the theory of thermal reactions.

Over time variants of the bulb experiment were developed with a view to measuring rates of fast reactions, as in a flow reactor at moderate temperatures, or in a shock tube at high temperatures. In either case such experiments consist of bringing reactant (usually at room temperature) into contact with excess inert gas at elevated temperature. Briefly, the initial conditions in these systems can be characterized as "cold" reactant and "hot" heat bath, involving energy transfer from heat bath to reactant, eventually followed by chemical reaction. A reaction that has been investigated by all of these methods is $C_2H_6 \rightarrow CH_3 + CH_3$: static [5], flow-reactor [6], and shock-tube [7] techniques have been applied, to mention only the more recent work.

Laser-photolysis or chemical-activation experiments (discussed below in Section 5.7) represent a somewhat different case in that initially an excited reactant is produced with a fairly narrow distribution of energies and angular momenta, dispersed generally in excess room-temperature inert gas; this excitation is then degraded over time by collisions and chemical reaction. Here we have a case of initially "hot" reactant and "cold" heat bath, followed by loss of energy from reactant to heat bath. Reactions that have been studied by laser photolysis include the photoisomerization of various cycloheptatrienes [8] and the fragmentation of substituted benzenes by C–C or C–H bond splits [9].

5.1.2 A model of a thermal system

The collisional systems mentioned above, with the laser-photolysis or chemical-activation system seemingly different from the others, are actually similar because in all of them the direction of the collisional energy transfer is always in the sense of the energy gradient, i.e. "hot" → "cold"; in technical language this is referred to as "relaxation." Thus the collisional aspects of these systems can be dealt with within the same formalism since, insofar as the reactant is concerned, it is only the direction of the collisional energy transfer that is different, i.e. gain or loss of energy.

A simple and somewhat idealized model suitable for describing the principal characteristics of thermal systems of the kind mentioned above will be represented here by molecules of reactant M, assumed to be a not-too-small polyatomic with a quasi-continuum of vibrational energy levels, highly dispersed in a chemically inert buffer gas X, which acts as a heat reservoir of constant temperature T. By a suitable excitation mechanism (bulb, flow system, shock tube, laser, chemical activation) a flow of energy between the molecules of reactant and heat bath is initiated, which is entirely due to binary collisions M + X on account of the high dilution.

In general, of course, molecules possess degrees of freedom of rotation as well. Treatment of both vibrations and rotations leads to consideration of vibrational–rotational energy transfer, which represents a not inconsiderable complication that is deferred until Section 5.8. For now it will be assumed that only vibrational degrees of freedom matter, a useful simplification.

5.2 The master equation

A general description of the problem of collisional energy transfer, in the sense that it allows consideration of all collisional events, is represented by the so-called master equation [10], the derivation and solution of which will be the principal subject of most of this chapter.

In previous chapters the symbol E was used for molecular energy. At present we have to distinguish between post- and pre-collision energies, which, for notational simplicity, will be represented by x and y, respectively. Only when this distinction is not necessary, as in Section 5.8, will we revert to use of the symbol E for energy.

5.2.1 The transition probability

The central role in the treatment of every gas-phase collisional process, and hence also in the master equation, is played by the transition probability $p(x, y)$, defined here as the probability, per collision and per unit energy (in the continuous formulation), that, if a molecule had internal energy y before the collision, it shall have energy x after the collision. Thus $p(x, y)$ is an intrinsic molecular property of the collision partners, and represents therefore the fundamental information about a given relaxing system. Since it is postulated that the properties of the heat bath are constant in time, $p(x, y)$ is time-independent.

There are two conditions that every transition probability must satisfy. The first is the condition of normalization

$$\int_0^\infty p(x, y)\,\mathrm{d}x = 1 \qquad \text{for every} \quad y \qquad (5.1)$$

which ensures that every energy transfer process from, or to, any level y is accounted for. The second is the condition of detailed balance, defined, with

$y > x$, by

$$p(y, x)_{\text{up}} B(x) = p(x, y)_{\text{down}} B(y) \tag{5.2}$$

in which $B(w)$ $(w = x$ or $y)$ is the Boltzmann (equilibrium) distribution per unit energy, appropriate to bath temperature T:

$$B(x) = \frac{1}{Q_{\text{v}}} N_{\text{v}}(x) \exp[-x/(\mathcal{k} T)] \tag{5.3}$$

where $N_{\text{v}}(x)$ is the vibrational density of states (dimension energy^{-1}) at energy x, $\mathcal{k}$ is Boltzmann's constant, and $Q_{\text{v}} = \int_0^\infty N_{\text{v}}(x) \exp[-x/(\mathcal{k} T)] \, dx$ is the vibrational partition function (Eq. (4.3)). Detailed balance is the consequence of the time-reversal symmetry of the underlying equations of motion [11].

An example of a transition probability that is fairly tractable is the exponential model [1]

$$p(x, y) = \begin{cases} C(y) \exp[-(y - x)/\alpha], & y > x \text{ (down)} \\ C(y) \exp[-(x - y)/\beta], & x > y \text{ (up)} \end{cases} \tag{5.4}$$

where $C(y)$ is the normalization constant and α is a parameter with the dimension of energy that measures the strength of the coupling to the heat bath. In the absence of definite information α is generally assumed to be temperature-independent. The up-form follows by application of detailed balance (Eq. (5.2)).

This purely empirical transition probability (Fig. 5.1(a)) assigns the highest probability to $x = y$, i.e. an elastic collision with no net energy transfer, which expresses

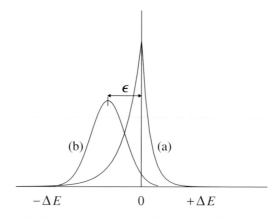

Fig. 5.1. (a) An example of the exponential transition probability of Eq. (5.4) as a function of $\Delta E = x - y$. This probability has a peak at $\Delta E = 0$ corresponding to an elastic collision, which transfers no energy. The probability is skewed toward negative ΔE. (b) An example of the offset Gaussian of Eq. (5.8). The maximum of this probability is shifted toward negative ΔE by offset ϵ.

the physically reasonable proposition that small energy transfers are more probable than large ones. It is therefore mostly used to model cases of inefficient energy transfer, generally involving rare gases, for which the coupling between reactant and heat bath is weak. Further details regarding the exponential model are the subject of Problem 5.1.

Recent experiments (for example, isomerization of cyclobutene into 1,3-butadiene, mediated by collisions with hexafluorobenzene [12]) have revealed the existence of so-called supercollisions, namely rare collisions that transfer large amounts of energy. Such cases may be modeled by a double-exponential model with the down-form of the probability given by [13]

$$p(x, y)_{\text{down}} = C(y)\{a \exp[-(y - x)/\alpha_1] + (1 - a) \exp[-(y - x)/\alpha_2]\} \quad (5.5)$$

where the first exponential term is the "ordinary" down-form, with "standard" α_1 and coefficient a close to unity. The second exponential term therefore is weighted by a very small coefficient $(1 - a)$ but large $\alpha_2 \gg \alpha_1$, which accounts for the supercollisions; $C(y)$ is the normalization constant. The result is that $p(x, y)_{\text{down}}$ decreases more slowly with $\Delta E = (y - x)$, as shown in Fig. 5.2, so that, even if $1 - a$ is very small, the effect of the second exponential term on average energy transfer is noticeable.

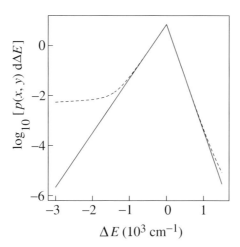

Fig. 5.2. A semi-logarithmic representation of the exponential transition-probability model. Solid line: the simple exponential model of Eq. (5.4) with $\alpha = 200\,\text{cm}^{-1}$ and $T = 300\,\text{K}$. Dashed line: the exponential supercollision model of Eq. (5.5), with $\alpha_1 = 200\,\text{cm}^{-1}$, $\alpha_2 = 5000\,\text{cm}^{-1}$, $a = 0.99$, and $T = 300\,\text{K}$. Comparison of the solid and dashed lines shows that supercollisions have little effect on small transfers of energy (up or down) but increase appreciably large down-transfers (negative ΔE).

An exponential-type transition probability that is based on theoretical arguments is, with $\delta = (y - x)$, and C_1, C_2, and C_3 explicit constants,

$$p(x, y)_{\text{down}} = \left(\delta^{4/3}/T^{1/6}\right) \exp\left[-C_1\left(\frac{\delta^2}{T}\right)^{1/3} + \frac{C_2\delta}{T} + \frac{C_3}{T}\right] \qquad (5.6)$$

which is known as the Schwartz–Slawsky–Herzfeld (SSH) formula [14]. It is derived for collision-induced transitions in a harmonic oscillator due to the repulsive part of the potential. It can be adapted to polyatomic molecules [15].

Another example of a simple model is the strong-collision probability (superscript "sc") which represents by contrast the strongest possible coupling between the reactant and the heat bath. Each collision is assumed to leave the reactant molecule in equilibrium with the heat bath at temperature T, so that the transition probability $p(x, y)^{\text{sc}}$ depends only on the final energy (x) and is independent of the initial energy (y):

$$p(x, y)^{\text{sc}} = B(x) \qquad (5.7)$$

where $B(x)$ is the equilibrium (Boltzmann) distribution at temperature T defined in Eq. (5.3). For interaction involving a reactant highly dispersed in a rare-gas heat bath this model is an extreme case, except that, if the reactant itself acts as its own heat bath (in the absence of any other bath gas), it often behaves as a strong collider. For gas–solid interaction strong-collision behavior is common [16], and is in fact the basis of the technique of very-low-pressure pyrolysis (VLPP) [17].

Another empirical, but more realistic, model is the off-set Gaussian, a simple form of which is defined by [1]

$$p(x, y) = \begin{cases} C \exp[-(y - x - \epsilon)^2/(4\epsilon \ell T)], & x < y \\ C \exp[-(x - y - \epsilon)^2/(4\epsilon \ell T)], & x > y \end{cases} \qquad (5.8)$$

$C = 2(\pi \epsilon \ell T)^{-1/2}$ is the normalization constant and ϵ is a constant of dimension of energy that determines the offset. The above transition probability, shown in Fig. 5.1(b), satisfies detailed balance for a quasi-diatomic system with constant density of states. By varying the offset ϵ the strength of the coupling to the heat bath can be varied. An off-set-type Gaussian probability can be derived from first principles for a specific model of reactant–bath-gas interaction, as for example in the biased-random-walk model [18].

In general, however, realistic models are neither simple nor tractable, in the sense that a numerical solution of the master equation is required.

5.2.2 Moments of the transition probability

A convenient means for characterizing a probability distribution is in terms of the moments. The first moment of $p(x, y)$ is defined by

$$\langle \Delta E(y) \rangle = \int_0^\infty (x - y) p(x, y) \, dx \tag{5.9}$$

which shows that $\langle \Delta E(y) \rangle$ is the average energy transferred per collision. It represents the outcome of *all* single-collision events (both up and down) and thus is an average *microscopic* property suitable for representing the intrinsic energy-transfer properties of a given relaxing M + X pair. For the strong-collision and simple-exponential model $\langle \Delta E(y) \rangle$ can be obtained analytically (Problems 5.2 and 5.3).

$\langle \Delta E(y) \rangle$ represents the average *total* energy transferred in a (single) collision, i.e. the algebraic sum of energy lost and gained. It still depends on the initial energy y, and as a result may be positive (mostly "up-collision") or negative (mostly "down-collision") [19]. This may be seen more clearly if Eq. (5.9) is written

$$\langle \Delta E(y) \rangle = - \int_0^y (y - x) p(x, y)_{\text{down}} \, dx + \int_y^\infty (x - y) p(x, y)_{\text{up}} \, dx$$

$$= - \langle \Delta E(y) \rangle_{\text{down}} + \langle \Delta E(y) \rangle_{\text{up}} \tag{5.10}$$

so that $\langle \Delta E(y) \rangle$ is now divided up between the average loss and average gain of energy per collision. If only the up-form is made temperature-dependent, the relative sizes of the down- and up-forms – and therefore the sign of $\langle \Delta E(y) \rangle$ (meaning that overall there will be loss or gain of energy by collision) – will depend on temperature. Higher moments of $p(x, y)$ can be defined similarly (Problem 5.4):

$$\langle \Delta E^k(y) \rangle = \int_0^\infty (x - y)^k p(x, y) \, dx \tag{5.11}$$

Note that detailed balance introduces asymmetry into the transition probability (cf. Fig. 5.1), which drops off more steeply on the up-side. If this were not the case, all odd moments, in particular $\langle \Delta E(y) \rangle$, would be zero.

5.2.3 The master equation

Let the collisional process be M + X, with M = reactant and X = buffer gas, and suppose that M with sufficient internal energy will decompose according to the unimolecular reaction M → products. "Sufficient energy" means energy in excess of threshold E_0. The master equation has the form of an ordinary kinetic equation that functions as a book-keeping device whereby the positive and negative terms, respectively, keep track of all rate processes that populate and depopulate a given energy level of the reactant M.

If x or y represent vibrational energies of the reactant molecule and rotational energy is for the moment ignored, let $[M(y, t)] \, dy$ be the concentration of reactant molecules M having internal energy $y > 0$ in the interval $y, y + dy$ at time t. Rather than working with concentrations as in an actual kinetic equation, it is more convenient to define a time-dependent population density $n(y, t) = [M(y, t)]/N_0$, where $N_0 = \int_0^\infty [M(y, 0)] \, dy$ is the total *initial* population of M of all energies. Thus $n(y, t)$ is the fractional population, i.e. the *probability*, per unit energy, that, at time t, a reactant molecule M shall have internal energy in the interval dy at y. The master equation then takes the form

$$\frac{\partial n(x, t)}{\partial t} = \omega \int_0^\infty p(x, y) \, n(y, t) \, dy - \int_0^\infty p(y, x)(\omega + k(x)) \, n(x, t) \, dy \quad (5.12)$$

where $k(x)$ is the microcanonical or specific-energy unimolecular rate constant for reaction (i.e. loss) of molecules M having energy $x \geq E_0$ (Eq. (3.28), with $x \equiv E$). This rate constant is of course zero for all $x < E_0$. The first integral represents collisional input from all energy levels to level x; the second term on the RHS represents population loss of level x to all other levels by collisions, plus the loss by reaction if $x \geq E_0$.

The second integral in Eq. (5.12), which involves only $p(y, x)$, is redundant if the transition probability is normalized according to Eq. (5.1). In the following we will always assume that this is the case, so the master equation will always be used in the form given below:

$$\frac{\partial n(x, t)}{\partial t} = \omega \int_0^\infty p(x, y) \, n(y, t) \, dy - (\omega + k(x)) \, n(x, t) \quad (5.13)$$

5.2.4 The collision frequency

The parameter ω is the collision frequency (dimension s^{-1}), so the product $\omega p(x, y)$ (or $\omega p(y, x)$) is the (first-order) *rate constant* for the transfer $x \leftarrow y$ (or $y \leftarrow x$), as of course it must be if the master equation is to be a kinetic equation defined in terms of concentrations. We may note in passing that only the transfer rate constant $\omega p(x, y)$ is the actual experimental observable in a relaxation from which $p(x, y)$ can be extracted if ω is defined.

In principle, the collision frequency ω can be expected to be energy-dependent:

$$\omega = Z(y)\,[X]\,\mathrm{s}^{-1} \tag{5.14}$$

where $Z(y)$ is a second-order energy-transfer rate constant and $[X]$ is the total (reactant plus buffer gas) concentration or pressure in the system. A useful approximation is to replace $Z(y)$ by the energy-independent constant Z, given by the collision rate for structureless particles:

$$Z(y) \approx Z = \pi \int_0^\infty b_c^2\, v\, f(v)\,\mathrm{d}v \tag{5.15}$$

where πb_c^2 is the collision cross-section, b_c is the critical impact parameter (cf. Problem 8.1 and Appendix 4), v is the relative velocity of $M + X$ and $f(v)$ is the distribution of v's (e.g. the Maxwell–Boltzmann distribution). The collision frequency ω thus becomes y-independent and can be taken outside the integrals in Eq. (5.12). The explicit expression for the cross-section depends on the interparticle potential, usually assumed to be of hard-sphere type, in which case b_c is a v-independent constant given by the mean molecular diameter. For more elaborate potentials, such as Lennard-Jones potentials for non-polar partners (Appendix 4) and Stockmayer potentials for polar partners [20], b_c is v-dependent.

5.2.5 Reaction equilibrium

As time increases without limit, an actual reaction will reach equilibrium, when the forward and reverse reactions proceed at equal rates. If equilibrium is totally on the side of products, as is often the case, the reaction is in effect a one-way reaction; the reaction ends with the total disappearance of M. Such a course of events is implicit in the formulation of the master equation given here in Eq. (5.12). If equilibrium is not entirely on the side of products, there will be a back reaction, which the above formulation of the master equation ignores.

In such cases Eq. (5.12) applies only to an initial stage before the back reaction becomes significant. This is not critical if the main focus of interest concerns events associated with collisional energy transfer before the intrusion of the back reaction becomes important. If the system is non-reactive ($k(x) = 0$ for all x), there is of course no such limitation and Eq. (5.12) applies at all times.

The above formulation of the master equation does not take into account the possibility of non-collisional input, which will be discussed later in Section 5.7.

5.2.6 Symmetrization

Insofar as the actual solution of Eq. (5.12) is concerned, it is advantageous if the equation is first symmetrized. Since the transition kernel $p(x, y)$ satisfies detailed balance, it is readily symmetrized by the similarity transformation

$$S(x, y) = p(x, y) \left(\frac{B(y)}{B(x)} \right)^{1/2} = S(y, x) \qquad (5.16)$$

where $B(x)$ is the Boltzmann distribution (Eq. (5.3)). Dividing Eq. (5.13) by $\omega[B(x)]^{1/2}$ defines a new population by $n'(x, t) = n(x, t)/[B(x)]^{1/2}$, and a new transition kernel $S'(x, y)$ by

$$S'(x, y) = S(x, y) - \left(1 + \frac{k(x)}{\omega} \right) \delta(x - y) \qquad (5.17)$$

where the δ-function signifies that the term in large parentheses affects only the diagonal of $S'(x, y)$. We thus obtain (with $n(y, t) = n'(y, t)B(y)^{1/2}$) the symmetrized form of Eq. (5.13) in terms of dimensionless time ωt:

$$\frac{\partial n'(x, t)}{\partial(\omega t)} = \int_0^\infty S'(x, y) n'(y, t) \, dy \qquad (5.18)$$

For the purpose of numerical solution this equation is used below in discrete form.

5.2.7 The discrete formulation

Except for special forms of the transition probability, such as simple exponential (Appendix 5) and strong-collision forms (vide infra), analytical solutions of the master equation are not possible, so a numerical solution is necessary. To this end, it is useful to first set up the master equation in Eq. (5.13) in terms of discrete energies. The vibrational-energy quasi-continuum of a polyatomic reactant molecule M is discretized into m states or "bins," and the solution is sought in matrix form. Using the ket $(| \ldots \rangle)$ and bra $(\langle \ldots |)$ notation, where ket represents a column vector and bra a row vector, the dimensionless matrix equivalent of Eq. (5.13) is

$$\frac{\partial}{\partial(\omega t)} |\mathbf{N}\rangle = \mathbb{J} \cdot |\mathbf{N}\rangle \qquad (5.19)$$

where the raised dot shall henceforth indicate matrix multiplication. $\mathbb{J}$ is the square $m \times m$ transport matrix $\mathbb{J} = \mathbb{P} - \mathbb{I} - \mathbb{K}$, where $\mathbb{P}$ is the $m \times m$ matrix of the per-collision probabilities p_{ij} for the transition $i \leftarrow j$ (i.e. the analog of $p(x, y)$). $\mathbb{I}$ is the diagonal unit matrix and $\mathbb{K}$ is the diagonal matrix of dimensionless rate constants $k(x_i)/\omega$; these diagonal matrices affect only the diagonal of $\mathbb{P}$ and serve the same purpose as the δ-function in Eq. (5.17). $|\mathbf{N}\rangle = [n(x_1, t), \ldots, n(x_m, t)]^{\mathrm{T}}$ (T denotes the transpose) is the m-element column vector of the time-dependent population distribution (i.e. the discrete analog of $n(x, t)$).

Since matrix $\mathbb{P}$ represents probabilities, it is normalized to unity (cf. Eq. (5.1)), i.e. every column sums to unity:

$$\sum_i p_{ij} = 1 \qquad \text{for all} \quad j \tag{5.20}$$

The p_{ij} satisfy detailed balance (cf. Eqs. (5.2) and (5.3)),

$$p_{ij} B(x_j) = p_{ji} B(x_i) \qquad \text{for all} \quad i, j \tag{5.21}$$

where $B(x_i)$ is the discrete Boltzmann distribution,

$$B(x_i) = W(x_i) \exp[-x_i/(\ell T)]/(pf) \tag{5.22}$$

and $pf = \Sigma_i W(x_i) \exp[-x_i/(\ell T)]$ is the partition function, where $W(x_i)$ is the number of states of reactant M at x_i. $W(x_i)$ is the discrete equivalent of the density of states $N(x)$ in a system with a continuum of energies (cf. Eq. (4.4)), and pf is the equivalent of Q_v in Eq. (5.3) (cf. Eq. (4.1)). We shall denote by $|\mathbf{B}\rangle$ the column vector of the equilibrium distribution $B(x_i)$.

5.3 Formal solution of the master equation

The formal solution of Eq. (5.19) is

$$\langle \mathbf{N}| = \langle \mathbf{N}_0| \cdot \exp(\mathbb{J}t) \tag{5.23}$$

where $\langle \mathbf{N}_0|$ is the row vector of the initial population. We shall consider an alternative solution in terms of an eigenfunction–eigenvector expansion, which is more instructive for the present purpose.

5.3.1 Eigenfunction–eigenvector expansion

It is useful first to summarize some known [21; 22, Chapter 6] general properties of the solution of Eq. (5.19). A numerically well-behaved eigenfunction–eigenvector expansion is obtained most readily if matrix $\mathbb{J}$ is first symmetrized by the same similarity transformation as was used above in connection with Eq. (5.16):

$$\mathbb{J}_s = \left(\text{diag } \mathbb{B}^{-1/2}\right) \cdot \mathbb{J} \cdot \left(\text{diag } \mathbb{B}^{1/2}\right) \tag{5.24}$$

where diag $\mathbb{B}^n$ is a diagonal matrix with elements $[B(x_i)]^n$. Thus $\mathbb{J}_s$ is the discrete analog of $S'(x, y)$ in Eq. (5.17) and the discretization is now applied to the solution of Eq. (5.18) in terms of dimensionless time ωt. Both $\mathbb{J}$ and the symmetrized matrix

$\mathbb{J}_s$, of size $m \times m$, will have the same set of m eigenvalues μ_k ($0 \leq k \leq m - 1$), understood to be dimensionless if the time is dimensionless. In general, algorithms for finding the eigenvalues of $\mathbb{J}_s$ are more robust than are algorithms for finding those of $\mathbb{J}$. The m associated eigenvectors of $\mathbb{J}_s$ form a matrix with columns $|\mathbf{f}_k\rangle$ that satisfy the eigenvalue equation (Problems 5.5 and 5.6)

$$\mathbb{J}_s \cdot |\mathbf{f}_k\rangle = \mu_k \cdot |\mathbf{f}_k\rangle \tag{5.25}$$

The eigenvectors of $\mathbb{J}_s$ (but not those of $\mathbb{J}$) corresponding to discrete eigenvalues are mutually orthonormal.

We shall assume that the eigenvalues are ordered in the sequence $\mu_0 \geq \mu_1 \geq \mu_2 \geq \ldots \geq \mu_{m-1}$. In terms of the eigenvalues μ_k and the eigenvector elements $f_k(x_i)$, the solution of Eq. (5.19) for elements $n(x_i, t)$ of $\mathbf{N}$ can be written

$$n(x_i, t) = \sum_{k=0}^{m-1} h_k \, \phi_k(x_i) \exp(\mu_k t) \tag{5.26}$$

where $\phi_k(x_i) = [B(x_i)]^{1/2} f_k(x_i)$. Note that $n(x_i, t)$ is the original population vector. The h_k are constants that depend on initial conditions (Problem 5.7):

$$h_k = \sum_i \frac{n(x_i, 0) f_k(x_i)}{[B(x_i)]^{1/2}} \tag{5.27}$$

Equation (5.26) is the basic result that will serve for the interpretation of principal characteristics of thermal systems. Note that the product $\mu_k t$ is always dimensionless; in particular, for t in seconds (s), the dimension of μ_k will be s^{-1}. In the following, μ_k will always be assumed to be dimensionless. For an explicit analytical form of the expansion in Eq. (5.26) in one particular case see Appendix 5.

5.3.2 Solution in a non-reactive system

It is worthwhile to deal first briefly with the principal features of the solution of the master equation in a non-reactive system since the presence of the reactive matrix $\mathbb{K}$ may be considered as basically a perturbation. In a non-reactive system, matrix $\mathbb{K} = 0$, so the system merely undergoes collisional energy transfer in the absence of reaction. The solution has the following properties.

(1) Matrix $\mathbb{J} = \mathbb{P} - \mathbb{I}$ is normalized to zero, with one eigenvalue equal to zero, all other ($m - 1$) eigenvalues (μ_k) being negative. Thus $\mu_0 = 0$, which ensures that, at long times, all terms in Eq. (5.26) except the first will vanish. Since the first eigenvector of the corresponding symmetrized matrix is $f_0(x_i) = [B(x_i)]^{1/2}$, and hence $h_0 = 1$, as is readily apparent from Eq. (5.27), it follows that, as $t \to \infty$, $n(x_i, t) \to B(x_i)$, meaning that, at long times, the population becomes the *thermal* equilibrium population.

(2) Since only energy transfer takes place, there is no loss of particles of M, so that the population $n(x_i, t)$ is normalized to unity at all times. This represents conservation of probability, which is reflected in the normalization to zero of matrix $\mathbb{J}$.

An exposition of the details of the eigenfunction–eigenvector expansion in a simple non-reactive system that has some pedagogic merit has been published [23].

5.3.3 Solution in a reactive system

The presence of a reactive channel modifies the solution of Eq. (5.26) in two ways.

(1) Since $\mathbb{K} \neq 0$, matrix $\mathbb{J}$ is no longer normalized to zero. The lack of normalization represents the loss probability owing to loss of particles of M due to reaction. In such a system $n(x_i, t)$ is normalized to unity only at $t = 0$, and at all other times

$$\sum_i n(x_i, t) < 1 \qquad \text{for all} \quad t > 0 \tag{5.28}$$

(2) Both $\mathbb{J}$ and the symmetrized matrix $\mathbb{J}_s$ have *all* eigenvalues (μ_k) negative (i.e. no eigenvalue is zero), so that, at long times ($t \to \infty$), all terms in Eq. (5.26) will vanish, meaning total disappearance of reactant molecules. An example of the distribution of eigenvalues and their dependence on pressure is shown in [24, Fig. 3.1].

A special case is the reversible isomerization reaction $A \leftrightarrow B$ in which the back reaction cannot be ignored. Here matrix $\mathbb{J}$ is normalized to zero since there is no net loss of particles; hence $\mu_0 = 0$. The rate therefore depends on higher eigenvalues in a complicated manner [25].

5.3.4 Steady state

Let $N(t) = \Sigma_i n(x_i, t)$ be the total fractional population of (intact) molecules M in a reactive system at time t; then the first-order rate constant for the disappearance of M is by definition $k(t) = -(N(t))^{-1} \, dN(t)/dt$, which is in principle time-dependent. If the smallest eigenvalue (in absolute value) $|\mu_0|$ is well separated from the others (the precise meaning of "well separated" is spelled out in Section 5.5), then, after a more or less short while, all terms in Eq. (5.26) except the first will die out and the expansion for $n(x_i, t)$ will reduce to the first term:

$$n(x_i, t) \to h_0\phi_0(x_i) \exp(\mu_0 t) = n(x_i, t)_{ss} \tag{5.29}$$

Let $h_0\phi_0(x_i) = n(x_i)_{ss}$, and then $n(x_i, t)_{ss} = n(x_i)_{ss} \exp(\mu_0 t)$, which means that the population relaxes in time with a simple exponential behavior. The period of time during which Eq. (5.29) is satisfied is called steady state; hence the subscript "ss." Since $N(t)_{ss} = \Sigma_i n(x_i)_{ss}$, it follows that $-(N(t))^{-1} \, dN(t)/d(\omega t)|_{ss} = -\mu_0$, so that $-\mu_0$ is in fact the (dimensionless) steady-state, and therefore time-independent, thermal rate constant. In dimension-bearing form it will be called henceforth k_{uni} (s^{-1}); thus $k_{uni} = -\omega\mu_0$. The higher eigenvalues μ_k correspond to relaxation modes that precede steady state, during which all properties of the system, including the rate constant (designated $k(t)$ above) are time-dependent.

5.3.5 The continuous formulation

If we drop the discrete index i, the picture is therefore that, as time progresses, the population density $n(x, t)$ becomes the steady-state population density $n(x, t)_{ss}$ (the continuous version of Eq. (5.29)), which is of particular interest in that its normalized form is time-independent. We can then define a time-*independent* (but pressure-dependent, via ω) steady-state distribution $P(x)_{ss}$:

$$P(x)_{ss} = \frac{n(x, t)_{ss}}{\int_0^\infty n(x, t)_{ss}\, dx} \tag{5.30}$$

This is in contrast with the non-reactive case, in which $n(x, t)$ evolves to the equilibrium population $B(x)$. Consequently, at steady state the rate constant $k(t)$ becomes independent of time and initial conditions. It is under steady-state conditions that $k(t)$, more commonly known as the overall or thermal rate constant k_{uni}, is identical to the rate constant $k_{uni} \equiv -\omega\mu_0$ defined above:

$$k_{uni} = \int k(x)P(x)_{ss}\, dx = -\omega\,\mu_0 \tag{5.31}$$

The steady-state distribution $P(x)_{ss}$ is in fact identical to the normalized eigenvector $\phi_0(x)$ corresponding to eigenvalue μ_0. The rate constant k_{uni} is thus obtainable from either the appropriate eigenvector or the eigenvalue of the transport matrix, which means that, for the determination of k_{uni}, a complete resolution of the master equation (all eigenvalues and/or eigenvectors) is not necessary.

5.3.6 The low-pressure limit

At the low-pressure limit the present model of a thermal unimolecular reaction amounts to a vibrational ladder with an absorbing barrier located at E_0, the reaction threshold. Collisions cause the reactant molecule M to execute the equivalent of a random walk up and down the ladder, until M reaches the absorbing barrier, at which point the molecule is counted as dissociated and assumed to be removed from the system. The rate is governed entirely by the energy-transfer process, so the rate is independent of the specific-energy rate constant $k(x)$. The master equation of

Eq. (5.12) therefore simplifies to

$$\frac{\partial n(x,t)}{\partial t} = \omega \left(\int_0^{E_0} p(x,y)\,n(y,t)\,\mathrm{d}y - n(x,t) \right) \tag{5.32}$$

where the upper limit on the integral is now E_0 instead of infinity.

In order to obtain the discrete equivalent of Eq. (5.32), one can take advantage of the fact that matrix $\mathbb{K}$ has elements that are zero below the threshold E_0. The transport matrix $\mathbb{J}$ can then be usefully partitioned at E_0 into four submatrices $\mathbb{J}_1, \mathbb{J}_2, \mathbb{J}_3$, and $\mathbb{J}_4$:

$$\mathbb{J} \Rightarrow \boxed{\begin{array}{c|c} \mathbb{J}_1 & \mathbb{J}_2 \\ \hline \mathbb{J}_3 & \mathbb{J}_4 \end{array}} \ldots E_0 \tag{5.33}$$

$$\vdots$$
$$E_0$$

Submatrix $\mathbb{J}_1$ involves only transitions below threshold, and $\mathbb{J}_4$ only those above threshold, while $\mathbb{J}_3$ involves activating transitions from below to above threshold, and $\mathbb{J}_2$ involves deactivating transitions from above to below threshold. Thus the random walk below threshold that describes the low-pressure limit corresponds to submatrix $\mathbb{J}_1$.

The consequence of the partitioning of matrix $\mathbb{J}$ is that submatrix $\mathbb{J}_1$ is not normalized to zero, although it does not contain the elements $k(x_i)$, so the lowest eigenvalue $|\mu_0|$ of $\mathbb{J}_1$ is not zero and the system is therefore reactive, as of course we expect. All properties of a system at the low-pressure limit can be obtained by application of the usual eigenfunction–eigenvalue expansion to matrix $\mathbb{J}_1$, which will have all the characteristics of a reactive system discussed in Section 5.3.3. In particular, the negative of the lowest eigenvalue of $\mathbb{J}_1$ (denoted $\mu_0(J_1)$) is the steady-state low-pressure rate constant k_0/ω (s^{-1}); thus $k_0 = -\omega\,\mu_0(J_1)$.

5.3.7 The incubation time

While the steady-state population $N(t)_{ss} = h_0 \exp(\mu_0 t)\Sigma_i \phi_0(x_i)$ decays exponentially with rate constant μ_0, at shorter times higher terms of the expansion in Eq. (5.26) start contributing, so that, when $-\ell_n N(t)_{ss}$ is extrapolated back to zero, it fails to reach $t = 0$ (Fig. 5.3) (recall that $N(t)$ is defined as the fractional population, so that $N(t = 0) = 1$). If we write $N(t)_{ss} = \exp[\mu_0(t - t_{inc})]$, where

$$\mu_0\, t_{inc} = -\ell_n \left[h_0 \sum_i \phi_0(x_i) \right] \tag{5.34}$$

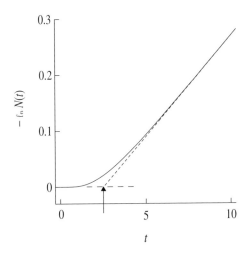

Fig. 5.3. A graphical representation (in arbitrary units) of the meaning of the incubation time. The dashed line is the extrapolation back to $N(t) = 1$ (the initial fractional population) of the steady-state population (the straight-line portion of the solid line). The arrow indicates the value of the incubation time.

we see that the onset of the exponential decay of the population is delayed by the amount t_{inc}, called the number, density incubation (or induction) time. Thus t_{inc} defines the "more or less short time" referred to above, which is necessary in order for the transients to die out.

Measuring incubation times (or indeed any pre-steady-state property) requires a difficult shock-tube experiment [26], and as a result there have been few experimental determinations for polyatomics. The data are usually given in the form of the ratio t_{inc}/μ_1, where μ_1 is the measured relaxation time, in principle the second-smallest eigenvalue. The ratio is found to decline with temperature: in the reaction $N_2O \rightarrow N_2 + O$, in the presence of argon, t_{inc}/μ_1 declines from ~ 8 at 2260 K to ~ 3 at 3580 K [27]; in the reaction norbornene (bicyclo [2,2,1]-hept-2-ene) $\rightarrow$ cyclopentadiene + ethylene, in the presence of krypton, t_{inc}/μ_1 declines from ~ 5 to ~ 2 in the range 900–1300 K [28]. More recently [29] the incubation time in the reaction $CH_3 + Ar \rightarrow CH + H_2 + Ar$ was measured. It was deduced that, at 2800 K, "cold" CH_3 needed about 10^4 collisions before decomposing.

5.3.8 Bulk properties

Once the population distribution $n(x_i, t)$ has been determined, it is possible to obtain all the *average* properties of the whole system at all times; these will be referred to as "bulk" properties for short, denoted by double angular brackets $\langle\!\langle \ldots \rangle\!\rangle$. They represent macroscopic observables, in principle time- and initial-condition-dependent, which, in a thermal system, are the only observables actually accessible to measurement.

Thus the observable or bulk-average energy is given by

$$\langle\langle y(t)\rangle\rangle = \int_0^\infty y\, n(y,t)\, dy \qquad (5.35)$$

and the macroscopic or bulk-average energy transfer by

$$\langle\langle \Delta E(t)\rangle\rangle = \int_0^\infty \langle \Delta E(y)\rangle\, n(y,t)\, dy \qquad (5.36)$$

which represents the average energy transferred per collision *in the bulk system* (Problem 5.8). In this sense the time-dependent rate constant $k(t)$ introduced above is actually a bulk-average rate constant:

$$k(t) \equiv \langle\langle k(t)\rangle\rangle = \frac{\int_0^\infty k(x)\, n(x,t)\, dx}{\int_0^\infty n(x,t)\, dx} \qquad (5.37)$$

At long times, via $n(x,t)$, bulk averages $\langle\langle y(t)\rangle\rangle$ and $\langle\langle \Delta E(t)\rangle\rangle$ tend to zero in a (one-way) reactive system, rather than to a constant, as was the case in a non-reactive system. Furthermore, it follows from Eq. (5.29) that, at steady state, these bulk averages relax exponentially according to $\exp(\mu_0 t) \equiv \exp(-k_{uni}t)$. For instance, we have for the bulk-average energy (Eq. (5.35))

$$k_{uni} = -\frac{1}{\langle\langle y(t)\rangle\rangle} \frac{d\langle\langle y(t)\rangle\rangle}{dt}\bigg|_{\text{at ss}} \qquad (5.38)$$

The definition of a bulk average shows clearly that, via $n(x,t)$ (or its derivative), a bulk average is a complicated function both of time and of initial conditions. In particular, there is a more or less distant relation [30] between $\langle\langle \Delta E\rangle\rangle_{ss}$ and the analogous microscopic property $\langle \Delta E(y)\rangle$ (Eq. (5.9)). The corollary is that, in the absence of other information, a bulk average offers in principle no access to microscopic properties of the system. This accounts for the difficulty of extracting $\langle \Delta E(y)\rangle$ from thermal data; in practice, therefore, $\langle \Delta E(y)\rangle$ is approximated by the y-independent $\langle \Delta E\rangle$ obtained from the collision efficiency β_c, as discussed later in Section 5.4.10.

5.4 Steady-state solutions

5.4.1 The temporal evolution of a thermal system

From the general features of the master-equation solutions discussed above we can now piece together the temporal sequence of events in a reactive system. Consider an idealized case in which at time zero the initial excitation takes place by sudden heating, as in a shock tube or by a laser flash. Immediately after, the system undergoes a more or less short transient during which all properties (e.g. the bulk-average energy $\langle\langle y \rangle\rangle$) are time-dependent as the system adjusts to the initial disturbance. This is because many, if not all, of the terms in Eq. (5.26) contribute significantly to the sum.

As time increases, the transient will usually die out quickly and the system will reach steady state (Eq. (5.29)), where it will remain for a period that is long compared with the transient, the result of only the term with the lowest eigenvalue effectively contributing to the sum in Eq. (5.26). In steady state, under which condition actual kinetic measurements are usually made, k_{uni} is independent of time. As time increases without limit the reaction ends with total disappearance of reactant M, which is implicit in the formulation of the master equation given in Eq. (5.12).

If the lowest eigenvalue is not well separated from others, the transient will not die out quickly enough, so there may be no measurable steady state.

5.4.2 The steady-state master equation

It is clear that full description of a thermal system from time zero onwards would require all eigenvalues and eigenvectors of matrix $\mathbb{J}$, a formidable task. If only steady-state properties of the reacting system are of interest, the lowest eigenvalue or the associated eigenvector are sufficient, which represents a simplification of the time-dependent master equation, provided that the reaction is not a multi-well sequence involving several intermediate structures.

At steady state, as we have seen, $n(x, t) \Rightarrow n(x) \exp(-k_{uni}t)$, using continuum analogs of quantities defined previously in connection with Eq. (5.29) in discrete terms and dropping the now-redundant subscript "ss." Thus $\partial n(x, t)/\partial t \Rightarrow -k_{uni} n(x) \exp(-k_{uni}t)$, so the steady-state version of the master equation of Eq. (5.13) is (Problem 5.9)

$$-k_{uni}\, n(x) = \omega \int_0^\infty p(x, y)\, n(y)\, dy - [\omega + k(x)]\, n(x) \qquad (5.39)$$

This is an eigenvalue equation, with k_{uni} the eigenvalue and $n(x)$ the corresponding eigenvector which, properly normalized, represents the steady-state population

$P(x)_{ss}$ (Eq. (5.30)). In dimensionless and symmetrized form the matrix equivalent of Eq. (5.39) is

$$-\frac{k_{uni}}{\omega}|\mathbf{N}\rangle = \mathbb{J}_s \cdot |\mathbf{N}\rangle \tag{5.40}$$

where $|\mathbf{N}\rangle$ is the column vector of the $n(x)$ and $\mathbb{J}_s$ is the same transport matrix as in Eq. (5.24). In fact this equation is the analog of Eq. (5.25) if $-k_{uni}/\omega$ is identified with μ_0 and $|\mathbf{N}\rangle$ with $|\mathbf{f}_0\rangle$, the corresponding eigenvector, so that k_{uni} is to be interpreted as the smallest eigenvalue in absolute terms, multiplied by ω. Thus, even though Eq. (5.39) represents no new information, it is interesting because it is solvable analytically for special forms of $p(x, y)$, as will be shown further below.

Using the steady-state form of $n(x, t)$ in Eq. (5.30), we then have from Eq. (5.31)

$$k_{uni} = \frac{\int_0^\infty k(x)\, n(x)\, dx}{\int_0^\infty n(x)\, dx} \tag{5.41}$$

which is an alternative means for determining the thermal rate constant k_{uni}. The general solution for arbitrary $p(x, y)$ has to be done numerically, in which case obtaining k_{uni} from the eigenvector as in Eq. (5.41) is preferable to the determination of k_{uni} as the eigenvalue, which is likely to be subject to numerical problems on account of the large size of the matrix $\mathbb{J}_s$. More on this in Section 5.5.4.

5.4.3 The high-pressure limit

When collisions are frequent enough to maintain an equilibrium population in the system, we have $n(x) \Rightarrow B(x)$ (Eq. (5.3)) regardless of the nature of the transition probability; then k_∞, the limiting high-pressure version of k_{uni}, is given quite generally by (Problem 5.10)

$$k_\infty = \lim_{\omega \to \infty} k_{uni} = \int_0^\infty k(x)B(x)\, dx \tag{5.42}$$

Since $k(x)$ is zero below the threshold E_0, the effective lower limit of the integral is E_0. On substituting for $k(x)$ from Eq. (3.28) and for $B(x)$ from Eq. (5.3), we have in terms of energy $x \equiv E$, and, with additional subscripts "v" to indicate that, at this stage, only vibrational degrees of freedom are involved,

$$k_\infty = \frac{\alpha}{h Q_v} \int_{E_0}^\infty G_v^*(E - E_0)\exp[-E/(kT)]\, dE$$

$$= \frac{\alpha k T}{h} \frac{Q_v^*}{Q_v}\exp[-E_0/(kT)] \tag{5.43}$$

Table 5.1. *Experimental high- and low-pressure parameters in selected reactions*

Reaction	$\log[A_\infty(s^{-1})]$	$E_{a\infty}$ (kcal mol^{-1})	E_{a0} (kcal mol^{-1})	Temperature range (K)	Ref.
Type 1					
$C_2H_6 \rightarrow CH_3 + CH_3$	17.3	90.8	83.3	300–2000	[a]
$CH_4 \rightarrow CH_3 + H$	$\{$ 18.0	96.4		1785–2325	[b]
	16.4	104.9	90.8	1000–3000	[a]
Type 2					
$C_2H_5Cl \rightarrow C_2H_4 + HCl$	13.8	57.4		1100–1400	[c]
$C_2H_5 \rightarrow C_2H_4 + H$	13.9	39.9	33.4	700–1100	[a]
Type 3					
Cyclopropane $\rightarrow$ propylene	14.9	63.9		690–750	[d]
Butene-2 $\rightarrow$ butene-2	14.5	65.0		742	[e]
(*cis*) (*trans*)					

[a] D. L. Baulch, C. J. Cobos, R. A. Cox, P. Frank, G. Hayman, Th. Just, J. A. Kerr, T. Murrells, M. J. Pilling, J. Troe, R. W. Walker and J. Warnatz, *J. Phys. Chem. Ref. Data* **23**, 847 (1994).
[b] D. F. Davidson, R. K. Hanson and C. T. Bowman, *Int. J. Chem. Kinet.* **27**, 305 (1995).
[c] H-L. Dai, E. Specht and M. R. Berman, *J. Chem. Phys.* **77**, 4494 (1982).
[d] U. Hohm and K. Kerl, *Ber. Bunsenges. Phys. Chem.* **94**, 1414 (1990).
[e] M. C. Lin and K. J. Laidler, *Trans. Faraday Soc.* **64**, 94 (1968).

using Eq. (4.11) to reduce the integral. Equation (5.43) is the standard transition-state-theory result for k_∞. The pre-exponential factor is thus given in terms of partition functions for the reactant and transition state (compare this with Kassel's indeterminate A (Eq. (1.24)) and Slater's pre-exponential factor (Eq. (2.10))).

The notion of "infinite" ω ($\omega \rightarrow \infty$) used to obtain the above expression for k_∞ is merely a mathematical convenience. Actually, at very high pressures k_{uni} goes through a maximum since eventually diffusion, rather than reaction, becomes rate-controlling, which causes k_{uni} eventually to decline (see Section 5.4.7 below). In practice k_∞ is obtained by extrapolation of experimental k_{uni} to pressures deemed sufficient to reach constant $k_{uni} \equiv k_\infty$ (generally a few bars, or less); this leaves some ambiguity associated with k_∞.

High-pressure experimental data are most often presented in the Arrhenius form

$$k_\infty = A_\infty \exp[-E_{a\infty}/(kT)] \tag{5.44}$$

where A_∞ and $E_{a\infty}$ are considered temperature-independent constants. Equation (5.43) shows that, in principle, A_∞ should be temperature-dependent, but such dependence is often lost in experimental error. A temperature dependence of $E_{a\infty}$ may signal a change of mechanism, or tunneling (Section 6.1.9). Table 5.1 lists experimental values of A_∞ and $E_{a\infty}$ for thermal reactions representative of the three Types. (The case of CH_4 in Table 5.1 also illustrates the fact that there is not always agreement regarding the values of the extrapolated parameters.)

5.4.4 The low-pressure limit

At the low-pressure limit collisions are infrequent and barely sufficient to raise the energy of the reactant to E_0, so the limiting low-pressure version of the steady-state master equation is

$$-k_0 \, n(x) = \omega \left(\int_0^{E_0} p(x, y) \, n(y) \, dy - n(x) \right) \tag{5.45}$$

obtained from Eq. (5.39) if $\omega \ll k(x)$ for all x. This is an eigenvalue equation for $-k_0$, in matrix form similar to Eq. (5.40), except that submatrix $\mathbb{J}_1$ is involved instead:

$$-\frac{k_0}{\omega} |\mathbf{N}\rangle = \mathbb{J}_1 \cdot |\mathbf{N}\rangle \tag{5.46}$$

Alternatively, Eq. (5.45) can be written

$$\left(1 - \frac{k_0}{\omega} \right) n(x) = \int_0^{E_0} p(x, y) \, n(y) \, dy \tag{5.47}$$

i.e.

$$\lambda_m |\mathbf{N}\rangle = \mathbb{P}_1 \cdot |\mathbf{N}\rangle \tag{5.48}$$

where $\mathbb{P}_1$ is the counterpart of $\mathbb{J}_1$ in a similarly partitioned matrix $\mathbb{P}$, λ_m is the *largest positive* eigenvalue, and $|\mathbf{N}\rangle$ is the corresponding eigenvector, the same as in Eq. (5.46); then $k_0/\omega = 1 - \lambda_m$.

In either case the physical picture is that of k_0 as essentially the rate of activation, with E_0 acting as a perfect sink, i.e. every molecule reaching E_0 from lower energy levels immediately dissociates and is removed from the system; hence Eq. (5.45) contains no reference to the specific rate constant $k(x)$. This is more obvious from an alternative expression, which is the subject of Problem 5.11.

5.4.5 Recombination reactions

Consider the reaction $M \rightarrow R_1 + R_2$, with rate constant k_{uni}, and its reverse, i.e. the recombination $R_1 + R_2 \rightarrow M$, with rate constant k_{rec}. At equilibrium $M \rightleftarrows R_1 + R_2$ the forward ($\rightarrow$) and reverse ($\leftarrow$) rates are equal, i.e. $k_{uni}[M] = k_{rec}[R_1][R_2]$, which defines the equilibrium constant K_c:

$$K_c = \frac{[R_1][R_2]}{[M]} = \frac{k_{uni}}{k_{rec}} \tag{5.49}$$

Here square brackets signify concentrations that obtain at equilibrium. Although the relation above was derived at equilibrium, the *ratio* k_{uni}/k_{rec} is equal to the equilibrium constant K_c irrespective of whether equilibrium obtains or not.

So here we have a way to convert forward and reverse rate constants into each other by means of the equilibrium constant. This is of some importance in Type-1 thermal reactions because dissociation rates (i.e. k_{uni}) are experimentally accessible only at elevated temperatures, owing to the appreciable magnitude of the critical energy E_0, whereas recombination rates (i.e. k_{rec}) are experimentally accessible also at low temperatures since recombinations of radicals have no formal critical energy. By means of Eq. (5.49) it is possible to combine data from several sources to characterize a given reaction over a large temperature domain. An example is the well-studied reaction $C_2H_6 \rightarrow 2CH_3$ and its reverse (note the wide temperature limit given for this reaction in Table 5.1).

5.4.6 The strong-collision steady-state solution

In order to obtain an analytical solution we invoke now the so-called "strong"-collision assumption (superscript "sc"), mentioned previously in Section 5.2, whereby each collision is assumed to leave the reactant molecule in equilibrium with the heat bath at temperature T. Thus $p(x, y) \Rightarrow B(x)$ (Eq. (5.7)); since $n(y)$ is normalized to unity, the integral on the RHS of Eq. (5.39) is simply $B(x)$, so the equation becomes

$$n(x)^{sc} = \frac{\omega B(x)}{\omega + k(x) - k_{uni}^{sc}} = P(x)_{ss}^{sc} \tag{5.50}$$

which is the strong-collision form of the properly normalized steady-state distribution introduced previously (Eq. (5.30)). If we now substitute the above form of $P(x)_{ss}^{sc}$ into Eq. (5.31), we have

$$k_{uni}^{sc} = \omega \int_{E_0}^{\infty} \frac{B(x) k(x) \, dx}{\omega + k(x) - k_{uni}^{sc}} \tag{5.51}$$

which may be considered as the "exact" form of the strong-collision unimolecular rate constant. The usual derivation neglects k_{uni}^{sc} in the denominator of the integral, which amounts to assuming implicitly that $\omega + k(x) \gg k_{uni}^{sc}$ for all ω and x. Such is the case of Kassel's (Eq. (1.29)) and Slater's (Eq. (2.8)) k_{uni}, which were derived by a simple kinetic argument. Neglecting k_{uni}^{sc} with respect to $\omega + k(x)$ is actually a very good approximation, except at very high temperatures for reactions with low activation energies [31, 32].

At the limiting high-pressure limit k_{uni}^{sc} of Eq. (5.51) reduces to k_∞ of Eq. (5.43), whereas at the low-pressure limit it reduces to

$$k_0^{sc} = \omega \int_{E_0}^{\infty} B(x)\,dx = \frac{\omega}{Q_v} \int_{E_0}^{\infty} N_v(x) \exp[-x/(kT)]\,dx \qquad (5.52)$$

It is shown in Problem 5.12 that the same result can be obtained by a different argument.

5.4.7 Fall-off

The customary representation of k_{uni} over a range of pressures is in the form of a double-logarithmic plot of $\log k_{uni}$ versus $\log p$ (p is the pressure or concentration). The slope of such a plot is

$$\text{Slope} = \frac{\partial \log k_{uni}}{\partial \log p} = \frac{p}{k_{uni}} \frac{\partial k_{uni}}{\partial p} \qquad (5.53)$$

a relation that holds for logarithms of any base. Taking for simplicity $k_{uni}^{sc} \ll \omega + k(x)$ in Eq. (5.51), $\omega = Zp$ (Eqs. (5.14) and (5.15)), and writing $K(x) = k(x)/[\omega + k(x)]$, we have

$$\text{Slope} = \frac{\int B(x)[K(x)]^2\,dx}{\int B(x)K(x)\,dx} \qquad (5.54)$$

As $\omega \ll k(x)$, $K(x) \to 1$, so at low pressure the slope tends to unity, whereas as $\omega \to \infty$, $K(x) \to k(x)/\omega$ and consequently the slope tends to zero. The order of reaction is given by $1 + \text{Slope}$, so the reaction approaches second order at low pressures, and first order at high pressures, as shown many years ago by Lindemann [33]. This behavior as a function of pressure or concentration is commonly referred to as "fall-off."

The low-pressure limit, i.e. the condition $\omega \ll k(x)$, is assumed to be realized experimentally when the reaction can be shown to be of second order, i.e. the fall-off plot has reached unit slope. The condition $\omega \to 0$ is a very special case considered in Section 5.10.

In principle every thermal reaction will exhibit fall-off, but the pressure domain where fall-off is actually measurable depends in essence on the complexity of the reactant, and to a lesser extent on the nature of the bath gas. Experiment shows that $\log p_{1/2}$, the pressure at which k_{uni} has fallen off to $\frac{1}{2}k_\infty$, decreases approximately linearly with the number of degrees of freedom of the reactant [34, Fig. 11.2]. Specifically, for large polyatomics fall-off is generally below 1 bar, which means

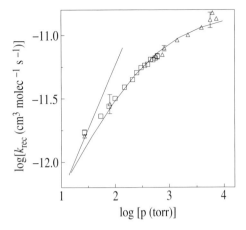

Fig. 5.4. The fall-off of k_{rec} for the recombination $CH_3 + NO$ in argon at 296 K. Squares are experimental data of J. W. Davies, N. J. B. Green and M. J. Pilling, *J. Chem. Soc. Faraday Trans.* **87**, 2317 (1991); triangles are those of E. W. Kaiser, *J. Phys. Chem.* **97**, 11 682 (1993). The continuous line was calculated from Eq. (5.40) as the lowest eigenvalue of a 186×186 exponential-model transition-probability matrix with $\langle \Delta E \rangle_{down} = 66 \, cm^{-1}$. At the indicated pressures the reaction is a long way from the low-pressure limit, as is evident on comparing experiment and calculation with the unit-slope straight line on the left. This is one of the few examples of a polyatomic recombination studied at pressures as high as 10^4 torr.

that k_∞ is accessible experimentally, but the second-order domain may be at such low pressures that the determination of k_0 is difficult. Conditions in the case of small molecules, and diatomics in particular, are exactly the reverse: fall-off is at such high pressures (e.g. several thousand bars for Br_2, see below) that only the second-order domain is accessible at atmospheric pressures.

In view of Eq. (5.49) it is obvious that recombination reactions will likewise be subject to fall-off, as illustrated in Fig. 5.4 for $CH_3 + NO$. Over a sufficiently large pressure domain the rate constant k_{rec} will exhibit a pressure dependence analogous to that of k_{uni}, that is, a rise to a maximum, followed by a slow decline at still higher pressures due to rate-controlling diffusion. The onset of diffusion control in $k_{rec,\infty}$ was demonstrated experimentally in the recombination $Br + Br \rightarrow Br_2$ in the presence of Ar and N_2 at $\sim$4000 bar [35], and in the recombination of CCl_3 radicals in N_2 at $\sim$300 bar [36].

5.4.8 Activation energy

It is customary, largely for historical reasons, to define the activation energy E_a as the slope of $\ln k_{uni}$ versus $1/T$, i.e. by

$$E_a = \frac{\partial \ln k_{uni}}{\partial (1/T)} = \mathit{k} T^2 \frac{\partial \ln k_{uni}}{\partial T} \qquad (5.55)$$

A fairly transparent expression for E_a can be obtained from the strong-collision expression for k_{uni}^{sc}. To this end, it is convenient to define a steady-state microcanonical rate constant $k(x)_{ss} = \omega k(x)/[\omega + k(x)]$, and a distribution $B'(x) = N(x)\exp[-x/(kT)]$, which is the numerator of the equilibrium distribution in Eq. (5.3). We then have, neglecting k_{uni}^{sc} in the denominator of Eq. (5.51),

$$k_{uni}^{sc} = \frac{1}{Q}\int\limits_0^\infty k(x)_{ss} B'(x)\,dx \qquad (5.56)$$

The temperature dependence of the collision frequency can be neglected, since it depends on temperature only as $T^{\mp 1/2}$, whereas the partition function Q and $B'(x)$ depend on temperature exponentially. Then, since $\partial B'(x)/\partial T = xB'(x)/(kT^2)$, Eq. (5.55) yields

$$E_a = \frac{\int_{E_0}^\infty x\,k(x)_{ss} B'(x)\,dx}{\int_{E_0}^\infty k(x)_{ss} B'(x)\,dx} - \langle E\rangle \qquad (5.57)$$

where $\langle E\rangle = kT^2\partial\ln Q/\partial T$. The RHS can be interpreted as the difference, at specified ω, between the average energy of molecules reacting per unit time, $\langle E^*\rangle$, and $\langle E\rangle$, the average energy of all molecules, reacting and non-reacting [37]; clearly $\langle E^*\rangle > \langle E\rangle$.

The limiting high-pressure activation energy follows directly from Eq. (5.43):

$$E_{a\infty} = kT^2\frac{\partial\ln k_\infty}{\partial T} = kT + \langle E^*\rangle - \langle E\rangle + E_0 \qquad (5.58)$$

$E_{a\infty}$ is therefore distinct from the critical energy E_0. This has some importance for the determination of bond energies, the accurate measure of which is E_0. If k_∞ is known from experiment only in the simple Arrhenius form as $k_\infty = A_\infty\exp[-E_{a\infty}/(kT)]$ (Eq. (5.44)), the experimental $E_{a\infty}$ is the best available estimate of E_0 in the absence of other information; it is exploited further in Section 5.6.1.

At low pressure $k(x)_{ss} \to \omega$, and the limiting low-pressure activation energy is similarly the difference between the average energy of all reacting molecules (not just per unit time) and the previously defined $\langle E\rangle$. With the help of the approximation used in Problem 5.12,

$$E_{a0} = \frac{\int_{E_0}^\infty xB'(x)\,dx}{\int_{E_0}^\infty B'(x)\,dx} - \langle E\rangle \qquad (5.59)$$

Thus $E_{a0} < E_{a\infty}$, so that the activation energy E_a, like k_{uni}, will exhibit a falloff with pressure (or concentration). Experimental determination of E_{a0} cannot be dissociated from the collision cross-section used, which in turn depends on the bath gas. Table 5.1 lists experimental values of E_{a0} for Type-1 and Type-2 reactions; the reactant itself was the bath gas for both Type-1 reactions, whereas ethane was the bath gas for the Type-2 dissociation of the ethyl radical. Table 5.1 shows that the difference $E_{a\infty} - E_{a0}$ is much less than what can be calculated classically (Problem 5.13).

5.4.9 The thermodynamic formulation

If in Eq. (5.43) we let $K_c^* = (Q_v^*/Q_v)\exp[-E_0/(kT)]$, K_c^* can be looked upon as the equilibrium constant for $M \rightleftharpoons M^*$, the pseudo-equilibrium between transition state M^* and reactant M, expressed in terms of partition functions. Equation (5.43) then becomes

$$k_\infty = \frac{\alpha k T}{h} K_c^* \tag{5.60}$$

From standard thermodynamics $K_c^* = \exp[(\Delta S^*/T) - (\Delta H^*/kT)]$, where ΔS^* and ΔH^* are the entropy and enthalpy of activation, respectively. Since $kT^2(\partial \ln K_c^*/\partial T) = \Delta H^*$, the activation energy corresponding to k_∞ of Eq. (5.60) is $kT + \Delta H^*$, while that corresponding to Eq. (5.44) is the experimental activation energy $E_{a\infty}$. On equating the two expressions we have $\Delta H^* = E_{a\infty} - kT$, so Eq. (5.60) takes the form (e = 2.718...)

$$k_\infty = \frac{\alpha e k T}{h} \exp(\Delta S^*/T)\exp[-E_{a\infty}/(kT)] \tag{5.61}$$

where the pre-exponential factor A_∞ in Eq. (5.44) is now expressed in terms of entropy of activation.

Since $\Delta S^* = S^*$ (transition state) $- S$(reactant), the interest of Eq. (5.61) is that it offers a qualitative interpretation of the structure of the transition state relative to the structure of the reactant. Numerically $ekT/h \sim 10^{13.75}$ s^{-1} at 1000 K (assuming that $\alpha = 1$ for simplicity). Therefore we can expect A_∞ to be of this magnitude if $\Delta S^* = 0$, i.e. if the structure of the transition state is similar to that of the reactant. Manifestly this is the case of Type-2 and -3 reactions in Table 5.1. Transition states that resemble the reactants are commonly called "tight."

The case of Type-1 reactions is different since, from Table 5.1, we see that $A_\infty > 10^{14}$ s^{-1}, so that clearly $\Delta S^* > 0$. If we look upon entropy as a measure of disorder, then the transition state of Type-1 reactions is clearly more "disordered," meaning that it has more quantum states than the reactant. Such "disordered" transition states are called "loose." It is shown in Section 5.8 that the additional quantum states come principally from incipient rotations of fragments as they begin to form.

By the same token, we might conclude that, if $A_\infty < 10^{13}$ s^{-1}, ΔS^* must be negative, i.e. a transition state that is more ordered than the reactant. However, the origin of such a low pre-exponential factor is quite different: it is the manifestation of a spin-forbidden reaction, as discussed in Chapter 6; see Section 6.6.2 and Table 6.1.

The thermodynamic approach is helpful in better characterizing the transition state, and hence $\langle E^* \rangle$, needed for obtaining a better value of E_0 from $E_{a\infty}$ in Eq. (5.58). This approach can be systematized for the estimation of reaction energetics as "Thermochemical Kinetics," which was pioneered by Benson [38]. For an extended summary and examples of the application of Benson's method see [22, Chapter 4].

5.4.10 Collision efficiency

The strong-collision assumption is dictated more by mathematical convenience than by physical reality, which suggests that – at least in the gas phase – most collisions are actually "weak," whereby the amount of energy transferred by collision ($\langle\langle\Delta E(y)\rangle\rangle$, Eq. (5.9)) is mostly small with respect to kT (however, see Eq. (5.5) and accompanying comments). An analytical solution for k_0 with a weak-collision transition probability, e.g. exponential (Appendix 5), is no longer simple.

In reality many collisions are necessary in order to transfer an amount of energy for which a hypothetical single "strong" collision would be sufficient. One way to account for the inefficiency of collisional energy transfer and still preserve the simplicity of a strong-collision-type solution is to reduce $Z(y)$, the second-order energy-transfer rate constant (Eq. (5.14)). Using the same simplification as before, we replace $Z(y)$ by the energy-independent Z, but now multiplied by a "collision-efficiency" factor $\beta_c < 1$, assumed to be a constant for a given combination of substrate M and collider X (β_c must not be confused with β in Eq. (5.4)).

This amounts to re-scaling pressures or concentrations in the sense that, with decreased Z, pressure or concentration has to be increased in the same proportion for the same fall-off. Physically this means that on average roughly $1/\beta_c$ collisions are necessary in order to transfer energy $\langle\Delta E\rangle$. Thus in Eq. (5.51) the original collision frequency ω is written $\omega\beta_c$, with the result that the strong-collision fall-off curve of k_{uni}^{sc} conserves its shape and is simply displaced to higher pressures by $|\log\beta_c|$ on the usual double-logarithmic plot.

The flip side is, of course, that, proceeding this way, we have no means to estimate β_c except by reference to experiment. Actually the situation is not much worse than if we had attempted a full weak-collision matrix solution using one of the transition-probability models (in Eqs. (5.4) and (5.8)), which each include a parameter (α and ϵ, respectively) that has to be obtained from other information; the same is true of the two expressions for β_c given below.

The most suitable experimental datum is the limiting low-pressure rate constant k_0 which is the "pure" result of collisional energy transfer across the activation barrier at E_0. Considering the experimental low-pressure rate constant for collider X as the weak-collision k_0^{wc}, the "experimental" collision efficiency is then

$$\beta_c = \frac{k_0^{wc}}{k_0^{sc}} \tag{5.62}$$

where k_0^{sc} is given by Eq. (5.52), provided that the two k_0 are compared at the same pressure or concentration (or both replaced by the slope $\partial k_0/\partial p$). In practice, β_c is obtained from the amount by which the calculated strong-collision fall-off curve has to be shifted along the pressure axis to higher pressures in order to coincide with experiment.

The most detailed investigation is that of Rabinovitch and co-workers [39], who measured essentially the equivalent of β_c for 109 different bath gases in the thermal isomerization of CH_3NC at 280 K. By and large, in most reactions, $\beta_c \sim 0.1$–0.2 for monotomic gases, 0.2–0.4 for diatomics, and takes higher values for polyatomics, up to $\beta_c = 1$ for the reactant itself.

Information of interest is how β_c translates into the average energy transferred per collision. It is shown in Appendix 5 that, for the simple-exponential transition-probability model,

$$\beta_c = \frac{k_0^{wc}}{k_0^{sc}} \cong \left(1 + \frac{\ell T}{\alpha}\right)^{-2} \tag{5.63}$$

In this model $\langle \Delta E \rangle = \beta - \alpha$, with $\beta = \alpha \ell T/(\alpha + \ell T)$, so that, on taking the absolute value, $|\langle \Delta E \rangle_{down}| = \alpha$ (cf. Problem 5.1), we have

$$\langle \Delta E \rangle = -\frac{\beta_c \ell T}{1 - \beta_c^{1/2}}; \qquad |\langle \Delta E \rangle_{down}| = \frac{\beta_c^{1/2} \ell T}{1 - \beta_c^{1/2}} \tag{5.64}$$

An alternative expression for β_c involving the second moment $\langle \Delta E^2 \rangle$ can be obtained by using for k_0^{wc} the result of Problem 5.14, from which, with the approximation of Problem 5.12 for k_0^{sc},

$$\beta_c \cong \frac{\langle \Delta E^2 \rangle}{2(\ell T)^2} \tag{5.65}$$

Equations (5.63) and (5.65) yield similar results for β_c at $\alpha \leq 100 \, \text{cm}^{-1}$ and elevated temperature; otherwise Eq. (5.65) yields a higher β_c for the same α and T. Extended tables of experimental values of $\langle \Delta E \rangle$ and $\langle \Delta E \rangle_{down}$ for polyatomics, based on modern techniques, may be found in [2]. The conclusion is that, at present, the exact energy and temperature dependences of $\langle \Delta E \rangle$ remains unknown.

While the above relations between β_c and $\langle \Delta E \rangle$ or $\langle \Delta E^2 \rangle$ were obtained from the exponential-transition-probability model, it turns out that, at least insofar as thermal reactions are concerned, all transition-probability models yield experimentally indistinguishable results if results are compared at the same $\langle \Delta E \rangle$. This can be understood by realizing that a thermal rate constant is an extensively averaged bulk property, so that much of the detail in the transition probability (which is a microscopic property) is washed out.

5.4.11 The Fokker–Planck equation

Another approximate analytical solution of the master equation of Eq. (5.12) can be attempted by performing an expansion [10], in which both $n(x, t)$ and $p(x, y)$ are expanded in a Taylor series about $x \approx y$. If collisions are "weak," the amount of energy transferred by collision ($\langle \Delta E(y) \rangle$, Eq. (5.9)) is small with respect to $k T$, and the transition probability $p(x, y)$ is sharply peaked about $x \approx y$ (as represented for instance by the exponential transition probability, Eq. (5.4)), so that only terms up to second order can be retained without serious error.

After some manipulation [40–42] the expansion thus truncated is obtained in the form

$$\frac{\partial n(x, t)}{\partial t} = -\frac{\partial}{\partial x}[v(x) n(x, t)] + \frac{\partial^2}{\partial x^2}[D(x) n(x, t)] - k(x) n(x, t) \qquad (5.66)$$

which is known as the Fokker–Planck (FP) equation. Here $v(x)$ is the so-called drift vector, which is given by

$$v(x) = \omega \int_0^\infty (x - y) p(y, x) \, dy = \omega \langle \Delta E(x) \rangle \qquad (5.67)$$

which will be recognized as involving the average energy transferred by collision (Eq. (5.9)). $D(x)$ is the so-called diffusion tensor defined by

$$D(x) = \frac{1}{2} \omega \int_0^\infty (y - x)^2 p(y, x) \, dy = \frac{1}{2} \omega \langle \Delta E^2(x) \rangle \qquad (5.68)$$

which is therefore related to the second moment of the transition probability defined previously (Eq. (5.11) with $k = 2$).

Assuming for the moment that the system is non-reactive, i.e. that $k(x) = 0$, the population in such a system reaches the Boltzmann distribution $B(x)$ at long times, as we have seen in Section 5.3. Thus, as $\partial n(x, t)/\partial t \Rightarrow 0$, $n(x, t) \Rightarrow B(x)$, so that, for $k(x) = 0$, Eq. (5.66) must satisfy

$$0 = -\frac{\partial}{\partial x}[v(x) B(x)] + \frac{\partial^2}{\partial x^2}[D(x) B(x)] \qquad (5.69)$$

On integrating, we have

$$v(x) = \frac{1}{B(x)} \frac{d}{dx}[D(x)B(x)] \tag{5.70}$$

The same relation can be derived somewhat less transparently from detailed balance [41]. Equation (5.70) shows that $v(x)$ and $D(x)$ are related, which implies in turn that, in order to correctly represent behavior at long times, either $v(x)$ or $D(x)$ will have to be re-defined with respect to the definitions given above in Eqs. (5.67) and (5.68).

5.4.12 Energy diffusion

If we prefer to keep the original form of $D(x)$ and choose $v(x)$ for re-definition, we rewrite Eq. (5.70) in the form

$$v(x) = D'(x) + D(x) \frac{d \ln B(x)}{dx} \tag{5.71}$$

where the prime indicates the derivative with respect to x. On substituting back into Eq. (5.66), the result of replacing $v(x)$ is [43–45]

$$\frac{\partial n(x,t)}{\partial t} = \frac{\partial}{\partial x} \left(D(x) \frac{\partial n(x,t)}{\partial x} - D(x)n(x,t) \frac{d \ln B(x)}{dx} \right) - k(x)n(x,t) \tag{5.72}$$

which may be termed the "energy-diffusion" equation. The familiar diffusion coefficient in coordinate space has the dimension length^2 s^{-1}, while the dimension of $D(x)$ is energy^2 s^{-1}, so that by analogy $D(x)$ and the equation it contains may be considered to represent diffusion in energy space.

5.4.13 The low-pressure solution

For the purpose of obtaining an expression for the limiting low-pressure rate constant, Eq. (5.72) is specialized to energies below threshold ($x < E_0$), where $k(x < E_0) = 0$. Then at steady state $n(x,t) \Rightarrow n(x)\exp(-k_0 t)$ (since now $k_{uni} \Rightarrow k_0$), so that Eq. (5.72) becomes (cf. Eq. (5.45))

$$-k_0 n(x) = \frac{d}{dx} \left(D(x) \frac{dn(x)}{dx} - D(x)n(x) \frac{d \ln B(x)}{dx} \right) \tag{5.73}$$

It is convenient to introduce the variable $h(x) = n(x)/B(x)$, which means that the steady-state population is now expressed as the fractional deviation from the equilibrium population. Equation (5.73) becomes

$$\frac{d}{dx} \left(D(x)B(x) \frac{dh(x)}{dx} \right) = -k_0 h(x)B(x) \tag{5.74}$$

Integrating this gives $D(x)B(x)[\mathrm{d}h(x)/\mathrm{d}x] = C_1$, with the integration constant given by $C_1 = k_0 \int_0^{E_0} h(x)B(x)\,\mathrm{d}x$ (since at this point C_1 is arbitrary, the sign does not matter). On integrating again, there appears another integration constant, C_2:

$$h(y) = C_1 \int\limits_y^{E_0} \frac{\mathrm{d}x}{D(x)B(x)} + C_2 \qquad (5.75)$$

At the low-pressure limit we must have $h(E_0) = 0$ since all particles reaching the threshold E_0 dissociate and are removed from the system; thus $C_2 = 0$. Furthermore, $n(x)$ is normalized below threshold, so that

$$\int\limits_0^{E_0} n(x)\,\mathrm{d}x = \int\limits_0^{E_0} h(x)B(x)\,\mathrm{d}x = 1 \qquad (5.76)$$

which means that $C_1 = k_0$. On multiplying both sides of Eq. (5.75) by $B(y)\,\mathrm{d}x$ and integrating with respect to y, we have, using Eq. (5.76) and adding the superscipt "wc" to k_0 to signify that weak collisions are involved,

$$1 = k_0^{\mathrm{wc}} \int\limits_0^{E_0} B(y)\,\mathrm{d}y \int\limits_y^{E_0} \frac{\mathrm{d}x}{D(x)B(x)} \qquad (5.77)$$

Thus the limiting weak-collision low-pressure rate constant $k_0^{\mathrm{wc}}(\mathrm{s}^{-1})$ in the diffusion model is obtained analytically in the form

$$k_0^{\mathrm{wc}} = \left(\int\limits_0^{E_0} \frac{\mathrm{d}x}{D(x)B(x)} \int\limits_0^x B(y)\,\mathrm{d}y \right)^{-1} \qquad (5.78)$$

where the integrals in Eq. (5.77) have been transformed using integration by parts (Problem 5.15).

5.4.14 *The general FP solution*

If we choose instead $D(x)$ to be re-defined, integrating Eq. (5.70) yields

$$D(x) = \frac{1}{B(x)} \int\limits_0^x v(y)B(y)\,\mathrm{d}y \qquad (5.79)$$

Substituting this into Eq. (5.66) leads to a FP equation that is not amenable to analytical reduction, so a numerical solution is necessary. In this case it need not be specialized to energies below threshold so that a general-pressure solution can be obtained, again on the assumption that the average energy transfer per collision is small. Solution by finite-differences leads to a tridiagonal matrix that is somewhat easier to diagonalize than a standard weak-collision matrix [34, pp. 210ff].

5.5 Numerical steady-state solutions

5.5.1 General considerations

Apart from the strong-collision transition probability, and a few other more or less artificial transition-probability models for which analytical results are possible (see [1, 2] for a discussion of some of these), an adequate treatment of a thermal system requires a numerical approach to deal with the transport matrix. As is apparent from Eq. (5.1), proper normalization and subsequent numerical manipulation would require, strictly speaking, the transport matrix $\mathbb{J}$ to be of infinite size, and the full description of the system according to the expansion in Eq. (5.26) would then require all the infinite number of eigenvalues and eigenvectors. This is of course not feasible, and in fact not necessary if the main interest is in the properties of the system at or near steady state.

Thus, if only the steady-state rate constant k_{uni} is required, substantial simplification is possible since, as we have seen, the smallest eigenvalue in absolute terms or its eigenvector is sufficient. Note that, if the transport matrix used in Eq. (5.19) is defined by $\mathbb{J} = \mathbb{P} - \mathbb{K}$, the (dimensionless) rate constant k_{uni}/ω will be equal to 1 minus the *largest positive* eigenvalue. Among a number of numerical routines designed to find either the smallest or the largest eigenvalue, we will mention here only those that are specifically adapted to the thermal rate problem. The choice is then either to obtain the eigenvalue of the transport matrix, or to obtain the corresponding eigenvector, i.e. the steady-state distribution P_{ss}, from which k_{uni} follows from Eq. (5.31).

A representative of the first approach is the mean first-passage time, and the reduced-matrix approximation is typical of the second. These will now be discussed in turn.

5.5.2 The mean first-passage time (MFPT)

An alternative measure of the progress of reaction can be obtained by measuring the times $t_1, t_2, t_3, \ldots$ it takes for the molecules numbered 1, 2, 3, $\ldots$ to react and be removed from the system. The mean first-passage time $\bar{t}$ is defined by [46]

$$\bar{t} = \frac{t_1 + t_2 + t_3 + \cdots}{N_0} \tag{5.80}$$

where N_0 is the total number of molecules at $t = 0$. In terms of the continuous formulation we have

$$N(t)/N_0 = \int_{\mathcal{R}} n(x, t)\, dx \tag{5.81}$$

where $n(x, t)$ is the fractional population of molecules with energy x at time t, defined previously in connection with Eq. (5.12), and the domain of integration covers the relevant energy space $\mathcal{R}$. $N(t)$ is the total population of molecules of all energies at time t. Thus

$$\bar{t} = \frac{1}{N_0} \int_0^{\infty} N(t)\, dt \tag{5.82}$$

so that $\bar{t}$ can be interpreted as the average lifetime. It is convenient to define $\tau(x)$ by

$$\tau(x) = \int_0^{\infty} n(x, t)\, dt \tag{5.83}$$

and then

$$\bar{t} = \int_{\mathcal{R}} \tau(x)\, dx \tag{5.84}$$

If the reaction is at the low-pressure limit, $\mathcal{R}$ corresponds to molecular energy levels below the reaction threshold E_0 which acts as the absorbing barrier (*vide supra*). Then $\bar{t}$ is a measure of the average time it takes for the molecules to pass across the barrier for the first time; hence the designation MFPT. In this sense the inverse of k_0 given by Eq. (5.78) can also be considered as another MFPT formula.

In the discrete formulation, and at arbitrary pressure, we can start by integrating Eq. (5.23) with respect to time ($\mathbb{J}$ is here assumed to be in principle a reactive transport matrix). With $\mathbf{N}_0$ as row vector $\langle \mathbf{N}_0 |$, the result is (row) vector $\mathbf{T}$, which is the analog of $\tau(x)$:

$$\mathbf{T} = \mathbf{N}_0 \int_0^{\infty} \exp(\mathbb{J}t)\, dt = -\mathbb{J}^{-1}\mathbf{N}_0 \tag{5.85}$$

where $\mathbb{J}^{-1}$ is the matrix inverse of $\mathbb{J}$. According to Eq. (5.84), $\bar{t}$ is the vector $\mathbf{T}$ summed:

$$\bar{t} = -\sum_i (\mathbb{J}^{-1}\mathbf{N}_0)_i \tag{5.86}$$

The same result can be obtained somewhat less directly from the eigenfuction– eigenvalue expansion (Problem 5.16).

If the system starts from the lowest level, $\langle \mathbf{N}_0 |$ becomes the row vector $\mathbf{N}_0(1)$ with elements $\{1\ 0\ 0\ 0\dots\}$. Then, to within a good approximation, the product $\mathbb{J}^{-1}\mathbf{N}_0(1)$ is

$$\mathbb{J}^{-1}\mathbf{N}_0(1) \cong \operatorname{diag} \mathbb{J}^{-1} = \operatorname{diag} \mathbb{J}_s^{-1} \tag{5.87}$$

since $\mathbb{J}$, $\mathbb{J}_s$, and their inverses have the same diagonal. If $\mathbb{J}$ is a non-reactive matrix, or if $\mathbb{J}$ is a reactive matrix with nearest-neighbor transitions only, then $\mathbb{J}^{-1}\mathbf{N}_0(1)$ equals $\operatorname{diag} \mathbb{J}^{-1}$ *exactly* [21, 47]. Equation (5.86) thus reduces to

$$\bar{t} \cong -\operatorname{Tr} \mathbb{J}^{-1} = -\operatorname{Tr} \mathbb{J}_s^{-1} \tag{5.88}$$

(Tr denotes the trace, i.e. sum of diagonal elements).

The same relation can be made to apply at the low-pressure limit by substituting for $\operatorname{Tr} \mathbb{J}^{-1}$ the trace of the inverse of submatrix $\mathbb{J}_1$ (Eq. (5.33)).

The obvious question concerns the relation of the above $\bar{t}$ to the "true" rate constant, the latter being given, as we know, by the lowest eigenvalue in absolute terms of $\mathbb{J}$ or $\mathbb{J}_s$. It can be shown, using linear-algebra theory, that the eigenvalues μ of any $m \times m$ symmetric matrix $\mathbb{M}$ are given quite generally by the solution of the characteristic equation

$$\mu^m + a_{m-1}\mu^{m-1} + \cdots + a_1\mu + a_0 = 0 \tag{5.89}$$

where the a_i are coefficients related to various elements of $\mathbb{M}$. If the lowest (in absolute terms) eigenvalue μ_0 is well separated from the others, i.e. $|\mu_0| \ll |\mu_k|$ ($k = 1, 2, \ldots, m - 1$), the characteristic equation reduces to the linear relation

$$a_1\mu_0 + a_0 \cong 0 \tag{5.90}$$

so that $\mu_0 \cong -a_0/a_1$. Now [48] $a_1 = -(\operatorname{Det} \mathbb{M}) \times (\operatorname{Tr} \mathbb{M}^{-1})$ (Det denotes the determinant), and $a_0 = \operatorname{Det} \mathbb{M}$ so that $|\mu_0| \cong -(\operatorname{Tr} \mathbb{M}^{-1})^{-1}$.

The answer, therefore, is that, if the lowest eigenvalue $|\mu_0|$ of matrix $\mathbb{J}$ or $\mathbb{J}_s$ is well separated from the others, $|\mu_0|$ is to a good approximation given by $1/\bar{t}$ of Eq. (5.88). We have seen above (Eq. (5.29)) that this corresponds to steady-state conditions under which the system relaxes exponentially; in that case $|\mu_0| \equiv k_{\text{uni}}$ and, taking $N(t) \cong N_0 \exp(-k_{\text{uni}}t)$, we obtain after substitution into Eq. (5.82) the expected result $\bar{t} \cong k_{\text{uni}}^{-1}$ [49].

The estimate of the steady-state rate constant as the inverse of $|\bar{t}|$ can be improved [50] by a generalization of Eq. (5.88) involving higher powers of the matrix inverse:

$$|\bar{t}| \cong (|\operatorname{Tr} \mathbb{J}^{-n}|)^{1/n} = (|\operatorname{Tr} \mathbb{J}_s^{-n}|)^{1/n} \tag{5.91}$$

Usually $n = 2$ provides a significant improvement, and with $n = 3$ the result for $|\bar{t}|$ will converge to within a few percent of $|\mu_0|^{-1}$.

5.5.3 The MFPT and incubation time

The MFPT is useful when experiment does not give access to the actual rate constant. Such is the case, for example, in multiphoton dissociation, for which experiment measures the reaction yield (Y) as a function of "fluence," i.e. essentially time. In terms of the population at time t, $N(t)/N_0 = 1 - Y(t)$. The complication here is that measurements are not made under steady-state conditions.

If a system is *not* in steady state, $\bar{t}$ will be a poor representation and an expansion in higher moments of the MFPT is necessary. If we consider Eqs. (5.83) and (5.84) as defining the first moment of $\bar{t}$, then the moment of order n is usually defined by

$$\bar{t}_n = n \int_0^\infty dt\, t^{n-1} \int_{\mathcal{R}} dx\, n(x, t) \tag{5.92}$$

Using a procedure similar to that used in deriving Eq. (5.86), the discrete analog is

$$\bar{t}_n = (-1)^n n! \sum_j (\mathbb{J}^{-n} \mathbf{N}_0)_j \tag{5.93}$$

Equation (5.91) can be considered a special case of $(\bar{t}_n/n!)^{1/n}$.

Typically, the yield $Y(t)$ in a multiphoton-dissociation experiment on a reactive system is [51]

$$Y(t) = \begin{cases} 0, & t < \theta \\ 1 - \exp[-\kappa(t - \theta)], & t > \theta \end{cases} \tag{5.94}$$

This means that the appearance of products is delayed by the time θ which is therefore an "incubation time." Note that θ is unrelated to the incubation time t_{inc} as defined in Eq. (5.34). In the present notation, $Y(t) = 1 - N(t)/N_0$; thus, for a dissociating system represented by matrix $\mathbb{J}$, we have [52] from Eq. (5.82) for the first moment of the MFPT $\bar{t}_1 = \theta + \kappa^{-1}$, while from Eq. (5.93) the second moment is $\bar{t}_2 = \theta^2 + (2/\kappa^2) + 2\theta/\kappa$; by elimination $\kappa^{-1} = \sqrt{\bar{t}_2 - \bar{t}_1^2}$, and $\theta = \bar{t}_1 - \kappa^{-1}$. The reaction yield given by Eq. (5.94) is thus described by two moments of the MFPT, but, if necessary, the fit can be refined by including higher moments [53].

5.5.4 The reduced-matrix approach

In this method the rate constant is obtained from Eq. (5.31) by computing the steady-state distribution first, that is, the eigenvector corresponding to the lowest eigenvalue, rather than the eigenvalue itself.

Examination of the population of dissociating molecules as a function of energy reveals that, depending upon the pressure, the population is basically an equilibrium population over a significant range of energies below the reaction threshold

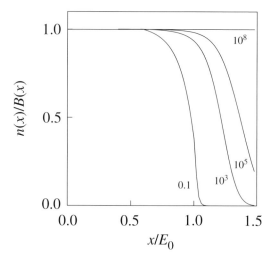

Fig. 5.5. An example of fractional deviation of the steady-state population $n(x)$ from the equilibrium distribution $B(x)$ as a function of dimensionless energy x/E_0 at several pressures, calculated for the dissociation of ethane at 2000 K using an off-set Gaussian transition probability ($\epsilon = 500\,\mathrm{cm}^{-1}$). The curves represent distributions at pressures of 0.1, 10^3, 10^5, and 10^8 torr, the latter as a horizontal line, meaning that the equilibrium population is maintained at all energies shown. E_0 is the reaction threshold.

(Fig. 5.5), which means that the effective size of the matrix necessary for computation may be reduced considerably.

Consider the master equation of Eq. (5.13) written in terms of the time-independent fractional population $n(y)$, i.e. at steady state. Let the reaction threshold be at energy E_0, and suppose that, up to energy $E_m \leq E_0$, the population is the equilibrium distribution $B(y)$ (Eq. (5.3)), so that the integral $\int_0^\infty p(x, y)\, n(y)\, dy$ can be split into two parts: one from 0 to E_m with population $n(y) \Rightarrow B(y)$, and another from E_m to ∞ with population $n(y)$. The master equation becomes

$$0 = \int_0^{E_m} p(x, y) B(y)\, dy + \int_{E_m}^\infty p(x, y) n(y)\, dy - \left(1 + \frac{k(x)}{\omega}\right) n(x) \qquad (5.95)$$

By virtue of detailed balance (Eq. (5.2)), $p(x, y) B(y) = p(y, x) B(x)$, so Eq. (5.95) becomes

$$B(x) \int_0^{E_m} p(y, x)\, dy = \left(1 + \frac{k(x)}{\omega}\right) n(x) - \int_{E_m}^\infty p(x, y) n(y)\, dy \qquad (5.96)$$

In discretized form let E_m correspond to level m; then Eq. (5.96) becomes, for level $i \geq m$,

$$B(x_i) \sum_{j \leq m} p_{ji} = n(x_i) \left(1 + \frac{k(x_i)}{\omega}\right) - \sum_{j > m} p_{ij} n(x_j) \qquad (5.97)$$

where $B(x_i)$ is an element of the discrete equilibrium population (Eq. (5.22)), $n(x_i)$ is an element of the time-independent population $n(x)$, and $k(x_i)$ is an element of $k(x)$.

The square in the diagram below represents the complete transition matrix $\mathbb{P}$, which is assumed to be of size $s \times s$. The summation on the LHS of Eq. (5.97) represents transitions that start above level m and end below level m (shown as an area labeled $\mathbb{P}_a$ in the diagram), while the summation on the RHS represents transitions that both start and end above level m (the area labeled $\mathbb{P}_b$).

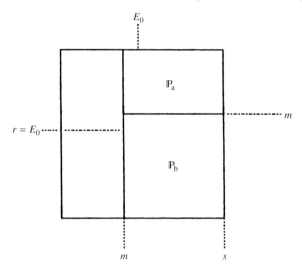

Thus Eq. (5.97) in matrix form, with raised dot indicating matrix multiplication, is

$$|\mathbf{V}\rangle = \mathbb{J}_b \cdot |\mathbf{N}\rangle \qquad (5.98)$$

where, using the notation introduced in Eq. (5.19), $\mathbb{J}_b = \mathbb{I} + \mathbb{K} - \mathbb{P}_b$ is the transport matrix, so that matrix $\mathbb{P}$ originally of size $s \times s$ is thus reduced for computation to matrix $\mathbb{P}_b$ of effective size $(s - m) \times (s - m)$. $\mathbb{J}_b$ has an obvious similarity to submatrix $\mathbb{J}_4$ in Eq. (5.33), except for sign and the fact that the partition is below threshold. $|\mathbf{V}\rangle$ is the column vector $B(x_i) \Sigma_{j \leq m} p_{ji}$ (Eq. (5.97)), and $|\mathbf{N}\rangle$ is the column vector of the populations $n(x_i)$, all of dimension $s - m$.

The formal solution of Eq. (5.98) is $|\mathbf{N}\rangle = \mathbb{J}_b^{-1} \cdot |\mathbf{V}\rangle$; however, in order to avoid computing the inverse of a likely ill-conditioned matrix $\mathbb{J}_b$, a re-scaling procedure is preferable [54]. If $B'(x_i)$ represents equilibrium populations for $i \geq m$, then $\mathbb{F} = \text{diag } B'(x_i)$ will be the corresponding diagonal matrix (of size $(s - m) \times (s - m)$). Now Eq. (5.98) can be written equivalently as $|\mathbf{V}\rangle = \mathbb{J}_b \cdot \mathbb{F} \cdot \mathbb{F}^{-1} \cdot |\mathbf{N}\rangle$; then the matrix product $\mathbb{J}_b \cdot \mathbb{F} = \mathbb{A}$ yields the symmetric matrix $\mathbb{A}$, while $\mathbb{F}^{-1} \cdot |\mathbf{N}\rangle = |\mathbf{Z}\rangle$ represents the column vector $|\mathbf{Z}\rangle$ with elements $z_i = n'(x_i)/B'(x_i)$, i.e. the steady-state population divided by the equilibrium population for $i \geq m$. Equation (5.98) thus becomes

$$|\mathbf{V}\rangle = \mathbb{A} \cdot |\mathbf{Z}\rangle \tag{5.99}$$

The immediate solution of this equation for vector $|\mathbf{Z}\rangle$ is obtainable, in principle, by matrix inversion: $|\mathbf{Z}\rangle = \mathbb{A}^{-1} \cdot |\mathbf{V}\rangle$. However, a stable solution of Eq. (5.99) may be obtained more easily in general by an iterative method, several of which have been described in the literature [22, 55–57].

A method that is not diffcult to implement and yields satisfactory results is a version of the Gauss–Seidel method [58] using the so-called "variable overrelaxation" for faster convergence. For the kth iterate of z_i we have [54, 58]

$$z_i^{k+1} = (1 - \gamma_i)z_i^k + \frac{\gamma_i}{a_{ii}}\left(v_i - \sum_{j=1}^{i-1}a_{ij}z_j^{k+1} - \sum_{j=i+1}^{s-m}a_{ij}z_j^k\right) \tag{5.100}$$

where the a_{ij} are elements of matrix $\mathbb{A}$ and the z_i have the same significance as before; v_i are the elements of $|\mathbf{V}\rangle$. γ_i is the relaxation parameter given by

$$\gamma_i = 1 + \frac{c - 1}{1 + k(x_i)/\omega} \tag{5.101}$$

c being an adjustable constant determined by trial for best results, generally $c \approx 1.5$. For $\gamma_i = 1$, Eq. (5.100) represents the standard Gauss–Seidel result. In the interest of rapidity of execution, the iteration starts with the initial estimate for z given by the deviation from the strong-collision steady-state distribution (Eq. (5.50)) in the form $z_i = (1 + k(x_i)/\omega)^{-1}$ (padded with $(r - m)$ 1's if the threshold is at level r), and is continued until convergence. Since matrix $\mathbb{A}$ is positive definite, with positive elements only on the diagonal, it can be shown [59] that the Gauss–Seidel iteration converges independently of the initial vector.

The full vector of steady-state populations is re-assembled as $|\mathbf{N}\rangle^\mathsf{T} = B(x_i)((1\,1\,1\ldots 1), z_i)$, the inside round bracket containing a string of m 1's. The rate constant k_{uni} is given by the discrete analog of Eq. (5.31) as $k_{\text{uni}} = \Sigma_i k(x_i)z_i B(x_i)$, where the counter i starts at $i = m + 1$ and runs until $i = s$.

It is evident from Fig. 5.5 that, for a given transition probability, the depopulation below threshold propagates to lower energies as pressure decreases, so that, if m is defined as the point at which the steady-state distribution begins to deviate from the equilibrium one, m shifts to lower energy levels with decreasing pressure. Some trial and error is therefore necessary to determine the appropriate value of m for a particular combination of transition-probability matrix and collision frequency ω.

5.5.5 *The power method*

At the low-pressure limit, as we have seen, only a partitioned matrix, $\mathbb{J}_1$, or $\mathbb{P}_1$ is involved (Eq. (5.33)), so the method of solution can be considered under the reduced-matrix heading. The idea is the following [60, 61]: starting with an arbitrary initial eigenvector $|\mathbf{V}^{(0)}\rangle$, a better approximation will be $\mathbf{V}^{(1)} = \mathbb{P}_1 \cdot \mathbf{V}^{(0)}$. This process is repeated, so that after n iterations we have

$$\mathbf{V}^{(n)} = \mathbb{P}_1 \cdot \mathbf{V}^{(n-1)} = \ldots = (\mathbb{P}_1)^n \cdot |\mathbf{V}^{(0)}\rangle \qquad (5.102)$$

It can be shown that, if $\mathbb{P}_1$ is a regular stochastic matrix, $\mathbf{V}^{(n)}$ will converge to the correct eigenvector after a sufficient number of iterations [62]. An additional iteration then produces the eigenvalue:

$$\mathbb{P}_1 \cdot \mathbf{V}^{(n)} = \lambda \mathbf{V}^{(n)} \qquad (5.103)$$

In this case λ is the *largest* positive eigenvalue (cf. Eq. (5.47)), so that the limiting low-pressure constant is $k_0/\omega = 1 - \lambda$. The corresponding steady-state population is obtained by normalizing $\mathbf{V}^{(n)}$. The disadvantage of the method is that, unless the starting vector $\mathbf{V}^{(0)}$ is close to the actual eigenvector, the number of iterations required is likely to be excessive.

5.6 Thermal rate as a Laplace transform

The Laplace transformation, as we have seen in the previous chapter (cf. also Appendix 2), provides a connection between E, the internal energy of a given molecular species (represented by the symbol x in the present chapter), a microcanonical or specific-energy property, and temperature T (via the transform parameter $s = 1/(kT)$), which is a canonical, i.e. bulk, property. The subject of thermal reactions was developed up to now on the assumption that the necessary microcanonical information is available, and proceeding from there bulk relations were obtained by Laplace-like averaging over all energies. Given the reciprocal relation between energy and temperature, it should be possible – under certain conditions – to proceed in the reverse direction and deduce specific-energy information from thermal bulk data. This is the subject of the present section.

5.6.1 *The case of dissociation*

If the $B(x)$ part in the expression for the limiting high-pressure rate constant k_∞ (Eq. (5.42)) is written out explicitly, the integral can be considered operationally as $1/Q$ times the Laplace transform of $k(x)N(x)$, with $s = 1/(kT)$ as the transform parameter (cf. Appendix 2):

$$k_\infty = \frac{1}{Q} \int_0^\infty k(x)\, N(x) \exp[-x/(\mathcal{k}\,T)]\, \mathrm{d}x = \frac{1}{Q} \mathcal{L}\{k(x)\, N(x)\} \qquad (5.104)$$

For clarity it is worthwhile to recall that k_∞ and $k(x)$ are the thermal and micro-canonical rate constants, respectively, for the dissociation reaction $M \longrightarrow$ products, which have implicitly been assumed so far; Q is the partition function for the internal degrees of freedom of reactant M involved in the density $N(x)$, i.e. principally vibrational.

If experiment provides k_∞ in the Arrhenius form (Eq. (5.44)) with A_∞ and $E_{a\infty}$ temperature-independent constants, then, on equating experimental and theoretical expressions for k_∞, we have

$$\mathcal{L}\{k(x)\, N(x)\} = A_\infty Q(s) \exp(-E_{a\infty}s) \qquad (5.105)$$

where $Q(s)$ is written for Q on the RHS of the equation above in recognition that Q is $\mathcal{k}\,T$- and hence s-dependent. By inversion

$$k(x)\, N(x) = A_\infty \mathcal{L}^{-1}\{Q(s) \exp(-E_{a\infty}s)\} \qquad (5.106)$$

and the inverse transform on the RHS is immediately recognized, by virtue of Section A2.2 in Appendix 2, as merely the zero-shifted inverse transform of $Q(s)$, i.e. the reactant density of states zero-shifted by $E_{a\infty}$. Thus [63]

$$k(x) = \begin{cases} \dfrac{A_\infty N(x - E_{a\infty})}{N(x)}, & x \geq E_{a\infty} \\ 0, & x < E_{a\infty} \end{cases} \qquad (5.107)$$

It is therefore possible, at least in principle, to recover the microcanonical rate constant $k(x)$ from the experimental thermal rate constant k_∞, the only other necessary information being the molecular parameters of the reactant for the calculation of the density of states $N(x)$. The qualification "in principle" is important, since the inversion process presupposes that ideally the experimental form of k_∞ is valid at all temperatures, whereas it is usually known over only a limited temperature range. Additionally, k_∞ is not directly experimentally observable and requires a more or less extensive extrapolation from lower pressures.

Despite this handicap, several tests [64, 65] have proved the practical usefulness of such inversion (see [24, Chapter 4] and [66, ref. 4]), for, if the only available items of information about a given reaction are experimental data in the form of Eq. (5.44) and molecular parameters of the reactant, the best approximation to the microcanonical constant $k(x)$ on the basis of such limited data is Eq. (5.107), with

$E_{a\infty}$ as the best available estimate of the threshold energy E_0. Note that this way $k(x)$ is obtained without any reference to a transition state.

Using this estimate of $k(x)$ it is then possible to reconstitute the entire fall-off of k_{uni} [67], which generally turns out to be not far from what can be obtained by more sophisticated methods based on better data.

With only minor complication the method can also be applied [68] when the pre-exponential factor A_∞ is found to be temperature-dependent in the form $A_\infty = A'_\infty(\ell T)^n$. Equation (5.106) then reads

$$k(x)N(x) = A'_\infty \mathcal{L}^{-1}\left\{\frac{Q(s)}{s^n}\exp(-E_{a\infty}s)\right\} \qquad (5.108)$$

where the zero-shifted inverse transform of $Q(s)/s^n$ is handled without difficulty by the method of steepest descents. The same method can deal with more compli-cated non-Arrhenius forms of k_∞ [66]. The consequence in these cases is a mod-ified energy dependence of the numerator in the expression for $k(x)$, Eq. (5.107) (Problem 5.17).

5.6.2 The case of recombination

When the inversion procedure is based on the high-pressure experimental rate con-stant for *dissociation* (Eq. (5.44)), as above, the disadvantage is that the activation energy $E_{a\infty}$ is generally large, and therefore likely to be subject to large experi-mental error in absolute terms, which perforce translates into a corresponding error in the microcanonical rate constant $k(x)$.

On the other hand, if the dissociation is of the special type XY $\rightarrow$ X + Y, where X and Y are radicals, the *recombination* rate constant has a temperature coefficient that is usually zero or slightly negative, and is therefore known with a correspond-ingly small absolute error. Most importantly, however, recombination rates are often available from experiment over a much larger temperature range than is the case for dissociation, so a similar inversion procedure involving the recombination rate constant is on somewhat more secure grounds and promises a more accurate $k(x)$ [69].

We now have to distinguish between dissociation (subscript d) and recombination (subscript r) rate constants, and between reactant (subscript XY) and products (subscripts X and Y). At limiting high pressure we have

$$k_{d,\infty} = k_{r,\infty}K_c \qquad (5.109)$$

K_c is the equilibrium constant

$$K_c = Q_t\frac{Q_X Q_Y}{Q_{XY}}\exp[-\Delta H_0/(\ell T)] \qquad (5.110)$$

where for convenience the translational partition function for relative motion $Q_t = (2\pi \mu \ell T)^{3/2}/h^3$ was factored out (μ is the reduced mass of the products X and Y), so that Q_{XY}, Q_X, and Q_Y are the respective *total* partition functions for degrees of freedom excluding translation (i.e. vibrational, rotational, and electronic), and ΔH_0 is the 0 K enthalpy of the reaction.

We assume that the experimental *recombination* rate constant is given by

$$k_{r,\infty} = A_{r,\infty} \exp[-E_{r,\infty}/(\ell T)] \tag{5.111}$$

i.e. an analog of Eq. (5.44), but usually with $E_{r,\infty} \approx$ zero or small. The theoretical high-pressure *dissociation* rate constant (with additional subscript "d") can be represented by (cf. Eq. (5.104))

$$k_{d,\infty} = \frac{1}{Q_{XY}^v} \mathcal{L}\{k(x)N(x)\} \tag{5.112}$$

where Q_{XY}^v is the *vibrational* partition function of reactant XY (emphasized with superscript "v"). From Eqs. (5.109)–(5.112) we have formally $k(x)N(x) = \mathcal{L}^{-1}(Q_{XY}^v k_{r,\infty} K_c)$. On substituting into this the explicit form of the equilibrium constant from Eq. (5.110), we see immediately that the *vibrational* part of the total partition function Q_{XY} drops out. On taking the inverse transform of the result we have, for $x \geq E_{r,\infty} + \Delta H_0$,

$$k(x)N(x) = \frac{A_{r,\infty}(2\pi \mu)^{3/2}}{h^3} \mathcal{L}^{-1}\left\{\frac{Q_X(s)Q_Y(s)}{s^{3/2}Q_{XY}^{rot}(s)} \exp[-(E_{r,\infty} + \Delta H_0)s]\right\} \tag{5.113}$$

where $Q_{XY}^{rot}(s)$ is the surviving rotational part of the partition function for the reactant. Here x is still the internal energy of the reactant and $N(x)$ its density of states, but the RHS is now (apart from a constant factor in front) a zero-shifted (by $E_{r,\infty} + \Delta H_0$) inverse transform of the partition-function ratio $Q_X(s)Q_Y(s)/Q_{XY}^{rot}(s)$, divided by $s^{3/2}$, which originates from the translational partition function Q_t.

$Q_{XY}^{rot}(s)$ and the rotational parts of $Q_X(s)Q_Y(s)$ each refer to a three-dimensional "external" rotation, the partition function of which is proportional to $s^{-3/2}$ (Appendix 1, Section A1.2.2) so that Eq. (5.113) simplifies to

$$k(x)N(x) = \frac{A_{r,\infty}(2\pi \mu)^{3/2}q_r}{h^3} \mathcal{L}^{-1}\left\{\frac{Q_X^v(s)Q_Y^v(s)}{s^3} \exp[-(E_{r,\infty} + \Delta H_0)s]\right\}$$

$$\tag{5.114}$$

where $Q_X^v(s)$ and $Q_Y^v(s)$ are now *vibrational* partition functions of the products and q_r is the common factor remaining after $s = 1/(\ell T)$ has been factored out of the rotational partition functions. The result of the inversion depends on ΔH_0, which is of the same order of magnitude as $E_{a\infty}$, but usually better known from the thermochemistry of the reaction.

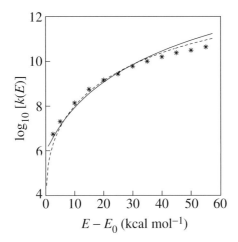

Fig. 5.6. An example of a microcanonical rate constant $k(E)$ obtained by inversion of thermal data for ethane. Solid line: by inversion of $k_{\infty,\text{diss}} = 10^{16} \exp[-30\,642\,\text{cm}^{-1}/(\not{k}\,T)]\,\text{s}^{-1}$. Dashed line: by inversion of $k_{\infty,\text{rec}} = 6 \times 10^{-11}\,\text{cm}^3\,\text{molec}^{-1}\,\text{s}^{-1}$, temperature-independent. Asterisks are data calculated by Wardlaw and Marcus (*J. Chem. Phys.* **83**, 3462 (1985)).

If $A_{\text{r},\infty}$ has a temperature dependence, positive or negative, this will increase or decrease the power of s in the denominator of the inverse transform. The inverse transform, apart from the different zero shift, is here basically the same as in Eq. (5.108), and is directly obtainable by the S-D method. Figure 5.6 compares the energy dependences of $k(x)$ in the reaction $C_2H_6 \leftrightarrow CH_3 + CH_3$ obtained by inversion of $k_{\text{r},\infty}$ and $k_{\text{d},\infty}$.

5.7 Non-collisional input

The energy necessary for reaction need not be uniquely the result of collisional interaction, as has been assumed so far in this chapter. Energy can initially be deposited in the reactant more directly, for example by a laser flash, photolysis, or chemical activation, whereby the source of the energy is radiation or the exothermicity of the reaction forming the reactant of interest. Subsequently this energy is degraded by the combined action of collisions and chemical reaction in much the same manner as described by the master equation (Eq. (5.13)), except for the addition of the source term $Rf(x)$:

$$\frac{\partial n(x,t)}{\partial t} = Rf(x) + \omega \int_0^\infty p(x,y)\,n(y,t)\,dy - (\omega + k(x))\,n(x,t) \qquad (5.115)$$

where $f(x)$ is the distribution function for activation, normalized to unity ($\int f(x)\,dx = 1$), i.e. the probability that an outside source shall furnish reactant M with internal energy x. If $n(x, t)$ is the fractional (i.e. dimensionless) population of M with internal energy x at time t, as in Eq. (5.13), R is a suitably normalized input flux in units of s^{-1}.

5.7.1 Chemical activation

Suppose that reactant M is formed by the bimolecular recombination X + Y. With energy zero at the ground state of M, as is appropriate for Eq. (5.115), the various energy parameters are shown in Fig. 5.7, from which it is clear that chemical activation requires $E_0' > E_0$. If X and Y are radicals, as is usually the case, there is no barrier to their recombination (i.e. $E_0' \cong \Delta H_0$), so the condition for chemical activation is then $\Delta H_0 > E_0$. Thus M will be activated for further reaction provided that the exothermicity of the activating reaction is larger than the threshold energy for decomposition of M.

Activated M, i.e. M with internal energy $x \geq E_0'$, represented in Fig. 5.7 and below by M(x), can then decompose to form products (D) with rate (not rate *constant*) ρ, or be collisionally deactivated (collision frequency ω) to form stabilized

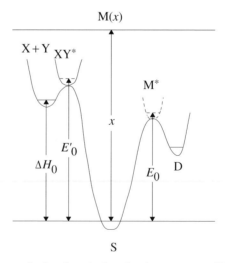

Fig. 5.7. Energy parameters in the chemical-activation process of Eq. (5.116). Energy zero is at the ground state of stabilized M (=S). E_0 is the critical energy for the process M(x) → D (D = decomposed M) via transition state M*. The entrance channel is the bimolecular reaction X + Y → M(x) via the transition state XY* with critical energy E_0' with respect to S, which produces M with internal energy x relative to its ground state (S). ΔH_0 is the exothermicity of the reaction X + Y → S.

M (S). The overall chemical activation scheme for production of M(x) is

$$X + Y \underset{k_d}{\overset{k_r}{\rightleftarrows}} M(x) \overset{\rho}{\longrightarrow} D \text{ (decomposed M)} \qquad (5.116)$$

$$\downarrow \omega$$

$$S \text{ (stablized M)}$$

The interest of this procedure is that recombinations of radicals proceed at low temperature, so M(x) will be formed with a distribution of energies that is much narrower than the high-temperature distributions (1000 K or more) produced in the usual kind of thermal reactions, which obliterate much of the detail of the dynamics.

If the non-collisional input R (the "source") is just balanced by collisions and reaction, the system can be considered to be in a steady state (more on this below), and, assuming no products to be present initially, Eq. (5.115) becomes for unit input ($R/\omega = 1$)

$$f(x) = \left(1 + \frac{k(x)}{\omega}\right) n(x) - \int_0^\infty p(x, y) \, n(y) \, dy \qquad (5.117)$$

where $k(x)$ is the microcanonical rate constant for the process M(x) $\rightarrow$ D. The matrix version of this equation reads

$$|F\rangle = -\mathbb{J} \cdot |N\rangle \qquad (5.118)$$

where $|F\rangle$ is the column vector of discretized $f(x)\,dx$, normalized to unity. The matrix $\mathbb{J} = \mathbb{P} - \mathbb{I} - \mathbb{K}$ is the same transport matrix as in Eq. (5.19), where $\mathbb{K}$ is the diagonal matrix with elements $k(x_i)/\omega$, and $|N\rangle$ is the column vector of the dimensionless populations $n(x_i)$. The solution for the steady-state population $|N\rangle$ is obtained by matrix inversion:

$$|N\rangle = (-\mathbb{J})^{-1} \cdot |F\rangle \qquad (5.119)$$

or, in terms of the elements $n(x_i)$ of $|N\rangle$, as $n(x_i) = [\mathbb{K} + \mathbb{I} - \mathbb{P}]_i^{-1} f_i$. Observe that the result for $|N\rangle$ does not assume any particular form of the transition probability $\mathbb{P}$.

In the above chemical activation scheme $R/\omega = k_r[X][Y]/\omega$, which has the dimension of concentration, so that the product $(R/\omega) n(x_i)$ are components of the steady-state concentration of chemically activated reactant M of internal energy x_i. The rate of formation of products is then $\rho = (R/\omega)\Sigma_i n(x_i) k(x_i)$.

The ratio ρ/R is dimensionless and represents the fraction of molecules M decomposed, i.e. using the notation of Eq. (5.116), $\rho/R = D/(D + S)$, which is sometimes referred to as the "reaction efficiency." The limiting low-pressure value of ρ/R is obtained by noting that, as $\omega \rightarrow 0$, elements $k(x_i)/\omega$ of $\mathbb{K}$ will dominate in matrix $\mathbb{J}$; thus the inverse $\mathbb{J}^{-1}$ contains mostly elements $\omega/k(x_i)$ and hence

$\lim(\rho/R)_{\omega\to 0} \to \Sigma_i f_i = 1$. The meaning is that, at low pressure, as might be expected, few molecules are stabilized since most decompose. At the other extreme, as ω increases without limit, matrix $\mathbb{K}$ will vanish, and so will the remaining elements $([\mathbb{I} - \mathbb{P}]_i^{-1} f_i)/\omega$, so that $\lim(\rho/R)_{\omega\to\infty} \to 0$, i.e. few molecules decompose since most are stabilized.

5.7.2 A special case

If the transition probability has the strong-collision form, $p(x, y) \Rightarrow B(x)$ (Eq. (5.7)), where $B(x)$ is the equilibrium Boltzmann distribution, Eq. (5.117) becomes

$$f(x) = \left(1 + \frac{k(x)}{\omega}\right) n(x) - B(x) \int_0^\infty n(y)\, dy \qquad (5.120)$$

In matrix form this equation means that matrix $-\mathbb{J}$ in Eq. (5.118) is now diagonal with elements $1 + (k(x_i)/\omega) - B(x_i)$. Since the inverse of a diagonal matrix is merely the inverse of the diagonal, we have immediately from Eq. (5.119) the strong-collision result

$$n(x_i)_{sc} = \frac{f(x_i)}{1 + (k(x_i)/\omega) - B(x_i)} \qquad (5.121)$$

Equation (5.118) is of the same type as Eq. (5.99), so the same Gauss–Seidel method can be applied to obtain a solution for the $n(x_i)$, as an alternative to direct matrix inversion. Equation (5.121) is then useful for the initial estimate of $n(x_i)$, and for the relaxation parameter, in the form $\gamma_i = 1 + (c - 1) n(x_i)_{sc}$.

5.7.3 The "sink"

Chemical-activation experiments are usually done at low temperature (<300 K), for which collisional energy transfer from below to above threshold is not important. This suggests [70] that only submatrix $\mathbb{J}_4$ (Eq. (5.33)) need be considered, instead of the full matrix $\mathbb{J}$, an approximation called the "steady-state sink method" [71] since energy levels below threshold are considered as a sink to which molecules are irreversibly deactivated.

A more detailed analysis [71] reveals that experimental conditions are reasonably represented by the steady state as defined above, on condition that the lowest eigenvalue of matrix $-\mathbb{J}$ is well separated from others. The approximation of the "sink method" worsens with rising temperature, because of the increasing probability of collisional re-activation of deactivated molecules, which can be remedied by placing the "sink" some levels below threshold. It is then found that, on placing the "sink" progressively at lower levels, the results tend toward a level-independent limit, which corresponds to the result obtained using the full matrix $(-\mathbb{J})^{-1}$ but from which the first row and the first column are deleted [72].

5.7.4 The distribution function

There remains the determination of the distribution function $f(x)$. In the case of room-temperature photochemical activation by monochromatic radiation (say of energy $\lambda \geq E_0$), the distribution function $f(x)$ will be, to a good approximation, simply the room-temperature equilibrium Boltzmann distribution $B(x)$ of the reactant shifted by λ to higher energies. Conditions in infrared multiphoton activation are considerably more complicated [73].

In the chemical activation scheme of Eq. (5.116) the necessary information can be obtained by assuming that the process $X + Y \rightleftarrows M(x)$ is in equilibrium, and then using microscopic reversibility to deduce the rate constants of interest. The explicit form of the source term for the formation of $M(x)$, with zero of energy at its ground state (Fig. 5.7), is thus ($x > E_0'$)

$$R f(x) = [X][Y] \, k_{\mathrm{r}}(x - \Delta H_0) \, B_{X+Y}(x - \Delta H_0) \tag{5.122}$$

where $B_{X+Y}(x - \Delta H_0)$ is the Boltzmann distribution (Eq. (5.3)) for the entrance channel system $X + Y$ at energy $x - \Delta H_0$ and temperature T:

$$B_{X+Y}(x - \Delta H_0) = \frac{N_{X+Y}(x - \Delta H_0) \exp[-(x - \Delta H_0)/(kT)]}{Q_X Q_Y} \tag{5.123}$$

The microcanonical rate constants for the "forward" and "reverse" activation processes at energy $x > E'$ are

$$k_{\mathrm{r}}(x - \Delta H_0) = \frac{G_{XY}^*(x - E_0')}{h N_{X+Y}(x - \Delta H_0)}; \qquad k_{\mathrm{d}}(x) = \frac{G_{XY}^*(x - E_0')}{h N_A(x)} \tag{5.124}$$

where $G_{XY}^*(x - E_0')/h$ are the fluxes through the transition state XY^*, which at equilibrium must be equal for $k_{\mathrm{r}}(x - \Delta H_0)$ and $k_{\mathrm{d}}(x)$. Thus, by virtue of microscopic reversibility,

$$k_{\mathrm{r}}(x - \Delta H_0) \, N_{X+Y}(x - \Delta H_0) = k_{\mathrm{d}}(x) \, N_A(x) \tag{5.125}$$

The equilibrium constant K_{c} for the process $X + Y \rightleftarrows M$ can be expressed either in terms of partition functions or in terms of rate constants:

$$K_{\mathrm{c}} = \frac{Q_A}{Q_X Q_Y} \exp[\Delta H_0/(kT)] = \frac{k_{\mathrm{r},\infty}}{k_{\mathrm{d},\infty}} \tag{5.126}$$

where $k_{\mathrm{r},\infty}$ and $k_{\mathrm{d},\infty}$ are the limiting high-pressure thermal rate constants for recombination and dissociation, respectively, the latter given by Eq. (5.42), but used here with the additional subscript "d" for clarity. On substituting Eqs. (5.123)–(5.126) into Eq. (5.122) and collecting terms, we have

$$Rf(x) = [X][Y]\,k_{r,\infty} \left(\frac{k_d(x)\,N_A(x)\,\exp[-x/(\hbar T)]\,dx}{Q_A\,k_{d,\infty}} \right) \tag{5.127}$$

where $[X][Y]k_{r,\infty} = R$, and thus $f(x)$ is the factor in large parentheses. Observe that $k_{d,\infty}$ is by definition the Boltzmann average of $k_d(x)$ (cf. Eq. (5.42)), so $k_{d,\infty} = \int_0^\infty k_d(x)B(x)\,dx$, where, in the present context, $B(x) = N_A(x)\exp[-x/(\hbar T)]\,dx/Q_A$, so $k_{d,\infty}$ is actually the normalization constant for the numerator in $f(x)$. Thus, if $k_{d,\infty}$ is written out explicitly, it is clear that $f(x)$ is a properly normalized distribution function, given by properties of the back-dissociating chemically activated reactant M:

$$f(x) = \frac{k_d(x)\,N_A(x)\,\exp[-x/(\hbar T)]\,dx}{\int_0^\infty k_d(x)\,N_A(x)\,\exp[-x/(\hbar T)]\,dx} \tag{5.128}$$

Moreover, if we substitute for $k_d(x)$ from Eq. (5.124), $N_A(x)$ drops out and $f(x)$ is given entirely in terms of properties of the transition state XY*. It is then obvious that the effective lower limit of the integral is E_0' since this is the threshold energy for $k_d(x)$.

It can be shown (Problem 5.18) that a good approximation to the distribution function $f(x)$ of Eq. (5.128) is a Boltzmann distribution of M, shifted to higher energies by $E_{a\infty}$, the temperature coefficient of $k_{d,\infty}$, if $E_{a\infty}$ is the only available estimate of E_0'. A comparison of several empirical forms of $f(x)$ shows [74] that all give comparable results, provided that the variance is the same.

To be useful, experimental studies using chemical activation require systems with known thermochemistry. Among the first were reactions H + butene-2 (*cis* or *trans*) $\rightarrow$ C$_4$H$_9^{\#}$ (*sec*-butyl radical) $\rightarrow$ CH$_3$ + C$_3$H$_6$ [75], and elimination of HY (Y = F, Cl, Br) from recombination of halogen-substituted methyl radicals, in reactions of the type 2CH$_2$Y $\rightarrow$ (YCH$_2$CH$_2$Y)$^{\#}$ $\rightarrow$ CH$_2$ = CHY + HY [76]. The symbol # indicates excitation, i.e. $x - E_0'$ in Fig. 5.7, which is 42 kcal mol^{-1} for C$_4$H$_9^{\#}$, and of the order of 50 kcal mol^{-1} for (YCH$_2$CH$_2$Y)$^{\#}$. Stabilization in the collisional part of the reaction sequence, i.e. the formation of S in Eq. (5.116), has been much used to study collisional energy transfer (see [2] for details).

More recently, chemical-activation studies were applied to multichannel reactions of atmospheric interest, e.g. NH$_2$ + NO $\rightarrow$ NH$_2$NO$^{\#}$ $\rightarrow$ etc., a process of importance in the formation and removal of NO$_x$ in combustion processes [77], and CH + C$_2$H$_2$ $\rightarrow$ C$_3$H$_3^{\#}$ $\rightarrow$ etc., a process implicated as the precursor to the formation of soot in combustion of hydrocarbons [78]. Details of these reaction channels are mentioned in Section 5.9.

5.8 Rotational effects

In the treatment of thermal reactions we have so far considered only a reactant with internal, mostly vibrational, energy. If the decomposing reactant has angular momentum, i.e. rotational energy, the fragmentation can be considered as proceeding under an effective potential, which was discussed in some detail in Section 3.4, and will be again in Chapter 8 from the point of view of the reverse association of fragments.

However, in the context of thermal reactions, which is the subject of the present section, the focus of interest is the effective potential insofar as it affects the *forward* decomposition of the reactant with angular momentum, or more precisely, as it modifies the height of the potential barrier separating the reactant from products.

5.8.1 The two-dimensional master equation

In a more realistic representation of collision events, it is necessary to consider that the outcome of a collision is likely to be the transfer both of vibrational and of rotational energy. The solution of the corresponding master equation is a considerably more difficult problem than that of the relatively simple master equation considered in previous sections.

To simplify, we shall consider at the outset only conditions at steady state, i.e. we shall work with a version of the master equation in the form of Eq. (5.39), suitably modified. The former transition probability $p(x, y)$, where x and y originally referred, respectively, to final and initial *vibrational* energies, will now be a function of final and initial *rotational* energies as well:

$$p(x, y) \Rightarrow p(x_v, y_v; x_r, y_r) \tag{5.129}$$

where, as before, y refers to *initial* energies, x refers to *final* (post-collision) energies, and subscripts v and r distinguish between vibrational and rotational energy, respectively. In this notation E_0 and E_0^* are particular values of x_v. The probability of Eq. (5.129) is assumed to be normalized according to (cf. Eq. (5.1))

$$\int_0^\infty \int_0^\infty p(x_v, y_v; x_r, y_r) \, dx_v \, dx_r = 1 \qquad \text{for every} \quad y_v, y_r \tag{5.130}$$

The former steady-state population $n(x)$ and the microcanonical constant $k(x)$ are now likewise functions of x_v and x_r:

$$n(x) \Rightarrow n(x_v, x_r); \qquad k(x) \Rightarrow k(x_v, x_r) \tag{5.131}$$

The population is normalized to unity:

$$\int_0^\infty \int_0^\infty n(x_v, x_r)\, dx_v\, dx_r = 1 \tag{5.132}$$

As a consequence, the steady-state master equation becomes two-dimensional:

$$-k_{uni}\, n(x_v, x_r) = \omega \int_0^\infty \int_0^\infty p(x_v, y_v; x_r, y_r)\, n(y_v, y_r)\, dy_v\, dy_r - [\omega + k(x_v, x_r)]\, n(x_v, x_r) \tag{5.133}$$

In order to simplify further, we shall assume that the transition probability p factorizes into vibrational (V) and rotational (R) parts:

$$p(x_v, y_v; x_r, y_r) = V(x_v, y_v) R(x_r, y_r) \tag{5.134}$$

each part being normalized separately to unity:

$$\int_0^\infty V(x_v, y_v)\, dx_v = 1; \qquad \int_0^{E_r^{\max}} R(x_r, y_r)\, dx_r = 1 \tag{5.135}$$

The upper limit $E_r^{\max}$ corresponds to rotational energy so high that the maximum in the effective potential reduces to an inflection point (Appendix 4, Section A4.5).

   The factorization amounts to assuming that vibrational and rotational energies are transferred independently of each other. It is found experimentally that, in general, rotational relaxation times are an order of magnitude faster than vibrational relaxation times, which suggests that rotational energy transfer occurs on a much shorter time scale and is therefore more efficient than the transfer of vibrational energy. We may therefore reasonably assume that the rotational transition probability will be of the strong-collision type, one such that each collision results in an equilibrium distribution of rotational energies corresponding to a heat-bath temperature T. The probability is thus a function of the final energy only and independent of the initial energy, in complete analogy with the vibrational strong-collision transition probability (cf. Eq. (5.7)):

$$R(x_r, y_r) \Rightarrow R(x_r) = \frac{\exp[-x_r/(\mathit{k}T)]}{\mathit{k}T\{1 - \exp[-E_r^{\max}/(\mathit{k}T)]\}} \tag{5.136}$$

5.8.2 The strong-collision solution

A more drastic approximation is to assume that the vibrational transition probability likewise has the strong-collision form:

$$V(x_v, y_v) \Rightarrow B(x_v) \tag{5.137}$$

where $B(x_v)$ is the Boltzmann distribution of vibrational energies x_v, corresponding to bath temperature T:

$$B(x_v) = N(x_v) \exp[-x_v/(k\,T)]/Q_v \qquad (5.138)$$

This is of course Eq. (5.3) with the additional subscript "v" on x. The master equation of Eq. (5.133) thus reduces to

$$-k_{\mathrm{uni}}^{\mathrm{sc}}\, n(x_v, x_r) = \omega \int_0^\infty \int_0^\infty B(x_v)\, R(x_r)\, n(y_v, y_r)\, \mathrm{d}y_v\, \mathrm{d}y_r - [\omega + k(x_v, x_r)]\, n(x_v, x_r)$$

$$(5.139)$$

Now both $B(x_v)$ and $R(x_r)$ factor out of the double integral, while the double integral over $n(y_v, y_r)$ is unity on account of the normalization in Eq. (5.132), so the steady-state distribution is

$$n(x_v, x_r) = \frac{\omega B(x_v) R(x_r)}{\omega + k(x_v, x_r) - k_{\mathrm{uni}}} \qquad (5.140)$$

Consequently we have

$$k_{\mathrm{uni}}^{\mathrm{sc}} = \omega \int \int \frac{B(x_v)\, R(x_r)\, k(x_v, x_r)\, \mathrm{d}x_v\, \mathrm{d}x_r}{\omega + k(x_v, x_r) - k_{\mathrm{uni}}^{\mathrm{sc}}} \qquad (5.141)$$

For simplicity, in the following we shall drop $k_{\mathrm{uni}}^{\mathrm{sc}}$ from the denominator, which should not affect the results significantly, given all the other approximations. This equation is the analog of Eq. (5.51), which was previously obtained in the rotation-less case.

In some cases it is convenient to obtain $k_{\mathrm{uni}}^{\mathrm{sc}}$ in two steps, by first integrating Eq. (5.141) over x_v at fixed x_r, which produces the r-dependent strong-collision $k_{\mathrm{uni}}^{\mathrm{sc}}(x_r)$, and then integrating over x_r. Thus, from Eqs. (5.140) and (5.138),

$$k_{\mathrm{uni}}^{\mathrm{sc}}(x_r) = \frac{\omega}{Q_v} \int \frac{k(x_v, x_r)\, N(x_v) \exp[-x_v/(k\,T)]\, \mathrm{d}x_v}{\omega + k(x_v, x_r)} \qquad (5.142)$$

followed by

$$k_{\mathrm{uni}}^{\mathrm{sc}} = \int k_{\mathrm{uni}}^{\mathrm{sc}}(x_r)\, R(x_r)\, \mathrm{d}x_r \qquad (5.143)$$

This procedure is used in the variational procedure of Chapter 7.

With an explicit form for $k(x_v, x_r)$ obtained in Chapter 3 it is now possible to deal further with Eq. (5.141), which will be done first for two special cases.

5.8.3 *Special case 1: high pressure*

This has been considered before in the vibrations-only case (Eq. (5.42)). In the present case the condition $\omega \gg k(x_v, x_r)$ in Eq. (5.141) leads to

$$k_\infty = \int \int B(x_v)\, R(x_r)\, k(x_v, x_r)\, dx_v\, dx_r \qquad (5.144)$$

Since there is no need now to distinguish between pre- and post-collision energies, we may revert to the notation of Chapter 3, noting the correspondence $x_v \equiv E_v$, $x_r \equiv E_r$. After substitution for $B(x_v)$ from Eq. (5.138), for $R(x_r)$ from Eq. (5.136), and for $k(x_v, x_r)$ from Eq. (3.38), it is convenient to change variables by writing $z = E - E_0^*$, using the general form of the barrier $E_0^* = E_0 - \mathfrak{F}(E_r)$ (cf. Section 3.4.3). After the density $N(E_v)$ has been canceled out in $B(E_v)$ and $k(E_v, E_r)$, the integral in Eq. (5.144) separates into two independent parts:

$$k_\infty = \frac{\alpha f_\infty \exp[-E_0/(kT)]}{hQ_v} \int_0^\infty G^*(z)\exp[-z/(kT)]\,dz \qquad (5.145)$$

where

$$f_\infty = \int_0^{E_r^{\max}} \frac{\exp\{[\mathfrak{F}(E_r) - E_r]/(kT)\}\,dE_r}{kT\{1 - \exp[-E_r^{\max}/(kT)]\}} \qquad (5.146)$$

The integral in Eq. (5.145) is the Laplace transform of $G^*(z)$, which we know from Eq. (4.11) is equal to kTQ_v^*, where Q_v^* is the partition function for the internal states of the transition state. Thus, in Eq. (5.145), $k_\infty = \alpha f_\infty (kT/h)(Q_v^*/Q_v)\exp[-E_0/(kT)]$, which is the standard transition-state result for zero angular momentum, multiplied by f_∞, which can be considered to represent the centrifugal correction factor that corrects for the neglect of rotations.

5.8.4 *Analytical results*

Equation (5.146) is the general result for f_∞ with an arbitrary potential that generally requires a numerical evaluation owing to the usually complicated form of the function $\mathfrak{F}(E_r)$.

However, an analytical solution is possible for a Type-1 reaction with a Lennard-Jones barrier of Eq. (3.32). $E_r^{\max}$ is determined by the condition $E_0^* = 0$, which yields in this case $E_r^{\max} = \frac{3}{2}E_0$ (cf. Fig. 3.3). Except for reactions with unusually low activation energies, this $E_r^{\max}$ is sufficiently high to allow one to extend the upper integration limit in f_∞ to infinity, and at the same time to drop the exponential in the denominator. If we let $a = [1/(kT)][2/(27E_0)]^{1/2}$, then, from Eq. (3.32) we have $[\mathcal{F}(E_r) - E_r]/(\ell T) = -aE_r^{3/2}$, so that Eq. (5.146) becomes

$$f_\infty \cong \frac{1}{\ell T} \int_0^\infty \exp\left(-aE_r^{3/2}\right) dE_r = \Gamma\left(\frac{2}{3}\right)\left(\frac{4E_0}{\ell T}\right)^{1/3} \qquad \text{(Type 1)} \quad (5.147)$$

The factor f_∞ thus has a weak negative temperature dependence and reduces the high-pressure activation energy $E_{a\infty}$ (Eq. (5.58)) by $\frac{1}{3}\ell T$.

For Type-2 and 3 reactions the barrier is given by Eq. (3.37), from which $E_r^{\max} = E_0/[1 - (r_e^2/r_0^2)]$ and $[\mathcal{F}(E_r) - E_r]/(\ell T) = -aE_r$, where $a = [1/(\ell T)](r_e/r_0)^2$. This $E_r^{\max}$ is again sufficiently high that Eq. (5.146) can be written with an infinite upper limit; thus

$$f_\infty \cong \frac{1}{\ell T} \int_0^\infty \exp(-aE_r) dE_r = \left(\frac{r_0}{r_e}\right)^2 \qquad \text{(Types 2 and 3)} \qquad (5.148)$$

In this case f_∞ is temperature-independent. Since the moment of inertia I is defined by $I = \mu r^2$ (μ is the reduced mass of the fragments), the above f_∞ can be defined equivalently as the ratio of the corresponding moments of inertia: $f_\infty = I_0/I_e$.

5.8.5 Special case 2: low pressure

For $\omega \ll k(E_v, E_r)$, Eq. (5.141) becomes the analog of Eq. (5.52):

$$k_0^{sc} = \omega \int \int B(E_v) R(E_r) dE_v dE_r$$

$$= \frac{\omega}{Q_v} \int_0^{E_r^{\max}} dE_r \int_{E_0^*}^\infty dE_v N(E_v) \exp[-E_v/(\ell T)] \frac{\exp[-E_r/(\ell T)]}{\ell T\{1 - \exp[-E_r^{\max}/(\ell T)]\}}$$

$$(5.149)$$

Since E_0^* is a function of E_r, this double integral does not separate into two independent parts. On proceeding to integrate by parts, and assuming $E_r^{\max}/(\ell T)$ to be high enough that $\exp[-E_r^{\max}/(\ell T)] \cong 0$, we obtain (Problem 5.19)

$$k_0^{sc} = \frac{\omega}{Q_v} \int_{E_0}^\infty dE_v N(E_v) \exp[-E_v/(\ell T)]$$

$$- \frac{\omega}{Q_v} \int_0^{E_r^{\max}} dE_0^* N_v(E_0^*) \exp[-(E_0^* + E_r)/(\ell T)] \qquad (5.150)$$

Now the first term is merely the low-pressure rate constant in the absence of rotations, which we may call $k_{0,\mathrm{nr}}^{\mathrm{sc}}$ ("nr" denotes "no rotations"). Note that, in the second term, $\mathrm{d}E_0^* = -\mathrm{d}\mathfrak{F}(E_\mathrm{r})$, so the term is actually positive, and consequently the above $k_0^{\mathrm{sc}} > k_{0,\mathrm{nr}}^{\mathrm{sc}}$, as one would expect.

Since there is no sufficiently accurate analytical formula for the density of states, the integrals in Eq. (5.150) have to be evaluated numerically. If we define a low-pressure centrifugal correction factor f_0, analogous to f_∞, by $f_0 = k_0^{\mathrm{sc}}/k_{0,\mathrm{nr}}^{\mathrm{sc}}$, and change the integration limits in $k_{0,\mathrm{nr}}^{\mathrm{sc}}$, the slightly simplified result for an arbitrary barrier is

$$f_0 = 1 + \frac{\displaystyle\int_\infty^{E_\mathrm{r}^{\max}} \mathrm{d}\mathfrak{F}(E_\mathrm{r})\, N_\mathrm{v}[E_0 - \mathfrak{F}(E_\mathrm{r})]\exp[\mathfrak{F}(E_\mathrm{r}) - E_\mathrm{r}]/(\ell T)}{\displaystyle\int_0^\infty \mathrm{d}y\, N_\mathrm{v}(E_0 + y)\exp[-y/(\ell T)]} \qquad (5.151)$$

Although this is not immediately obvious, the factor f_0 also has a negative temperature dependence, as shown in Fig. 5.8, which compares the temperature dependences of f_∞ and f_0. In general the temperature dependence of f_0 is comparable to that of f_∞, so there is little effect on the fall-off in activation energy.

In the case of a harmonic diatomic molecule, the density of states is a constant. If, in addition, $E_\mathrm{r}^{\max} \to \infty$ and the reaction is of Type 2 or 3, then, using

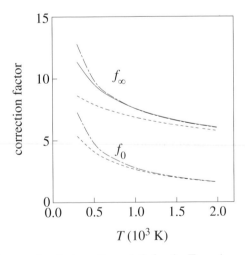

Fig. 5.8. Centrifugal correction factors f_∞ and f_0 for the Type-1 reaction $C_2H_6 \to CH_3 + CH_3$ between 300 and 2000 K. Solid line: analytical f_∞, Eq. (5.147). The other f_∞ were calculated from Eq. (5.146), and f_0 from Eq. (5.151): dashed line, Morse potential; dot–dashed line, Lennard-Jones potential.

$[\mathfrak{F}(E_r) - E_r]/(\ell T) = -aE_r$, we get

$$f_0 \cong \left(\frac{r_0}{r_e}\right)^2 \qquad \text{(Types 2 and 3, diatomic)} \qquad (5.152)$$

In this special case there is no pressure (and no temperature) dependence since $f_\infty = f_0$.

5.8.6 Intermediate pressures

To consider rotational effects at intermediate pressures, it is in principle necessary to go back to Eq. (5.141) and take an average over rotational energies, which amounts to a more complicated proposition [80] even in the strong-collision version, which uses equilibrium distributions both of vibrational and of rotational energies:

$$\langle k_{\text{uni}}\rangle = \frac{1}{\ell T Q_v} \int_0^\infty dE_r \exp[-E_r/(\ell T)]$$

$$\times \int_{E_0^*}^\infty dE_v \frac{\omega k(E_v, E_r) N(E_v) \exp[-E_v/(\ell T)]}{\omega + k(E_v, E_r)} \qquad (5.153)$$

where Q_v is the vibrational partition function of the reactant and $E_r^{\max}$ is assumed to be sufficiently high to allow using infinity as the upper limit of the integral with respect to E_r. Equivalently, the average can be written as a sum over an equilibrium distribution of rotational quantum numbers J:

$$\langle k_{\text{uni}}\rangle = \frac{1}{Q_v Q_r} \sum_{J=0}^{J_{\max}} (2J + 1) \exp[-J(J + 1)B_e/(\ell T)]$$

$$\times \int_{E_0^*(J)}^\infty dE_v \frac{\omega k(E_v, J) N(E_v) \exp[-E_v/(\ell T)]}{\omega + k(E_v, J)} \qquad (5.154)$$

where $Q_r = \ell T/B_e$ is the partition function for the two-dimensional external rotation of the reactant and $J_{\max}$ is the integer nearest $(E_r^{\max}/B_e)^{1/2}$. The critical energy is now written $E_0^*(J)$ to emphasize its J-dependence through E_r.

A simple way to avoid the somewhat complicated algebra involved in the above $\langle k_{\text{uni}}\rangle$ is by performing a simple interpolation between the high- and low-pressure forms of $k_{\text{uni}}^{\text{sc}}$ using centrifugal correction factors (Eqs. (5.146) and (5.151)). This is accomplished by writing the microcanonical rate constant of Eq. (3.28) in the form

$$\langle k(E_v)\rangle \cong \frac{\alpha f_\infty}{f_0} \frac{G^*(E_v - E_0)}{h N(E_v)} = \frac{f_\infty}{f_0} k(E_v) \qquad (5.155)$$

which can be considered as $\langle k(E_v) \rangle$ averaged over rotational energies, defined as the standard (no-rotation) $k(E_v)$ multiplied by the ratio f_∞/f_0. If the strong-collision k_{uni}^{sc} of Eq. (5.51), which represents the no-rotation case, is now modified to read (taking note of the correspondence $x \to E_v$ and neglecting k_{uni}^{sc} in the denominator)

$$\langle k_{uni}^{sc} \rangle = \omega f_0 \int_{E_0}^{\infty} \frac{B(E_v)\,(f_\infty/f_0)\,k(E_v)\,dE_v}{\omega + (f_\infty/f_0)\,k(E_v)} \tag{5.156}$$

it can be readily verified that, at the limits $\omega \to \infty$ and $\omega \to 0$, this equation yields $\langle k_\infty^{sc} \rangle = f_\infty k_{\infty,nr}^{sc}$ and $\langle k_0^{sc} \rangle = f_0 k_{0,nr}^{sc}$, respectively. This interpolation formula for $\langle k_{uni}^{sc} \rangle$ produces results that are experimentally indistinguishable from the treatment of [80].

5.9 Multichannel reactions

In a number of cases the same reactant can give rise to two or more distinct sets of products; for example, in the case of the excited species mentioned in Section 5.7.4,

$$C_3H_3^{\#} \to \begin{cases} H + C_3H_2 \\ H_2 + C_3H \end{cases} \quad \text{and} \quad NH_2NO^{\#} \to \begin{cases} OH + N_2H \\ H_2O + N_2 \end{cases}$$

In other cases, new channels may appear in an otherwise "simple" reaction as temperature is raised, e.g.

$$\begin{aligned} CH_3 + CH_3 &\to C_2H_6 && \text{below 2000 K} \\ &\to C_2H_5 + H \\ &\to CH_4 + {}^1CH_2 \end{aligned} \Bigg\} \text{ in flames above 2000 K}$$

If, prior to decomposition, energy is randomized in the excited reactant, according to the fundamental assumption of the present treatment of unimolecular reactions (cf. Chapter 3), then we can expect that decomposition into each product channel will take place independently, each with its own transition state [81, 82].

Consider the following general scheme (Fig. 5.9):

$$M(E) \to \begin{cases} M_1^* \xrightarrow{k_1(E)} P_1 \\ M_2^* \xrightarrow{k_2(E)} P_2 \end{cases} \tag{5.157}$$

where reactant M with internal energy E yields product(s) P_1 via transition state M_1^* via rate constant $k_1(E)$, specified below, and similarly for product(s) P_2. The eigenvalue equation for k_{uni}^t, the rate constant for total disappearance of $M(E)$ in a thermal reaction with arbitrary transition probability, is still given by the master equation of Eq. (5.39) for k_{uni}, and by Eq. (5.45) for k_0 (noting the correspondence

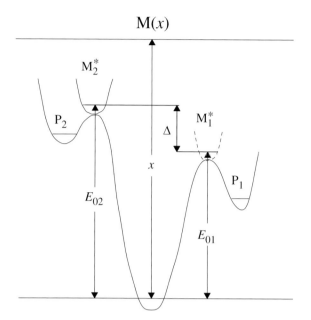

Fig. 5.9. Energy parameters in a two-channel reaction $M \rightarrow P_1$ and P_2, with x being the internal excitation energy of M. Channel 2, assumed to be the high-energy channel, proceeds through transition state M_2^* with critical energy E_{02}, while channel 1 proceeds through transition state M_1^* with critical energy E_{01}. Δ is the difference $E_{02} - E_{01}$. The drawing represents both channels as being reactions of Type 2, but the relations would be similar if one or both channels were of Type 1.

$x \rightarrow E$), except that $k(E)$ in both equations is now replaced by $k_1(E) + k_2(E)$, the total microcanonical rate. The eigenvalues then give the total rates k_{uni}^t and k_0^t, respectively.

5.9.1 The strong-collision solution

Proceeding as before in Section 5.4, we obtain immediately the strong-collision form of k_{uni}^t, assuming that channel 1 is the low-energy channel:

$$k_{uni}^{sc,t} = \omega \int_{E_{01}}^{\infty} \frac{B(E)\,[k_1(E) + k_2(E)]\,dE}{\omega + k_1(E) + k_2(E) - k_{uni}^{sc,t}} \tag{5.158}$$

which is the analog of Eq. (5.51). Neglecting $k_{uni}^{sc,t}$ in the denominator, this relation amounts to

$$k_{uni}^{sc,t} = \omega \int_{E_{01}}^{\infty} \frac{B(E)\,k_1(E)\,dE}{\omega + \sum k_i(E)} + \omega \int_{E_{02}}^{\infty} \frac{B(E)\,k_2(E)\,dE}{\omega + \sum k_i(E)} \tag{5.159}$$

where $\Sigma k_i(E) = k_1(E) + k_2(E)$. The high-pressure limit is then obviously $k_\infty^t = k_\infty(1) + k_\infty(2)$, so that in principle the high-pressure thermal rate constant of each channel is obtainable individually. At the low-pressure limit

$$\frac{k_0^{sc,t}}{\omega} = \int_{E_{01}}^{\infty} \frac{B(E)k_1(E)\,dE}{\Sigma k_i(E)} + \int_{E_{02}}^{\infty} \frac{B(E)k_2(E)\,dE}{\Sigma k_i(E)} = \int_{E_{01}}^{\infty} B(E)\,dE \qquad (5.160)$$

so that the total low-pressure rate separates for the two channels in proportion to their branching ratio. Generalization of these results to more than two channels should be obvious.

5.9.2 Rotational effects

It has been considered implicitly so far that only vibrational energy (E) is involved. If we wish to consider rotational effects, the simple procedure involving the centrifugal correction factors (within the strong-collision approximation) can be applied approximately if one of the channels is "tight," i.e. with $f_\infty \cong f_0 \cong 1$. Suppose that this channel is channel 2, while channel 1 is "loose" with $f_{\infty 1} > f_{01} \neq 1$. Equation (5.158) is then, with $f_r = f_{\infty 1}/f_{01}$, and neglecting $k_{uni}^{sc,t}$ in the denominator,

$$\langle k_{uni}^{sc,t} \rangle = \omega f_{01} \int_{E_{01}}^{\infty} \frac{B(E)[f_r k_1(E) + k_2(E)/f_{01}]\,dE}{\omega + f_r k_1(E) + k_2(E)/f_{01}} \qquad (5.161)$$

This formulation gives the correct high-pressure limits for the two channels:

$$k_\infty(1) = \int_{E_{01}}^{\infty} B(E)f_\infty k_1(E)\,dE; \qquad k_\infty(2) = \int_{E_{02}}^{\infty} B(E)k_2(E)\,dE \qquad (5.162)$$

For the total rate at the low-pressure limit we get

$$\frac{\langle k_0^t \rangle}{\omega} = f_{01} \int_{E_{01}}^{\infty} B(E)\,dE \qquad (5.163)$$

but this does not separate into the correct contributions from each channel as one would expect from the analog of Eq. (5.160).

For a better (and more complicated) treatment of rotational effects it is necessary to go back to Eq. (5.154) and write the microcanonical rate constant for each channel as a function both of vibrational energy and of angular momentum J, and then average the thermal k_{uni}^{sc} over all J.

5.10 Activation by black-body radiation

Back in the 1920s when thermal unimolecular reactions began to be studied in bulb experiments, the difficulty of how collisional activation, a second-order process, could give rise to a first-order reaction, was explained (for a while) by the hypothesis that perhaps activation was due to black-body radiation from the hot surface of the reactor. This hypothesis was laid to rest by Lindemann's collisional scheme (e.g. Section 1.2.3) that is the basis of every contemporary treatment of thermal unimolecular reactions *at finite pressure*.

5.10.1 The ion cyclotron

The invention of the ion cyclotron gave rise about fifteen years ago to a new field in mass spectrometry called Fourier-transform ion-cyclotron resonance (FT-ICR) spectrometry. Here, in a highly evacuated chamber ($<10^{-8}$ torr), ions are subjected to a combination of a weak electric field and a strong magnetic field, which creates an electromagnetic "bottle" where ions may be trapped and observed for long times (of the order of several seconds) in a virtually collision-free environment. It can be shown [83] that, under these conditions, ions can dissociate by absorption of black-body infrared radiation from the heated surroundings, or from a low-intensity infrared laser. The latter is not fundamentally different from standard infrared multiphoton activation, except for the substantially collision-free environment. Thus the original radiation hypothesis has regained credibility, albeit under conditions not experimentally realizable in the 1920s.

The first demonstration of black-body infrared radiative dissociation (BIRD) was for small cluster ions of the type [84]

$$(H_2O)_4H^+ \longrightarrow (H_2O)_3H^+ + H_2O$$

and was later extended to investigation of properties of large ions and ionic complexes of biological interest. These ionic proton-bound clusters are fundamental to many biological processes such as protein folding and bond cleavage in peptides. As shown below, BIRD provides access to the energetics of these processes, which is difficult to obtain otherwise.

5.10.2 Zero-pressure kinetics

The virtual absence of collisions in a FT-ICR experiment opens the way to unimolecular kinetics *at actual zero pressure*, unlike the "low-pressure limit" of Section 5.4.4, where Eq. (5.45) predicts $k_0 \to 0$ for $\omega \to 0$ since it is based on the assumption that collisions are the only source of activation.

The zero-pressure k_{uni} can be considered as given by Eq. (1.28) of Lindemann's scheme in which rate constants for collisional activation ($k_1[X]$) and deactivation ($k_2[X]$) are replaced by rate constants for radiative absorption (k_1^r) and radiative emission (k_2^r) of infrared photons; thus, in energy-dependent form,

$$k_{uni} = \int \frac{k_1^r(E) k_3(E)}{k_2^r(E) + k_3(E)} \, dE \qquad (5.164)$$

where $k_3(E)$ is the microcanonical rate constant for decomposition of the activated ionic cluster, which is given in principle by the RRKM $k(E)$ of Chapter 3.

When ions are confined in a heated FT-ICR trap, they not only absorb but also emit infrared photons. One can then envisage different regimes concerning k_{uni} of Eq. (5.164).

5.10.3 The Boltzmann limit

If ions absorb and emit infrared photons much faster than they decompose, i.e. if $k_2^r(E) \gg k_3(E)$, the decomposition will be the rate-limiting process. Under such conditions, ions of large biomolecules with very large numbers of degrees of freedom will achieve equilibrium with the black-body radiation field inside the heated trap [85]. Then $k_1^r(E)/k_2^r(E)$, the distribution of internal energies of ions, will be given by the Boltzmann distribution $B(E)$ (Eq. (5.3)), so that $k_{uni} = \int k_3(E) B(E) \, dE$, which is in fact the definition of k_∞ in a standard thermal reaction (cf. Eq. (5.42) with $x \equiv E$).

Paradoxically, then, a condition equivalent to the "high-pressure limit" is achieved at zero pressure. This has been demonstrated for the BIRD kinetics of the ions of ubiquitin (relative molecular mass 8600), a protein consisting of 76 amino-acid residues [86]. Infrared multiphoton dissociation by low-power CW CO_2 laser can provide similar results [87].

The interest of these manipulations is that, by changing the temperature, the above k_{uni} can be expressed in Arrhenius form, just like k_∞ (Eq. (5.44)), and thus provide information about dissociation energetics of large ions that is easier to obtain in this way than by the master-equation treatment mentioned below.

5.10.4 "Truncated" Boltzmann

For ions that are not large enough the condition $k_2^r(E) \gg k_3(E)$ is not satisfied and consequently equilibration with the radiation field is not achieved. The ionic population will no longer obey a Boltzmann distribution due to preferential loss of ions with energies above threshold, coupled with more or less severe depletion of the ionic population below threshold. The situation is thus analogous to Fig. 5.5, except that the drop of population below the Boltzmann limit $(n(x)/B(x) = 1)$ in the vicinity of the threshold E_0 can be shown to be a function of decreasing ionic size rather than decreasing pressure. Similarly, the activation energy, as determined from Eq. (5.55), will drop off with decreasing ionic size. Under these conditions useful information about the energetics can be extracted by applying a correction factor to a Boltzmann distribution truncated at threshold [83].

5.10.5 The master equation

On a more detailed level, the zero-pressure kinetics can be modeled using the master-equation approach. The basic equation is Eq. (5.12), slightly modified such that the collisional rate constants $\omega p(x, y)$ and $\omega p(y, x)$ are replaced in the present instance by the radiative constants k^r for black-body absorption and emission of photons. For actual computation the master equation is discretized as in Section 5.2.7, with the transport matrix $\mathbb{J}$ now defined by $\mathbb{J} = \mathbb{K}^r - \mathbb{K}$, where $\mathbb{K}^r$ is the matrix of radiative rate constants k^r_{ij} and $\mathbb{K}$ is the matrix of microcanonical RRKM rate constants k_i. The rate constant k_{uni} is then determined as the lowest eigenvalue of the transport matrix, in complete analogy with the standard thermal case [88].

The radiative rate constants are formulated in terms of a harmonic-oscillator model using Einstein coefficients for radiative emission and absorption, the Planck distribution for the black-body radiation density [89], and a term based on harmonic-oscillator transition probabilities [85, 88]. The complication is that few properties of the larger biomolecules have been measured, which therefore are obtainable only from *ab initio* calculations, such as the vibrational frequencies and transition dipole moments necessary for obtaining the radiative constants. The master-equation approach is useful for investigating the influences of various parameters on the energetics of a given reaction [85].

References

[1] W. Forst, in *Advances in Chemical Kinetics and Dynamics*, Vol. 2B. J. R. Barker, Ed. (JAI Press, Greenwich, CT, 1995), pp. 427–479.
[2] I. Oref and D. C. Tardy, *Chem. Rev.* **90**, 1407 (1990).
[3] R. G. Gilbert, *Aust. J. Chem.* **48**, 1787 (1995).
[4] F. W. Schneider and B. S. Rabinovitch, *J. Am. Chem. Soc.* **84**, 4215 (1962); B. S. Rabinovitch, P. W. Gilderson and F. W. Schneider, *J. Am. Chem. Soc.* **87**, 158 (1965); F. J. Fletcher, B. S. Rabinovitch, K. W. Watkins and D. J. Locker, *J. Phys. Chem.* **70**, 2823 (1966); Y. N. Lin and B. S. Rabinovitch, *J. Phys. Chem.* **72**, 1726 (1968).
[5] A. N. Trenwith, *J. Chem. Soc. Faraday Trans. 1*, **81**, 745 (1985).
[6] P. D. Pacey and J. H. Wimalasena, *J. Phys. Chem.* **88**, 5657 (1984); *Can. J. Chem.* **62**, 293 (1984).
[7] D. F. Davidson, M. D. DiRosa, R. K. Hanson and C. T. Bowman, *Int. J. Chem. Kinet.* **25**, 969 (1993).

[8] D. Astholz, J. Troe and W. Wieters, *J. Chem. Phys.* **70**, 5107 (1979).

[9] U. Brand, H. Hippler, L. Lindemann and J. Troe, *J. Phys. Chem.* **94**, 6305 (1990).

[10] See, for example, I. Oppenheim, K. E. Shuler and G. H. Weiss, *Stochastic Processes in Chemical Physics: The Master Equation* (MIT Press, Cambridge, MA, 1977); N. G. van Kampen, *Stochastic Processes in Physics and Chemistry* (North-Holland, Amsterdam, 1981), Ch. V.

[11] R. K. Boyd, *Chem. Rev.* **77**, 93 (1977).

[12] A. Pashtutski and I. Oref, *J. Phys. Chem.* **92**, 178 (1988); I. Oref, in *Advances in Chemical Kinetics and Dynamics*, Vol. 2B. J. R. Barker, Ed. (JAI Press, Greenwich, CT, 1995), pp. 285–298.

[13] J. Troe, *J. Chem. Phys.* **97**, 288 (1992).

[14] D. Rapp and T. Kassal, *Chem. Rev.* **69**, 61 (1969).

[15] J. R. Barker, *Ber. Bunsenges. Phys. Chem.* **101**, 566 (1997).

[16] W. Yuan and B. S. Rabinovitch, *J. Chem. Phys.* **80**, 1687 (1984).

[17] R. G. Gilbert, B. J. Gaynor and K. D. King, *Int. J. Chem. Kinet.* **11**, 317 (1979).

[18] D. L. Clarke, I. Oref, R. G. Gilbert and K. F. Lim, *J. Chem. Phys.* **96**, 5983 (1992).

[19] W. Forst, *Chem. Phys. Lett.* **131**, 209 (1989).

[20] J. O. Hirschfelder, C. F. Curtiss and R. B. Bird, *Molecular Theory of Gases and Liquids* (Wiley, New York, 1967).

[21] E. W. Montroll and K. E. Shuler, *Adv. Chem. Phys.* **1**, 361 (1958).

[22] R. G. Gilbert and S. C. Smith, *Theory of Unimolecular and Recombination Reactions* (Blackwells, Oxford, 1990), Ch. 6.

[23] W. Forst, *J. Chem. Educ.* **66**, 142 (1989).

[24] H. O. Pritchard, *The Quantum Theory of Unimolecular Reactions* (Cambridge University Press, Cambridge, 1984), p. 29, Fig. 3.1.

[25] N. J. B. Green, P. J. Marchant, M. J. Perona, M. J. Pilling and S. H. Robertson, *J. Chem. Phys.* **96**, 5896 (1992).

[26] W. Tsang and J. H. Kiefer, in *The Chemical Dynamics and Kinetics of Small Radicals*, K. Liu, A. Wagner, Eds. (World Scientific, Singapore, 1995), Ch. 3.

[27] J. E. Dove, W. S. Nip and H. Teitelbaum, in *15th Symposium on Combustion (Int'l)* (Combustion Institute, Pittsburgh, PA, 1974), p. 903.

[28] J. H. Kiefer, S. S. Kumaran and S. Sundaram, *J. Chem. Phys.* **99**. 3531 (1993).

[29] R. A. Eng, A. Gebert, E. Goos, H. Hippler and C. Kachiani, *Phys. Chem. Chem. Phys.* **3**, 2258 (2001).

[30] W. Forst, *Chem. Phys. Lett.* **157**, 374 (1989).

[31] S. Nordholm and H. W. Schranz, *Chem. Phys.* **62**, 459 (1981).

[32] I. Oref and D. C. Tardy, *J. Chem. Phys.* **91**, 205 (1989).

[33] F. A. Lindemann, *Trans. Faraday Soc.* **17**, 598 (1922).

[34] K. A. Holbrook, M. J. Pilling and S. H. Robertson, *Unimolecular Reactions*, 2nd Ed. (Wiley, Chichester, 1996).

[35] H. Hippler, V. Schubert and J. Troe, *J. Phys. Chem.* **81**, 3931 (1984).

[36] K. Luther, K. Oum and J. Troe, *J. Phys. Chem. A* **105**, 5535 (2001).

[37] R. C. Tolman, *J. Amer. Chem. Soc.* **42**, 2506 (1920).

[38] S. W. Benson, *Thermochemical Kinetics*, 2nd Ed. (Wiley, New York, 1976).

[39] S. C. Chan, B. S. Rabinovitch, J. T. Bryant, L. D. Spicer, T. Fujimoto, Y. N. Lin and S. P. Pavlou, *J. Phys. Chem.* **74**, 3160 (1970).

[40] I. Oppenheim, K. E. Shuler and G. H. Weiss, *Adv. Mol. Relax. Proc.* **1**, 13 (1967).

[41] P. Hänggi and H. Thomas, *Phys. Rep.* **88**, 207 (1982).

[42] P. Hänggi, P. Talkner and M. Borkovec, *Rev. Mod. Phys.* **62**, 251 (1990).

[43] M. Borkovec and B. J. Berne, *J. Chem. Phys.* **82**, 794 (1985).

[44] J. Troe, *J. Chem. Phys.* **77**, 3485 (1982).

[45] N. J. B. Green, S. H. Robertson and M. J. Pilling, *J. Chem. Phys.* **100**, 5259 (1994).

[46] S. K. Kim, *J. Chem. Phys.* **28**, 1057 (1958); B. Widom, *J. Chem. Phys.* **31**, 1387 (1959).

[47] K. E. Shuler and G. H. Weiss, *J. Chem. Phys.* **38**, 505 (1963).

[48] B. P. Demidovich and I. A. Marron, *Computational Mathematics* (Mir Publishers, Moscow, 1981), p. 376.

[49] W. Forst, *Theory of Unimolecular Reactions* (Academic Press, New York, 1973), p. 9.

[50] N. S. Snider, *J. Chem. Phys.* **65**, 1800 (1976).

[51] J. R. Barker, *J. Chem. Phys.* **72**, 3686 (1980).

[52] G. Yahav, Y. Haas, B. Carmeli and A. Nitzan, *J. Chem. Phys.* **72**, 3410 (1980).

[53] B. Carmeli and A. Nitzan, *J. Chem. Phys.* **76**, 5321 (1982).

[54] S. H. Kang and K.-H. Jung, *Chem. Phys. Lett.* **131**, 496 (1986).

[55] D. C. Tardy and B. S. Rabinovitch, *J. Chem. Phys.* **45**, 3720 (1966); *J. Chem. Phys.* **48**, 1282 (1968).

[56] B. J. Gaynor, R. G. Gilbert and K. D. King, *Chem. Phys. Lett.* **55**, 40 (1978); R. G. Gilbert and K. D. King, *Chem. Phys. Lett.* **49**, 367 (1980).

[57] H. W. Schranz and S. Nordholm, *Chem. Phys.* **62**, 459 (1981).

[58] C. E. Fröberg, *Introduction to Numerical Analysis* (Addison-Wesley, Reading, MA, 1965), Ch. 4.

[59] A. Ralston and P. Rabinowitz, *A First Course in Numerical Analysis* (McGraw-Hill, Auckland, 1986), Section 9.7–2.

[60] R. C. Bhattacharjee and W. Forst, in *15th Combustion Symposium* (*Int'l*) (Combustion Institute, Pittsburgh, PA, 1974), p. 681.

[61] J. Troe, *J. Chem. Phys.* **66**, 4745 (1977).

[62] W. Feller, *An Introduction to Probability Theory and Its Applications*, Vol. 1, 2nd Ed. (Wiley, New York, 1957), p. 357.

[63] W. Forst, *J. Phys. Chem.* **76**, 342 (1972).

[64] W. Forst, *J. Phys. Chem.* **86**, 1771 (1982).

[65] A. W. Yau and H. O. Pritchard, *Can. J. Chem.* **57**, 2458 (1979).

[66] C. Schoenenberger and W. Forst, *J. Comput. Chem.* **6**, 455 (1985). See also P. K. Venkatesh, R. W. Carr, M. H. Cohen and A. M. Dean, *J. Phys. Chem.* A **102**, 8104 (1998).

[67] B. Dill and H. Heydtmann, *Int. J. Chem. Kinet.* **9**, 321 (1977).

[68] W. Forst, *J. Phys. Chem.* **83**, 100 (1979).

[69] J. W. Davies, N. J. B. Green and M. J. Pilling, *Chem. Phys. Lett.* **129**, 373 (1986).

[70] M. Hoare, *J. Chem. Phys.* **38**, 1630 (1963).

[71] H. W. Schranz and S. Nordholm, *Chem. Phys.* **87**, 163 (1984).

[72] H. O. Pritchard and A. Lakshmi, *Can. J. Chem.* **57**, 2793 (1979).

[73] M. Quack, *Ber. Bunsenges. Phys. Chem.* **83**, 757 (1979).

[74] R. J. McCluskey and R. W. Carr, Jr, *Int. J. Chem. Kinet.* **10**, 171 (1978).

[75] B. S. Rabinovitch and R. W. Diesen, *J. Chem. Phys.* **30**, 735 (1959); V. D. Knyazev and W. Tsang, *J. Phys. Chem.* A **104**, 10747 (2000).

[76] D. W. Setser, in *Chemical Kinetics*, MTP International Review of Science Vol. 9 (Butterworths, London, 1972), p. 1.

[77] E. W. Diau, T. Yu, M. A. G. Wagner and M. C. Lin, *J. Phys. Chem.* **98**, 4034 (1994); E. W. G. Diau and S. C. Smith, *J. Phys. Chem.* **99**, 6589 (1995).

[78] L. Vereecken and J. Peeters, *J. Phys. Chem.* A **103**, 5523 (1999).

[79] W. Forst, *J. Phys. Chem.* **93**, 3145 (1989).

[80] S. C. Smith and R. G. Gilbert, *Int. J. Chem. Kinet.* **20**, 307 (1988).

[81] T. Just and J. Troe, *J. Phys. Chem.* **84**, 3068 (1980).
[82] N. Snider, *J. Phys. Chem.* **90**, 4366 (1986).
[83] R. C. Dunbar, *J. Phys. Chem.* **98**, 8705 (1994); R. C. Dunbar and T. B. McMahon, *Science* **279**, 194 (1998).
[84] D. Thölmann, D. S. Tonner and T. B. McMahon, *J. Phys. Chem.* **98**, 2002 (1994).
[85] W. D. Price and E. R. Williams, *J. Phys. Chem. A* **101**, 8844 (1997).
[86] W. D. Price, P. D. Schnier, R. A. Jockusch, E. F. Strittmatter and E. R. Williams, *J. Am. Chem. Soc.* **118**, 10 640 (1996).
[87] R. A. Jockusch, K. Paech and E. R. Williams, *J. Phys. Chem. A* **104**, 3188 (2000).
[88] R. C. Dunbar and R. C. Zaniewski, *J. Chem. Phys.* **96**, 5069 (1992); W. D. Price, P. D. Schnier and E. R. Williams, *J. Phys. Chem. B* **101**, 664 (1997).
[89] See, for example, H. Eyring, J. Walter and G. E. Kimball, *Quantum Chemistry* (Wiley, New York, 1963), pp. 114ff.

Problems

Problem 5.1. For a diatomic molecule the normalized Boltzmann distribution per unit energy (Eq. (5.3)) is $B(x) = (kT)^{-1} \exp[-x/(kT)]$. Using the exponential transition probability of Eq. (5.4), show that
(a) application of detailed balance using this diatomic-like distribution yields $\beta = \alpha kT/(\alpha + kT)$; and
(b) with normalization according to Eq. (5.1), the normalization constant is $C(y) = \beta + \alpha[1 - \exp(-y/\alpha)]$. If the lower limit of the normalization integral (Eq. (5.1)) is extended for mathematical convenience to $-\infty$, the normalization constant is $C = (\alpha + \beta)^{-1}$. This represents the so-called "simple" exponential model. Such normalization is likely to cause problems only at such low energies that $\exp(-y/\alpha) \approx 1$.

Problem 5.2. Derive the following results using the strong-collision transition probability:

$$\langle \Delta E(y) \rangle = \langle x \rangle_{\text{eq}} - y; \qquad \langle \Delta E^2(y) \rangle = \langle x^2 \rangle_{\text{eq}} - 2y \langle x \rangle_{\text{eq}} + y^2$$

where $\langle \ldots \rangle_{\text{eq}}$ signifies the equilibrium average (i.e. average over $B(x)$). In particular (cf. Section 4.4.4)

$$\langle x^2 \rangle_{\text{eq}} = kT^2 C_v + \langle x \rangle_{\text{eq}}^2 \qquad (C_v = \text{specific heat})$$

Show that, for a *diatomic* Boltzmann distribution, we have quite generally

$$\langle x^n \rangle_{\text{eq}} = n kT \langle x^{n-1} \rangle_{\text{eq}}$$

Problem 5.3. (a) Using results of the diatomic-like exponential model from Problem 5.1(b), show that

$$\langle \Delta E(y) \rangle = \{\beta^2 + \alpha^2 [\exp(-y/\alpha)(1 + y/\alpha) - 1]\}[C(y)]^{-1}$$

For sufficiently high energies this reduces to $\langle \Delta E \rangle = \beta - \alpha$, where $\langle \Delta E \rangle_{\text{down}} = -\alpha$, $\langle \Delta E \rangle_{\text{up}} = \beta$. This can be obtained directly from the "simple" exponential model.
(b) Show that, if $\langle \Delta E \rangle_{\text{down}}$ is normalized separately, the result is

$$\langle \Delta E \rangle_{\text{down}} = -\alpha + \frac{y \exp(-y/\alpha)}{1 - \exp(-y/\alpha)}$$

Problem 5.4. Using results of the diatomic-like exponential model from Problem 5.1(b), show that

$$\langle \Delta E^2(y) \rangle = 2 \left\{ \beta^3 + \alpha^3 \left\{ 1 - \exp(-y/\alpha) \left[\frac{1}{2} \left(\frac{y}{\alpha} \right)^2 + \frac{y}{\alpha} + 1 \right] \right\} \right\} [C(y)]^{-1}$$

At sufficiently high energies, this gives for the "simple" exponential model

$$\langle \Delta E^2 \rangle = 2(\alpha^3 + \beta^3)/(\alpha + \beta)$$

Problem 5.5. Equation (5.25) states that a *column* eigenvector is the *right* eigenvector of a matrix. Show that, for an arbitrary square matrix $\mathbb{M}$, this equation can also be written in terms of the *row* eigenvector $\langle \mathbf{f}_k |$

$$\langle \mathbf{f}_k | \mathbb{M}^T = \mu_k \langle \mathbf{f}_k |$$

where $\mathbb{M}^T$ is the matrix transpose. Thus a *row* eigenvector is the *left* eigenvector of the *transposed* matrix. The distinction is moot for a symmetric matrix, which is its own transpose.

Problem 5.6. Show that $|\mathbf{B}\rangle$ is a right eigenvector of $\mathbb{P}$ with eigenvalue 1.

Problem 5.7. Using Eq. (5.26) at $t = 0$, and exploiting the orthonormality of the eigenvectors $\mathbf{f}_k(x_i)$, derive Eq. (5.27).

Problem 5.8. Derive the following results using the strong-collision transition probability:

$$\langle\langle \Delta E \rangle\rangle = 0; \qquad \langle\langle \Delta E^2 \rangle\rangle = 2 k T^2 C_v$$

(cf. Problem 5.2).

Problem 5.9. Show that equation (5.39) is also obtainable from Eq. (5.13) by invoking Eq. (5.29).

Problem 5.10. Starting with the steady-state master equation (Eq. (5.39)), show that when $n(x) \Rightarrow B(x)$, k_{uni} is given by Eq. (5.42).

Problem 5.11. Show that, at the low-pressure limit,

$$k_0 = \omega \int_{E_0}^{\infty} dx \int_{0}^{E_0} dy \, p(x, y) \, n(y)$$

which may be interpreted as representing an "upward flow," i.e. flow of particles from below threshold to above threshold.

Problem 5.12. Show that, for the strong-collision transition probability ($p(x, y) \rightarrow B(x)$), k_0 in Problem 5.11 reduces to k_0^{sc} of Eq. (5.52), which may be further approximated by

$$k_0^{sc} \approx \frac{\omega}{Q} k T \, N(E_0) \exp[E_0/(kT)]$$

Problem 5.13. Using the result of Problem 1.5 in Chapter 1 for k_0, and Eq. (5.43) for k_∞, show that classically the difference between the high- and low-pressure activation energies for a molecule with v vibrational degrees of freedom is

$$E_{a\infty} - E_{a0} \approx (v - 1) k T \pm \frac{1}{2} k T$$

where the $+$ sign applies for the collision frequency ω in pressure units and the $-$ sign for ω in concentration units.

Problem 5.14. Using the result of Problem 5.4, show that Eq. (5.78) yields for the "simple" (i.e. quasi-diatomic) exponential model

$$k_0^{\mathrm{wc}} \cong \frac{\omega(\alpha^3 + \beta^3)}{(kT)^2(\alpha + \beta)} \exp[-E_0/(kT)] = \frac{\omega \langle \Delta E^2 \rangle}{2(kT)^2} \exp[-E_0/(kT)]$$

For polyatomics this result can be improved by re-defining E_0 as the effective threshold using the approximation in Problem 5.12 whereby $\exp[E_0/(kT)] \Rightarrow kTN(E_0)\exp[-E_0/(kT)]/Q$, so that

$$k_0^{\mathrm{wc}} \approx \frac{\omega \langle \Delta E^2 \rangle \, kT \, N(E_0) \exp[-E_0/(kT)]}{2Q(kT)^2}$$

Problem 5.15. Show how Eq. (5.78) is obtained from Eq. (5.77).

Problem 5.16. Obtain Eq. (5.86) from Eq. (5.26).

Problem 5.17. If experiment gives $k_\infty = A'_\infty (kT)^n \exp[-E_{a\infty}/(kT)]$, show that classically one obtains by inversion

$$k(x) = A'_\infty \frac{(x - E_{a\infty})^{v+n-1}}{x^{v-1}} \frac{\Gamma(v)}{\Gamma(v+n)}$$

assuming that the reactant molecule has v vibrational degrees of freedom.

Problem 5.18. If the only available information about a given chemical activation system is $k_{\mathrm{d},\infty}$, known from experiment in the form $k_{\mathrm{d},\infty} = A_\infty \exp[-E_{a\infty}/(kT)]$ (Eq. (5.44)), an approximation to $k_{\mathrm{d}}(x)$ can be obtained by Laplace inversion (Eq. (5.107)). Show that the distribution function $f(x)$ is then

$$f(x) = \frac{N_A(x - E_{a\infty}) \exp[-(x - E_{a\infty})/(kT)] \, dx}{Q_A}$$

which is a Boltzmann distribution zero-shifted by $E_{a\infty}$, the best available estimate of E'_0 under the circumstances.

Problem 5.19. Derive Eq. (5.150).

6

Non-classical effects near barrier maximum

It has been emphasized in Chapter 3 that, although RRKM is a theory that views reaction dynamics in basically classical terms, it nevertheless lends itself to incorporating certain quantum (that is, non-classical) effects into the actual fragmentation process. Two such effects, previously only alluded to in Chapter 3, will now be the subject of the present chapter. The discussion will be limited to features that can readily be incorporated into statistical unimolecular rate theory.

6.1 Tunneling

Tunneling, i.e. the passage through a potential-energy barrier of a particle with insufficient energy to surmount the barrier classically, is a standard quantum-mechanical problem [1], which is well documented for bimolecular hydrogen-transfer reactions [2]. In unimolecular reactions, tunneling is a phenomenon analogous to the bimolecular case, which is similarly limited in practice mostly to reactions involving (atomic or molecular) hydrogen, e.g. $H_2CO \rightarrow H_2 + CO$. The theoretical treatment of such reactions has traditionally received less attention since much of the ancillary information became available only fairly recently, from quantitatively useful quantum-mechanical calculations [3] and highly resolved experiments [4].

Calculation of the actual tunneling probability, as well as the relevant information (barrier height and shape, transition-state frequencies and moments of inertia) is strictly speaking the province of quantum mechanics, and therefore outside the scope of this treatise. In the following it is assumed that such data are available, and the discussion is limited to those aspects directly connected with the unimolecular rate constant.

6.1.1 The barrier

For a given particle, the tunneling probability depends on the height and thickness of the potential barrier, i.e. its shape. In the present context, by "barrier" is meant here a cut of the potential-energy surface along a one-dimensional reaction coordinate, on the assumption that the motion along the reaction coordinate is separable from other degrees of freedom. The problem is thus reduced to one dimension, so that barrier penetration can readily be incorporated into standard unimolecular theory as set out in Chapter 3.

6.1.2 The Eckart barrier

A barrier proposed by Eckart [5] can model a variety of physically reasonable shapes and admits an analytical solution of the corresponding Schrödinger equation. The Eckart barrier in its most general unsymmetrical form is given by [1]

$$V(r) = -\frac{Ay}{1-y} - \frac{By}{(1-y)^2} \qquad (6.1)$$

where y is a dimensionless variable defined by $y = -\exp(2\pi r/L)$. The variable r may be thought of as representing a distance along the reaction coordinate, while L is a characteristic length, which determines basically the width or thickness of the barrier.

The constants A and B (dimension: energy) determine the overall shape of the Eckart barrier. For a barrier as shown in Fig. 6.1 with $V_2 > V_1$ the potential in the forward direction ($\rightarrow$) is $V_1 = (A - B)^2/(4B)$, while that in the reverse direction ($\leftarrow$) is $V_2 = (A + B)^2/(4B)$ (Problem 6.1). The inverse relation is $A = V_2 - V_1$, $B = (V_1^{1/2} + V_2^{1/2})^2$, which can readily be verified. As $A \rightarrow 0$, the

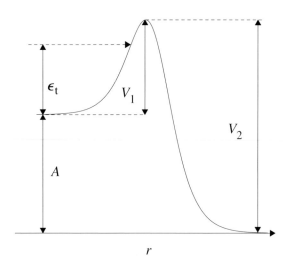

Fig. 6.1. An example of the asymmetrical Eckart barrier of Eq. (6.1), with $A = 1$, $B = 5$, and $L = 0.4$, in arbitrary units. Here $V_1 = (A - B)^2/(4B) = 0.8$ and $V_2 = (A + B)^2/(4B) = 1.8$.

barrier becomes symmetric ($V_1 = V_2$), whereas as $B \to 0$, the barrier becomes a simple S-shaped function of height $V_2 = A$ with no maximum in the forward direction ($V_1 = 0$). The steepness of the S-shape increases with decreasing L, and, in the limit $L \to 0$, becomes a step function (Eq. (6.14) below).

L is related to the curvature C (dimension: energy/length^2) at potential maximum by

$$C - \left(\frac{\partial^2 V(r)}{\partial r^2}\right)_{r=0} = -\frac{\pi^2}{2L^2}\frac{(A^2 - B^2)^2}{B^3} \tag{6.2}$$

if the potential maximum is placed at $r = 0$. Just as a particle moving in a potential *minimum* will give rise to a *real* vibrational frequency, a particle moving with respect to a potential *maximum* – which is the present case of barrier penetration – will give rise to an *imaginary* frequency that is routinely obtained in quantum-mechanical calculations of transition-state frequencies. If v^* is known from such a calculation, the relation to the curvature C is

$$v^* = \frac{1}{2\pi}\left(\frac{C}{\mu^*}\right)^{1/2} \tag{6.3}$$

so L may be defined alternatively in terms of v^*. This equation is in fact analogous to the expression for the classical frequency of a harmonic oscillator of mass μ^* and quadratic force constant C (dimension: force/length = energy/length^2), except that here C is negative, so v^* is imaginary; μ^* may be interpreted as the effective mass of the particle undergoing tunneling. In principle, this effective mass is not the same as the reduced mass, for it depends on geometric parameters of the reacting system and the reaction path [6].

The superscript asterisks are in line with the usual notation adopted throughout to signify properties related to the transition state, which is the case in the present context.

6.1.3 Tunneling probability

For the purpose of determining the tunneling probability $p(\epsilon_t)$ for a particle of effective mass μ^* approaching an unsymmetrical Eckart barrier (Fig. 6.1) from the left with translational energy ϵ_t, it is convenient to define variables a, b, c, and d:

$$a = 2\pi c \epsilon_t^{1/2}; \quad b = 2\pi c(\epsilon_t - A)^{1/2}; \quad d = \pi(4c^2 B - 1)^{1/2} \tag{6.4}$$

where

$$c = \frac{L}{h}(2\mu^*)^{1/2} = \frac{A^2 - B^2}{2h|v^*|B^{3/2}} \tag{6.5}$$

in terms of which $p(\epsilon_t)$ is given by [1, p. 146]

$$p(\epsilon_t) = 1 - \frac{\cosh(a-b) + \cosh d}{\cosh(a+b) + \cosh d} = \frac{2 \sinh a \sinh b}{\cosh(a+b) + \cosh d} \tag{6.6}$$

For a symmetric barrier, $a = b$, and

$$p(\epsilon) = \frac{\cosh(2a) - 1}{\cosh(2a) + \cosh d} \tag{6.7}$$

If $4c^2 B - 1$ is negative, d is imaginary and the $\cosh d$ term in Eqs. (6.6) and (6.7) becomes $\cos d$. Under certain conditions the hyperbolic functions in Eq. (6.6) may cause overflow problems, which are avoided by replacing hyperbolic functions by exponentials and canceling out positive exponentials top and bottom:

$$p(\epsilon_t) = \frac{1 - \exp(-2a) - \exp(-2b) + \exp(-2u)}{1 + \exp(-2u) + \exp[-(u-d)] + \exp[-(u+d)]} \tag{6.8}$$

where $u = a + b$.

These equations make it clear that the tunneling probability depends exponentially on the effective mass μ^* of the particle, so that tunneling will produce an isotope effect [2], meaning a substantially reduced tunneling probability for the heavier isotope, e.g. deuterium with respect to hydrogen, for which case the difference in mass is largest.

For a given potential and effective particle mass μ^*, the tunneling probability drops rapidly as the incident energy ϵ_t becomes less than E_0, the reaction threshold that represents the barrier height (Fig. 6.2). Hence, in fitting an analytical barrier, the shape, i.e. curvature, of the actual reaction barrier (if it is known from other

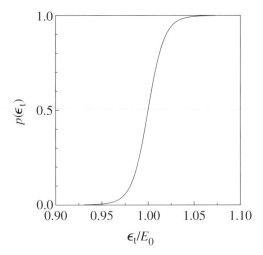

Fig. 6.2. The energy dependence of the tunneling probability $p(\epsilon_t)$ (Eqs. (6.6) and (6.8)) calculated for the molecular channel in the decomposition of formaldehyde $H_2CO \rightarrow H_2 + CO$ [10] using the Eckart barrier. Note that $p(\epsilon_t) = \frac{1}{2}$ at $\epsilon_t = E_0$.

information), is important only near the top. Note that, *at* the top of the barrier, the probability of tunneling is $p(\epsilon_t) = \frac{1}{2}$, and it reaches unity only at some energy *above* the barrier. In principle there is therefore barrier penetration at energies below the barrier, and reflection above the barrier.

6.1.4 Parabolic barriers

Since the barrier-penetration probability rapidly approaches zero as energy drops below threshold, even an unsymmetrical barrier may sometimes be approximated by an inverted parabola, for which the solution is simpler. The important proviso is that the parabola must have the proper curvature near the top. For a barrier represented by the inverted parabola

$$V(r) = E_0 - \frac{1}{2}Cr^2 \tag{6.9}$$

of height $V_{\max} = E_0$, curvature C at the top, and half-width at the base $w = (2E_0/C)^{1/2}$, the tunneling probability for a particle of effective mass μ^* approaching with energy ϵ_t is [7]

$$p(\epsilon_t) = \{1 + \exp[D(1 - \epsilon_t/E_0)]\}^{-1} = \frac{1}{2}\left\{1 + \tanh\left[\frac{1}{2}D\left(\frac{\epsilon_t}{E_0} - 1\right)\right]\right\} \tag{6.10}$$

where D is the dimensionless parameter

$$D = \frac{\pi w}{h}(2\mu^* E_0)^{1/2} = \frac{2\pi E_0}{h|v^*|} \tag{6.11}$$

using the definition of v^* in Eq. (6.3).

Figure 6.3 shows a parabolic barrier fitted to the asymmetrical Eckart barrier of Fig. 6.1, a rather unfavorable case. The fitting was done by equating curvatures at

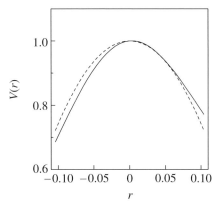

Fig. 6.3. The top of the unsymmetrical Eckart barrier of Fig. 6.1 (solid line) fitted with a parabolic barrier (Eq. (6.9), dashed line) having the same curvature at the top. $V(r)$ and r are in arbitrary units.

potential maximum. For a symmetric Eckart barrier the fit would be substantially better.

Standard texts on quantum mechanics [1, p. 162; 8] show that, for an arbitrary barrier $V(r)$, the tunneling probability $p(\epsilon_t)$ can be obtained from the Wentzel–Kramers–Brillouin (WKB) approximation in the form

$$p(\epsilon_t) \approx \exp\left(-\frac{2}{\hbar}\int_{r_1}^{r_2}\{2\mu[V(r) - \epsilon_t]\}^{1/2}\,dr\right) \tag{6.12}$$

where r_1 and r_2 are the classical "turning points." This approximation is accurate only if the integrand is large, that is, when $p(\epsilon_t)$ is small. However, in the vicinity of the reaction threshold (i.e. potential maximum), which is of importance for reaction rates, the tunneling probability is large ($\approx \frac{1}{2}$, cf. Fig. 6.2), and therefore Eq. (6.12) is unsuitable for kinetics calculations.

6.1.5 The rate constant and tunneling

In the usual formulation of the unimolecular rate constant $k(E)$, where E is the internal energy of the reactant, $G^*(E - E_0)$ refers to the number of ways available energy in the transition state can be shared among internal states of the transition state and ϵ_t, the relative translational energy of fragments (cf. Section 3.2).

If the barrier height separating reactant from fragments is E_0 (exclusive of zero-point energies of reactant and transition state), in the absence of tunneling fragments do not form unless $E \geq E_0$, and hence $k(E) = 0$ for all $E < E_0$. If tunneling operates, fragments can appear in principle for all energies below E_0, although with vanishing probability for E much below E_0.

Consequently, if $N^*(E)$ is the density of states for the transition state at energy $E > 0$ (*not* necessarily $E \geq E_0$), and $p(\epsilon_t)$ is the probability that fragments will appear with relative translational energy ϵ_t, the total number of ways of distributing energy E between the transition state and the relative translational energy of fragments can be represented semi-classically by the convolution [9, 10]

$$G^*(E) = \int_0^E p(\epsilon_t)\,N^*(E - \epsilon_t)\,d\epsilon_t, \qquad E > 0 \tag{6.13}$$

where $p(\epsilon_t)$ is the tunneling probability given by Eq. (6.6) or (6.10).

In the absence of tunneling, $p(\epsilon_t)$ becomes a step-function:

$$p(\epsilon_t) = \begin{cases} 0, & E < E_0 \\ 1, & E \geq E_0 \end{cases} \tag{6.14}$$

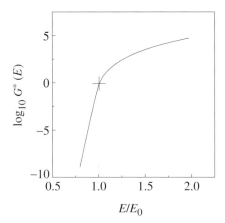

Fig. 6.4. The energy dependence of $G^*(E)$ calculated for the molecular channel of the reaction $H_2CO \rightarrow H_2 + CO$ using $p(\epsilon_t)$ of Fig. 6.2. The solid line shows $G^*(E)$ with tunneling (Eq. (6.13)); the cross (+) at $E/E_0 = 1$ and $G^*(E) = 1$ indicates where $G^*(E)$ begins in the absence of tunneling (Eq. (6.15)).

so Eq. (6.13) reduces to the integral in Eq. (3.20):

$$G^*(E) = \int_{E_0}^{E} N^*(E - \epsilon_t)\,d\epsilon_t = \int_{0}^{E-E_0} N^*(E - E_0 - \epsilon_t)\,d\epsilon_t, \qquad E \geq E_0 \quad (6.15)$$

The effect of tunneling on $G^*(E)$, as represented by Eq. (6.13), is to cause the disappearance of the sharp threshold at $E = E_0$ (Fig. 6.4) which characterizes $G^*(E)$ in the absence of tunneling (Eq. (6.15)); this has consequences for the temperature dependence of the thermal rate constant.

   Equation (6.13) thus extends the "standard" $G^*(E)$ of Eq. (6.15), and hence also the microcanonical rate constant $k(E)$, to energies below the classical threshold E_0, as shown in Fig. 6.4, and therefore may be considered a generalization of Eq. (3.21).

6.1.6 Experimental evidence

The experimental evidence for tunneling is therefore the appearance of products at energies below the threshold E_0. In the well-documented reaction $H_2CO \rightarrow H_2 + CO$, for example, products begin to form from ground-state formaldehyde some $7\,\text{kcal mol}^{-1}$ below threshold [3, 4]. Similarly, calculations show that $k(E)$ for the isomerization HCN $\rightarrow$ HNC can be as high as $10^5\,\text{s}^{-1}$ at $8\,\text{kcal mol}^{-1}$ below threshold [3]. Another candidate for tunneling is the *cis–trans* isomerization of HONO, which has been confirmed by calculations [11] but detected experimentally only at low temperature in a solid-nitrogen matrix [12], since at room temperature there exists a complex equilibrium in the gas phase among NO, NO_2, N_2O_3, N_2O_4, H_2O, and HNO_3 [13]. The possibility of tunneling has also been considered in the isomerization of the methoxy radical $CH_3O \rightarrow CH_2OH$ [14].

An example of tunneling that is well known to spectroscopists is the inversion of ammonia, the rate of which can be calculated semi-classically [15]. NH_3 has a pyramidal structure, the three hydrogens forming the base and the nitrogen atom oscillating between above and below the base by tunneling through the double-well potential hill separating the two conformations.

An interesting *a priori* candidate for tunneling would seem to be the Type-1 reaction $CH_4 \rightarrow CH_3 + H$. However, the effective mass is fairly substantial and the potential separating reactant and products is an effective potential of large width, so tunneling is negligible. On the other hand, in the reaction $CH_4^+ \rightarrow CH_3^+ + H$ some products are formed with a very small rate constant (so-called "metastables"), which has been attributed to tunneling below the top of the centrifugal barrier [16]. Fragmentations in which one of the products is a light atom have other interesting features, which are discussed in Section 8.10.

6.1.7 Tunneling and the thermal rate

The limiting high-pressure unimolecular rate constant is the Boltzmann average of the specific-energy (microcanonical) rate constant $k(E)$. As shown in Eq. (5.104), this can be written as the Laplace transform

$$k_\infty = \frac{1}{Q} \mathcal{L}\{k(E)\, N(E)\} \tag{6.16}$$

where (assuming that $\alpha = 1$), $k(E) = G^*(E)/[h N(E)]$ and Q is the partition function of the reactant. Using Eq. (6.13), Eq. (6.16) becomes

$$k_\infty = \frac{1}{hQ} \mathcal{L}\left\{ \int_0^E p(\epsilon_t)\, N^*(E - \epsilon_t)\, d\epsilon_t \right\} \tag{6.17}$$

Now, if we write $s = 1/(\ell T)$ for the Laplace transform parameter, and if

$$\mathcal{L}\{p(\epsilon_t)\} = P(s) \tag{6.18}$$

it follows from standard Laplace-transform theory [17] and Appendix 2, Section A2.2 that

$$\mathcal{L}\left\{ \int_0^E p(\epsilon_t)\, N^*(E - \epsilon_t)\, d\epsilon_t \right\} = P(s)\, Q^*(s) \tag{6.19}$$

remembering that $\mathscr{L}\{N^*(E)\} = Q^*(s)$ (Eq. (4.8)). The rate constant k_∞ of Eqs. (6.16) and (6.17) then becomes simply [10]

$$k_\infty = \frac{P(s)}{h}\frac{Q^*(s)}{Q(s)} \tag{6.20}$$

which incorporates tunneling, and the associated isotope effect, through the function $P(s)$. This tunneling isotope effect will be in addition to the isotope effect on the partition functions $Q^*(s)$ and $O(s)$ arising from the mass dependence of vibrational frequencies and moments of inertia. In the literature, $\Gamma^* = sP(s)\exp(E_0 s)$ is referred to as the thermal correction factor for tunneling [18], in terms of which Eq. (6.20) becomes

$$k_\infty = \Gamma^*\frac{\ell T}{h}\frac{Q^*}{Q}\exp[-E_0/(\ell T)] \tag{6.21}$$

In the absence of tunneling, $p(E)$ is the step-function mentioned above, the Laplace transform of which is $P(s) = \exp(-sE_0)/s$ which, when it is substituted into Eq. (6.20), yields (with $s = 1/(\ell T)$) the standard transition-state-theory result $k_\infty = (\ell T/h)(Q^*/Q)\exp[-E_0/(\ell T)]$, of which Eq. (6.21) may be considered to be a generalization.

6.1.8 The Laplace transform of $p(\epsilon_t)$

If the tunneling probability is given by the parabolic-barrier formula of Eq. (6.10), the explicit form of the Laplace transform in Eq. (6.18) is

$$\mathscr{L}\{p(\epsilon_t)\} = \int_0^\infty \frac{\exp(-s\epsilon_t)\,d\epsilon_t}{1 + \exp[D(E_0 - \epsilon_t)/E_0]} = P(s) \tag{6.22}$$

The substitution $\beta = D/E_0$, $x = \exp[s(E_0 - \epsilon_t)]$ brings it into the form

$$P(s) = \frac{\exp(-E_0 s)}{s}\int_0^{\exp(E_0 s)} \frac{dx}{1 + x^{\beta/s}} \tag{6.23}$$

where $\beta/s = D\ell T/E_0$ is now a dimensionless parameter. This integral is solvable analytically, but only for certain [19] values of β/s; however, for arbitrary values it can be fairly well approximated by [7, Appendix C; 20, p. 22]

$$P(s) \cong \begin{cases} \dfrac{E_0\exp(-D)}{sE_0 - D}, & \beta/s < 1 \\[4mm] \dfrac{\pi\exp(-sE_0)}{\beta\sin(\pi s/\beta)}, & \beta/s > 1 \end{cases} \tag{6.24}$$

For $\beta/s = 1$ Eq. (6.23) gives the exact result $P(s) = E_0 \exp(-E_0 s)$ if the upper limit in Eq. (6.23) is extended to infinity, as can reasonably be expected.

Another useful substitution in Eq. (6.22) is $a = \exp(D)$, $b = D/E_0$, $y = a \exp(-b\epsilon_t)$, with the result

$$P(s) = \frac{a^{-s/b}}{b} \int_0^a \frac{y^{-(1-s/b)} \, dy}{1+y} \tag{6.25}$$

For large D such that $a \cong \infty$, this integral has the analytical solution [21, p. 213, item 856.02]

$$\int_0^\infty \frac{y^{-(1-s/b)} \, dy}{1+y} = \pi \operatorname{cosec}\left(\frac{\pi s}{b}\right) \tag{6.26}$$

Using the expansion [22]

$$\operatorname{cosec} z = \frac{1}{z} + \frac{z}{6} + \frac{7z^3}{360} + \cdots \tag{6.27}$$

and replacing D in terms of $h\nu^*$ from Eq. (6.11), the parabolic $P(s)$ becomes

$$P(s) = kT \exp[-E_0/(kT)] \left[1 + \frac{1}{24}\left(\frac{h\nu^*}{kT}\right)^2 + \frac{7}{5760}\left(\frac{h\nu^*}{kT}\right)^4 + \cdots \right] \tag{6.28}$$

where the first two terms of the series represent an oft-quoted result that goes back to Wigner [23]. In general, however, the terms beyond the second are negligible only at a sufficiently high temperature.

6.1.9 Activation energy in tunneling

From Eq. (6.20) we have for the corresponding activation energy at the high-pressure limit

$$E_{a\infty} = -\frac{d\ln k_\infty}{ds} = -\frac{d\ln P(s)}{ds} + \langle E^* \rangle - \langle E \rangle \tag{6.29}$$

where $\langle E^* \rangle$ and $\langle E \rangle$ are the average energies of the transition state and the reactant, respectively (cf. Section 5.4.8.) From Eq. (6.24) the logarithmic derivatives of $P(s)$ are

$$-\frac{d\ln P(s)}{ds} \cong \begin{cases} \dfrac{E_0}{(E_0/kT) - D}, & \beta/s < 1 \\[3mm] E_0 + \dfrac{\pi E_0}{D} \cot\left(\dfrac{\pi E_0}{DkT}\right), & \beta/s > 1 \end{cases} \tag{6.30}$$

which means that, for $\beta/s < 1$ (i.e. for $T < E_0/Dk$), the logarithmic derivative

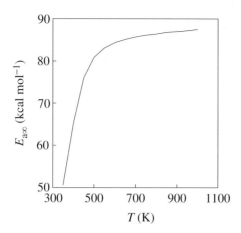

Fig. 6.5. The temperature dependence of $E_{a\infty}$ in the tunneling reaction $H_2CO \rightarrow H_2 + CO$, calculated for a symmetric Eckart barrier, with $E_0 = 87 \, \text{kcal mol}^{-1}$ and $D = 94.4$. From Eqs. (6.30), $T_c \approx E_0/(D\ell) = 464 \, \text{K}$.

of $P(s)$ increases roughly linearly with T, whereas for $\beta/s > 1$ it is virtually constant at $\cong E_0$; the separation between the two regimes occurs at $\beta/s = 1$. This means a strong temperature dependence of $E_{a\infty}$ below $T_c \cong E_0/(D\ell)$, and virtually constant $E_{a\infty}$ above T_c, inasmuch as $\langle E^* \rangle$ and $\langle E \rangle$ have only a modest temperature dependence. The consequence is a curvature in the Arrhenius plot of k_∞ at low temperature (cf. [10, Fig. 2]), in complete analogy with a similar effect of tunneling in bimolecular reactions [24].

As an example, Fig. 6.5 shows the temperature dependence of $E_{a\infty}$ in the reaction $H_2CO \rightarrow H_2 + CO$. Note that the overall shape of the curve is similar to the energy dependence of $G^*(E)$ in Fig. 6.4, indicating the reciprocal relationship between temperature and energy which is the basis of the Laplace-inversion method (Section 4.1).

6.2 Non-adiabatic transition

The standard RRKM expression for the microcanonical rate constant $k(E)$ is predicated on the assumption that the unimolecular fragmentation occurs throughout on a potential-energy surface of the same electronic multiplicity. This is not always the case, as illustrated by the thermal decomposition $N_2O(^1\Sigma) \rightarrow N_2(^1\Sigma) + O(^3P)$ (Fig. 6.6), one of the better-known examples of a non-adiabatic (spin-forbidden) unimolecular reaction [25–36]. For the decomposition to occur on the singlet surface, the oxygen-atom product would have to be in the ^1D state, which is an excited state. In a thermal reaction only the lowest-energy channel is open, which leads to the lower-energy ^3P state of the oxygen atom. Similar conditions obtain in the thermal decompositions $CO_2(^1\Sigma) \rightarrow CO(^1\Sigma) + O(^3P)$, $CS_2(^1\Sigma) \rightarrow CS(^1\Sigma) + S(^3P)$, and $COS(^1\Sigma) \rightarrow CO(^1\Sigma) + S(^3P)$.

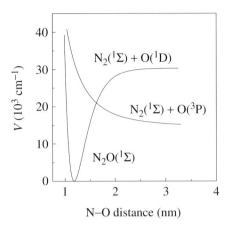

Fig. 6.6. A schematic diagram of the decomposition of collinear N_2O (N—N—O) into singlet- and triplet-oxygen product. The potential energy V is shown as a function of the N—O distance. The crossing point of the two potential-energy curves is roughly $21\,000\,\mathrm{cm}^{-1}$ above the ground state.

6.2.1 Reduction to one dimension

Figure 6.6 represents the non-adiabatic transition as a one-dimensional event, which would apply to a diatomic molecule dissociating into two atomic fragments. In the case of a polyatomic molecule, conditions are more complicated in that a non-adiabatic transition involves a crossing between two potential-energy surfaces that define a *seam* instead of a crossing *point*. In the spirit of simple transition-state theory [37], the problem is reduced to a quasi-diatomic case by separating out one particular degree of freedom chosen to be the reaction coordinate along which the non-adiabatic transition takes place, e.g. the N—O distance in the case of N_2O (Fig. 6.6). The other degrees of freedom are assumed to act merely as a reservoir of energy, in a fashion analogous to the RRKM prescription. The analog of the transition state is then the structure located at the crossing point of the two potential-energy surfaces, which is assumed to be located at energy E_c, which plays the role of the critical energy E_0 in standard transition-state theory.

In principle a non-adiabatic transition is likely only near the crossing point of the two potential-energy curves, and the likelihood will decrease with increasing translational energy ϵ_t in the reaction coordinate. In its simplest form the non-adiabatic transition probability can be expected to have the form [33] $\wp(\epsilon_t) \approx 1 - a/\epsilon_t^{1/2}$, where a is a parameter indicating the strength of non-adiabatic coupling (script $\wp$ is used here to distinguish it from the tunneling probability p). The translational energy ϵ_t defined here is completely analogous to the one-dimensional translational energy in the transition-state-theory reaction coordinate (Eq. (3.11)).

6.2.2 Calculation of $p(\epsilon_t)$

A more explicit form of $p(\epsilon_t)$ requires detailed information about the potential-energy curves near their crossing point, which in the most general case is not a simple matter even for a diatomic [38]. The problem is simplified by assuming that the potential-energy curves are linear in the vicinity of the crossing point (i.e. have constant slopes of opposite sign), and that the transition is induced by a weak spin–orbit interaction (which is the case of N_2O used here for illustration). Not considered is the case of conical intersections, to which the present development does not apply.

Let F_1 and F_2 be the slopes of the potential-energy curves (dimension: energy/distance) at the crossing point, E_c is the energy at the crossing point, and let $\Delta F = |F_1 - F_2|$ and $F = (|F_1 F_2|)^{1/2}$, H_{12} is the spin–orbit-coupling constant (dimension: energy), and μ is the reduced mass of the system. It can then be shown [38] that the quantum-mechanical $p(\epsilon_t)$ is given in terms of the square of the Airy function $\mathrm{Ai}(x)$ [22, p. 446] with a negative argument:

$$p(\epsilon_t) = \pi^2 \beta^{4/3} \, \mathrm{Ai}^2\!\left(-\epsilon \beta^{2/3}\right) \tag{6.31}$$

where ϵ and β are the dimensionless variables

$$\beta = \frac{4H_{12}^{3/2}}{\hbar}\left(\frac{\mu}{F\,\Delta F}\right)^{1/2}, \quad \epsilon = (\epsilon_t - E_c)\epsilon_0, \quad \epsilon_0 = \frac{\Delta F}{2F H_{12}} \tag{6.32}$$

The Airy function arises from an approximation to the overlap of vibrational wave functions. Figure 6.7 is a plot of Eq. (6.31) that shows $p(\epsilon_t)$ decaying exponentially to zero for negative values of $\epsilon_t - E_c$, indicating tunneling at energies below the crossing point. At energies above the crossing point $p(\epsilon_t)$ oscillates due to interference between two possible paths.

Another expression for $p(\epsilon_t)$ can be derived from time-dependent perturbation theory. It is known as the double-passage Landau–Zener formula [39; 40, p. 330]:

$$p(\epsilon_t) = 2\{1 - \exp\left[-\pi\beta/\left(4\epsilon^{1/2}\right)\right]\} \tag{6.33}$$

where β and ϵ have the same significance as in Eq. (6.32). The factor of 2 in front accounts for the assumption that a non-adiabatic transition can occur both in the "forward" and in the "reverse" motion of the system, which is also the case of $p(\epsilon_t)$ given by the Airy function in Eq. (6.31). Figure 6.7 compares $p(\epsilon_t)$ of Eq. (6.33) with that of Eq. (6.31), which shows that the Landau–Zener formula fails at energies in the vicinity of the energy of the crossing point, and does not account at all for tunneling at energies below E_c.

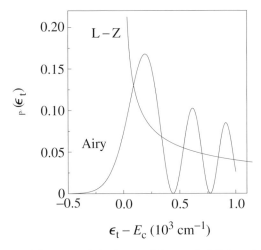

Fig. 6.7. A comparison of the non-adiabatic transition probabilities given by the Airy function (Eq. (6.31), curve labeled "Airy") and by the Landau–Zener formula (Eq. (6.33) or (6.34), curve labeled "L–Z"). At $\epsilon_t = E_c$, the L–Z curve goes off to infinity. The calculation is for the transition of singlet N_2O to its bent triplet, assuming equal slopes at the crossing point of the diabatic curves ($F_1 = 0.5\,\text{eV nm}^{-1} = -F_2$), the spin–orbit-coupling constant $H_{12} = 46\,\text{cm}^{-1}$, and $E_c = 14\,135\,\text{cm}^{-1}$. For the linear triplet conditions are estimated to be slightly different: $F_1 = 2.52\,\text{eV nm}^{-1}$, $F_2 = -1.28\,\text{eV nm}^{-1}$, and $E_c = 20\,560\,\text{cm}^{-1}$ [36].

6.2.3 Special cases

The argument of the exponential in the Landau–Zener formula of Eq. (6.33) is small for transitions induced by spin–orbit coupling, so it can be expanded to first order, which leads to

$$p(\epsilon_t) \approx \frac{\pi \beta}{2\epsilon^{1/2}} \tag{6.34}$$

Now, for $\epsilon \beta^{2/3} \gg 1$, $p(\epsilon_t)$ of Eq. (6.31) goes over asymptotically into [38, p. 54, Eq. (4.9)]

$$p(\epsilon_t) \approx \frac{\pi \beta}{\epsilon^{1/2}} \sin^2\left(\frac{2}{3}(\epsilon^{3/2}\beta) + \frac{\pi}{4}\right) \tag{6.35}$$

On replacing $\sin^2[\frac{2}{3}(\epsilon^{3/2}\beta) + \pi/4]$ by $\frac{1}{2}$, i.e. by its average, Eq. (6.35) is seen to reduce to Eq. (6.34). Thus the high-energy tail of the Airy-function formula of Eq. (6.31) coincides with the Landau–Zener equation (6.34).

At the other extreme, i.e. for $\epsilon \beta^{2/3} \ll -1$, $p(\epsilon_t)$ of Eq. (6.31) reduces to [38, loc. cit.]

$$p(\epsilon_t) \approx \frac{\pi \beta}{4|\epsilon|^{1/2}} \exp\left(-\frac{4}{3}|\epsilon|^{3/2}\beta\right) \tag{6.36}$$

6.2.4 The Laplace transform of $p(\epsilon_t)$

The probability of non-adiabatic transition $p(\epsilon_t)$ will affect the sum of states $G^*(E)$ of the transition state in a manner analogous to Eq. (6.13):

$$G^*(E) = \int_0^E p(\epsilon_t) N^*(E - \epsilon_t) d\epsilon_t, \qquad E > 0 \tag{6.37}$$

In terms of the Laplace-transform notation introduced previously in connection with Eq. (6.16), the above equation represents the convolution of $p(\epsilon_t)$ with $N^*(E)$. Consequently the limiting high-pressure thermal rate constant k_∞ can be obtained by a procedure similar to that used in Eqs. (6.17)–(6.19).

To this end we need the Laplace transform of $p(\epsilon_t)$. With $s = 1/\Bbbk T$, the transform of Eq. (6.31) is formally

$$\mathcal{L}\{p(\epsilon_t)\} = \mathcal{P}(s) \tag{6.38}$$

If we let $\epsilon_t - E_c = x$ and $a = \beta^{2/3}\epsilon_0$, then

$$\mathcal{P}(s) = \exp(-E_c s) \int_{-E_c}^{\infty} p(ax) \exp(-xs) \, dx \tag{6.39}$$

Consider now splitting the domain of integration into two parts, one from $-E_c$ to 0, and the other from 0 to ∞, such that $\mathcal{P}(s) = \mathcal{P}_1(s) + \mathcal{P}_2(s)$, where

$$\mathcal{P}_1(s) = \exp(-E_c s) \int_{-E_c}^{0} p_1(ax) \exp(-xs) \, dx \tag{6.40}$$

$$\mathcal{P}_2(s) = \exp(-E_c s) \int_{0}^{\infty} p_2(ax) \exp(-xs) \, dx \tag{6.41}$$

Replacing $p_1(ax)$ by the approximation of Eq. (6.36), the result is (Problem 6.2)

$$\mathcal{P}_1(s) \cong \frac{1}{4}\pi\beta \left(\frac{\pi\epsilon_0}{s}\right)^{1/2} \exp\left(-E_c s + \frac{1}{12\beta^2}(\epsilon_0 s)^3\right) \tag{6.42}$$

where the symbols have the same meaning as in Eq. (6.32). The positive term in the exponential represents a lowering of the "critical energy" E_c due to tunneling. Similarly replacing $p_2(ax)$ by the approximation of Eq. (6.34), the result is

$$\mathcal{P}_2(s) \cong \frac{1}{2}\pi\beta \left(\frac{\pi\epsilon_0}{s}\right)^{1/2} \exp(-E_c s) \tag{6.43}$$

which is essentially Eq. (6.42) minus the positive term in the exponential. An

Table 6.1. *Experimental high-pressure parameters for spin-forbidden reactions*

Reaction	$\log[A_\infty \ (s^{-1})]$	$E_{a\infty}$ (kcal mol^{-1})	Temperature range (K)	Ref.
$N_2O(^1\Sigma) \rightarrow N_2(^1\Sigma) + O(^3P)$	11.4	60.7	900–2000	(a)
$CO_2(^1\Sigma) \rightarrow CO(^1\Sigma) + O(^3P)$	12.9	129.8	3000–3700	(b)
$CS_2(^1\Sigma) \rightarrow CS(^1\Sigma) + S(^3P)$	12.6	87.0	1950–2800	(c)
$COS(^1\Sigma) \rightarrow CO(^1\Sigma) + S(^3P)$	11.6	68.0	1500–3100	(d)

(a) W. D. Breshears, *J. Phys. Chem.* **99**, 12 529 (1995). (Literature survey.)
(b) H. G. Wagner and F. Zabel, *Ber. Bunsenges. Phys. Chem.* **72**, 705 (1974).
(c) H. A. Olschewski, J. Troe and H. G. Wagner, *Ber. Bunsenges. Phys. Chem.* **70**, 1060 (1966).
(d) H. G. Schecker and H. G. Wagner, *Int. J. Chem. Kinet.* **1**, 541 (1969).

analogous piecewise integration can be used in the convolution of Eq. (6.37) for the calculation of $G^*(E)$.

The rate constant k_∞ then follows in a manner analogous to that used in Section 6.1, that is, from the analog of Eq. (6.20) with $\mathcal{P}(s)$ replacing $P(s)$:

$$k_\infty = \frac{\mathcal{P}(s)}{h} \frac{Q^*(s)}{Q(s)} \tag{6.44}$$

For the case of collinear decomposition of N_2O, the simplest example, the limiting high-pressure thermal rate constant, calculated from $\mathcal{P}_1(s) + \mathcal{P}_2(s)$ (Eqs. (6.42) and (6.43)), using data in the legend of Fig. 6.7), is $k_\infty \approx 3 \times 10^{-3} \ s^{-1}$ at 900 K, if transition-state frequencies are approximated by those of N_2O with the N–O stretch deleted as the reaction coordinate. This k_∞ is roughly the same as that obtained in most other collinear calculations (e.g. [30]). However, the particular case of N_2O is more complicated in that the excited triplet of N_2O is bent, so that a multi-dimensional approach is in principle necessary [36]. In addition, the actual mechanism of the thermal reaction is not simple and experimental data from different sources are not always in agreement [41].

Table 6.1 lists experimental high-pressure parameters for three other non-adiabatic reactions, all of which were obtained from shock-tube experiments. Note in particular the low pre-exponential factors compared with the "normal" $A_\infty \approx e k T/h = 10^{14} \ s^{-1}$ at 2000 K (cf. Section 5.4.9). Another non-adiabatic reaction is $HN_3 \rightarrow N_2(^1\Sigma) + HN(^3\Sigma)$, which has received both experimental [42] and theoretical [43] attention.

References

[1] D. Rapp, *Quantum Mechanics* (Holt, Rinehart and Winston, New York, 1971).

[2] See, for example, *Isotope Effect in Chemical Reactions*, C. J. Collins and N. S. Bowman, Eds. ACS Monograph 167 (Van Nostrand-Reinhold, New York, 1970).

[3] S. K. Gray, W. H. Miller, Y. Yamaguchi and H. F. Schaefer, *J. Chem. Phys.* **73**, 2733 (1980); S.K. Gray, W. H. Miller, Y. Yamaguchi, and H. F. Schaefer, *J. Am. Chem. Soc.* **103**, 1900 (1981); Y. Osamura, H. F. Schaefer, S. K. Gray and W. H. Miller, *J. Am. Chem. Soc.* **103**, 1904 (1981).

[4] J. W. Keister, T. Baer, R. Thissen, C. Alcaraz, O. Dutuit, H. Audier and V. Troude, *J. Phys. Chem.* **A102**, 1090 (1998).

[5] C. Eckart, *Phys. Rev.* **35**, 1303 (1930).

[6] C. A. Gonzales, T. C. Allison and F. Louis, *J. Phys. Chem.* **A105**, 11034 (2001).

[7] R. P. Bell, *The Tunnel Effect in Chemistry* (Chapman and Hall, London, 1980).

[8] See, for example, A. S. Davydov, *Quantum Mechanics*, translated by D. ter Haar (Pergamon Press, Oxford, 1965), p. 81.

[9] W. H. Miller, *J. Am. Chem. Soc.* **101**, 6810 (1979).

[10] W. Forst, *J. Phys. Chem.* **87**, 4489 (1983).

[11] Y. Qin and D. L. Thompson, *J. Chem. Phys.* **100**, 6445 (1994).

[12] P. A. MacDonald and J. S. Shirk, *J. Chem. Phys.* **77**, 2355 (1982).

[13] G. E. McGraw, D. E. Bernitt and I. C. Hisatsune, *J. Chem. Phys.* **45**, 1392 (1966).

[14] S. M. Colwell and N. C. Handy, *J. Chem. Phys.* **82**, 1281 (1985).

[15] F. B. Brown, S. C. Tucker and D. G. Truhlar, *J. Chem. Phys.* **83**, 4451 (1985).

[16] A. J. Illies, M. F. Jarrold and M. T. Bowers, *J. Am. Chem. Soc.* **104**, 3587 (1982).

[17] D. V. Widder, *The Laplace Transform* (Princeton University Press, Princeton, NJ, 1971).

[18] H. S. Johnston and D. Rapp, *J. Am. Chem. Soc.* **83**, 1 (1961). Approximate closed-form expressions for Γ^* applicable to an unsymmetrical Eckart barrier are given by H. Shin, *J. Chem. Phys.* **39**, 2934 (1963).

[19] I. S. Gradshteyn and I. M. Ryzhik, *Table of Integrals, Series and Products*, 4th. Ed., translated by A. Jeffrey (Academic Press, New York, 1965), pp. 63ff.

[20] E. E. Nikitin, *Theory of Elementary Atomic and Molecular Processes in Gases*, translated by M. J. Kearsley (Oxford University Press, Oxford, 1974).

[21] H. B. Dwight, *Tables of Integrals and Other Mathematical Data*, 4th Ed. (Macmillan, New York, 1965).

[22] M. Abramowitz and I. A. Stegun, *Handbook of Mathematical Functions* (NBS Applied Mathematics Series 55, Washington, 1972).

[23] E. P. Wigner, *Z. Physik. Chem.* **B19**, 203 (1932).

[24] P. D. Pacey, *J. Chem. Phys.* **71**, 2966, (1979).

[25] A. E. Stearn and H. Eyring, *J. Chem. Phys.* **3**, 778 (1935).

[26] S. Glasstone, K. J. Laidler and H. Eyring, *The Theory of Rate Processes* (McGraw-Hill, New York, 1941), pp. 333–337.

[27] E. K. Gill and K. J. Laidler, *Can. J. Chem.* **36**, 1570 (1958).

[28] H. A. Olschewski, J. Troe and H. G. Wagner, *Ber. Bunsenges. Physik. Chem.* **70**, 450 (1966).

[29] R. G. Gilbert and I. G. Ross, *Aust. J. Chem.* **24**, 1541 (1971).

[30] G. Gebelein and J. Jortner, *Theoret. Chim. Acta* **25**, 143 (1972).

[31] E. R. Fisher and E. Bauer, *J. Chem. Phys.* **57**, 1966 (1972).

[32] J. B. Delos, *J. Chem. Phys.* **59**, 2365 (1973).

[33] J. C. Tully, *J. Chem. Phys.* **61**, 61 (1974).

[34] G. E. Zahr, R. K. Preston and W. H. Miller, *J. Chem. Phys.* **62**, 1127 (1975).

[35] A. W. Yau and H. O. Pritchard, *Can. J. Chem.* **57**, 1731 (1979).

[36] A. J. Lorquet, J. C. Lorquet and W. Forst, *Chem. Phys.* **51**, 253 (1980).

[37] J. C. Lorquet and B. Leyh-Nihant, *J. Phys. Chem.* **92**, 4778 (1988).

[38] E. E. Nikitin, in *Chemische Elementarprozesse*, H. Hartmann, Ed. (Springer-Verlag, Berlin, 1968), pp. 43–77.

[39] B. Desouter-Lecomte and J. C. Lorquet, *J. Chem. Phys.* **71**, 4391 (1979).

[40] H. Eyring, J. Walter and G. E. Kimball, *Quantum Chemistry* (Wiley, New York, 1963).

[41] W. Forst, *J. Phys. Chem.* **86**, 1776 (1982); W. D. Breshears, *J. Phys. Chem.* **99**, 12 529 (1995).

[42] B. R. Foy, M. P. Casassa, J. C. Stephenson and D. S. King, *J. Chem. Phys.* **92**, 2782 (1990).

[43] D. R. Yarkony, *J. Chem. Phys.* **92**, 320 (1990).

Problems

Problem 6.1. Show that (i) the maximum of the Eckart potential (Eq. (6.3)) is at $y_{max} = (A + B)/(A - B)$, from which $V_{max} = (A + B)^2/(4B)$; (ii) the curvature at potential maximum is $\partial^2 V/\partial y^2 = -(A - B)^4/(2B)^3$; (iii) as $r \to \infty$, $V(x) \to 0$, and as $r \to -\infty$, $V(x) \to A$.

Problem 6.2. Derive Eq. (6.42), starting with the approximation of Eq. (6.36).

7

A variational transition-state theory

Type-1 reactions present a special case in that, if the transition state is located at the maximum of effective potential regardless of excitation energy E, as was assumed in Section 3.4, standard transition-state theory often fails to reproduce the generally observed negative temperature coefficient for recombinations of radicals at high pressure (cf. Table 7.2 below).

On a more fundamental level, when the first computed trajectory results on the dissociation of triatomics were analyzed, it soon became evident that some trajectories "turn back," that is, recross the dividing surface that separates reactant from products (Section 3.1.4) to re-form the reactant, whereas standard transition-state theory assumes that this does not occur. Consequently the calculated rate constant is at best only an upper limit to the "true" rate constant.

To correct for this effect, Bunker and Pattengill [1] suggested that the system should be allowed to "find its own transition state" by minimizing the flux from reactant to products. This amounts to a variational approach, which was originally suggested by Horiuti [2] and is now generally recognized as the more realistic approach.

7.1 General aspects

7.1.1 The variational criterion

With E^* the internal energy of the transition state (Fig. 7.1), and using the angular-momentum quantum number J rather than the rotational energy E_r as the rotational variable, the microcanonical variational criterion is defined [3] as the local minimum in the total number of states $G^*(E^*, J, r)$ along the reaction coordinate r (taken to be the breaking bond):

$$\frac{\partial G^*(E^*, J, r)}{\partial r} = 0 \qquad (7.1)$$

If, at a given E^* and J, the solution of Eq. (7.1) for the local minimum is at $r = r^*$ (cf. Figs. 7.1 and 7.3 later), the variational equivalent of Eq. (3.38) using the variable J is (with additional asterisks)

$$k^*(E, J) = \frac{\alpha G^*(E^*, J, r^*)}{h\, N(E)} \tag{7.2}$$

where the numerator is understood to be evaluated at $r = r^*$. $N(E)$ is the density of internal states of the reactant at energy E defined with respect to the ground state of the effective potential (cf. Fig. 7.1).

In the implementations of Eq. (7.1) some vibrational modes of the reactant – so-called "transitional" modes – undergo a substantial change of character during the decomposition, in that they become fragment rotations, i.e. essentially vibrations of zero frequency. The vibrational frequency of a transitional mode is generally

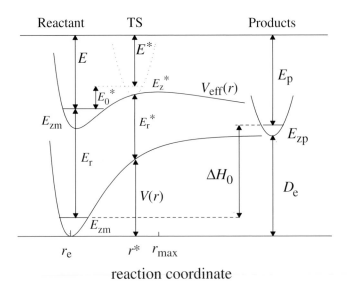

reaction coordinate

Fig. 7.1. A schematic representation of a section through the potential-energy surface in a Type-1 bond-fission reaction along the (one-dimensional) reaction coordinate r. The internal energy of the reactant molecule is E and its zero-point energy is E_{zm}. For zero angular momentum ($J = 0$) the attractive part of the potential $V(r)$ has no maximum (lower curve, bare barrier height D_e); the critical energy for decomposition is the enthalpy of reaction at $0\,K$, $\Delta H_0 = D_e - E_{zm} + E_{zp}$. For $J > 0$ (upper curve), the rotational energy of the molecule is E_r (short for $E_r(J)$), and the effective potential $V_{eff}(r)$ has a maximum at r_{max}. The variational transition state (dotted line, parameters denoted by asterisks) is shown located at $r^* < r_{max}$, since the variational routine generally finds transition states at distances shorter than r_{max}. $E_r^*(<E_r)$ is the rotational energy of the transition state, its zero-point energy is E_z^*, and its internal energy is E^*. At the particular r^* and J the minimum (critical) energy for decomposition is E_0^*. The zero-point energy of products is E_{zp}.

Table 7.1. *Reactant–product correlation in the reaction*
$$CX_3CX_3 \rightarrow CX_3 + CX_3$$

Reactant CX_3CX_3	Products $CX_3 + CX_3$
Conserved:	
C–C stretch	$\rightarrow$ reaction coordinate
6 C–X stretches	$\rightarrow (3 + 3)$ C–X stretches in $CX_3 + CX_3$
6 CX_3 deformations	$\rightarrow (3 + 3)$ CX_3 deformations in $CX_3 + CX_3$
Transitional:	
4 CX_3 rockings	$\rightarrow (2 + 2)$ CX_3 rotations of $CX_3 + CX_3$
1 X_3C–CX_3 internal torsion	$\rightarrow$ 1 CX_3 rotation
Rotations:	
1 K-rotor rotation	$\rightarrow$ 1 CX_3 rotation
2 J-rotor rotations	$\rightarrow$ 2 orbital rotations
Total 21 DOF including the reaction coordinate	

assumed to depend on the bond distance r according to the relation

$$v(r) = v_e \, S(r) \tag{7.3}$$

in which $v(r)$ is the r-dependent mode frequency and v_e is its equilibrium value in the reactant molecule. $S(r)$ is a switching function such that $S(r_e) = 1$ and $S(r = \infty) = 0$, where r_e is the equilibrium value of the reaction coordinate (Fig. 7.1).

Equation (7.3) describes the progressive "softening" of a transitional mode as it transforms into a rotation. The remaining modes which are not directly involved in the reaction are termed "conserved" modes since their frequencies change little, if at all, on going from reactant to products, and therefore have little influence on the r-dependence of $G^*(E^*, J, r)$.

As an example, a reactant $\rightarrow$ product correlation for all 21 degrees of freedom (DOF) in the decomposition of a molecule of the type CX_3CX_3 is shown in Table 7.1. Compare this with the more primitive correlation shown in Table 3.1 for $C_2H_6 \rightarrow CH_3 + CH_3$.

An alternative approach is to model the transitional modes by means of a two-dimensional hindered rotor, as in the fragmentation $CH_4 \rightarrow CH_3 + H$ (see Section 4.3.1 and Problem 7.1). The hindering barrier V_0 is made dependent on the interfragment separation r and Eq. (4.35) thus becomes $V(r, \theta) = V_0(r) \sin^2 \theta$. The dependence of $V_0(r)$ on r can be approximated in principle by an appropriate (but not necessarily explicitly the same) switching function [4] as in Eq. (7.3) above, which will progressively decrease the barrier from V_0 to zero:

$$V_0(r) = V_0 \, S(r) \tag{7.4}$$

The corresponding r-dependent sum of states then follows directly from Eqs. (4.40) and (4.41) merely on replacing V_0 by $V_0(r)$. A similar procedure may be adopted with respect to V_0 in the hindering potential of a one-dimensional rotor (Eq. (4.44)).

7.1.2 VTST

A rigorous variational transition-state theory (VTST) must correctly describe how, in the course of a bond fission, a transitional vibrational mode of a reactant evolves first into a hindered rotor, and then into a free rotor in the product. This is no simple matter; it requires that attention be paid to structural changes of the reactant in the course of the dissociation, on the basis of detailed knowledge of the relevant potential-energy surface and associated properties. Because of the difficulty and expense of a full potential-energy-surface calculation, such an energy surface is not readily available for polyatomic reactants of practical interest.

Given the complexity of rigorous VTST calculations [5], it is of some interest to propose a simplified approach that will preserve some of the original simplicity of the transition-state theory and offer physical insight without sacrificing the essence of the variational treatment. The approach is based on Eq. (7.3), which provides a useful point of departure for a fairly general treatment.

7.1.3 A simplified VTST

Within the context of Eq. (7.2), the obvious difficulty with Eq. (7.3) is that, as $r \to \infty$, $v(r) \to 0$, so that, if $G^*(E^*, J, r)$ is interpreted as the total number of *vibrational* states, to be calculated by one of the methods of Chapter 4, this number would in principle go off to infinity at large r as transitional mode frequencies approach zero. However, transitional modes in bond-fission reactions connect with fragment rotations (cf. Table 7.1), the number of states of which, while large, is not infinite.

This difficulty is obviated in a simplified VTST described below, which applies the steepest-descents (S-D) inversion routine of Chapter 4 to the variational problem [6]. An r-dependent partition function for the transition state is constructed by interpolation between total *partition functions* (rather than frequencies) for the reactant and product, using an r-dependent switching function $S(r)$. Inversion of this interpolated transition-state partition function yields an r-dependent state count $G^*(E^*, J, r)$ for the transition state, which is then used in Eq. (7.1) to find the variational minimum and the corresponding minimized rate constant. This approach ensures that $G^*(E^*, J, r)$ will smoothly converge to the correct value for the sum of states of the products.

The procedure makes use only of information that is readily available, namely basic molecular properties of the reactant and products, sufficient for the calculation of partition functions. Since the object is to calculate a dynamical property (the rate constant) from static properties (partition functions), a reference point from experiment is necessary, which translates into the appearance of an adjustable parameter, a feature in common with all other variational procedures.

7.1.4 The effective potential

The reaction will be assumed to proceed under the effective potential as described in Section 3.4.1 (Eq. (3.30)), using the approximation that angular momentum in the products is mostly orbital, i.e. that the requirement for conservation of angular momentum reduces to $J \approx L$ (but see the end of Section 9.2).

It will be further assumed that a Type-1 reactant decomposing according to $M \rightarrow R_1 + R_2$ is a prolate symmetric top with rotational constants $A > B = C$ (Appendix 1, Section A1.4). In the course of the decomposition the separating $R_1 \cdots R_2$ fragments will maintain this configuration, which becomes more and more prolate as R_1 and R_2 move apart, i.e. as the interfragment distance increases from r_e to $r \rightarrow \infty$ (see Section 9.2.1 for additional discussion on this point). The rotational constant A which refers to rotation about the $R_1 \cdots R_2$ axis is not significantly affected, but the doubly degenerate rotational constant B, which refers to rotation about the two axes perpendicular to the $R_1 \cdots R_2$ axis, becomes smaller (because of the increasing moment of inertia), and therefore also the J-dependent rotational energy $E_r(J)$ (the "J-rotor") becomes smaller. $E_r(J)$ is the non-randomizable (adiabatic) part of the rotational energy which is responsible for the creation of a centrifugal barrier.

The quantum number K, on the other hand, is assumed for simplicity *not* to be conserved because in a symmetric top the corresponding energy $E_r(K)$ (the "K-rotor") contributes to the density of internal states (Section 3.3.2). As an approximation, the one-dimensional K-rotor may be considered as uncoupled from the J-rotor, in which case the restriction $-J \leq K \leq +J$ is ignored. However, at the price of only a slight complication of the S-D routine, this restriction can be lifted (Sections 4.5.9 and 4.5.10).

The question of the effective potential is re-examined in Chapter 9 from a more fundamental point of view.

7.1.5 The vibrational potential

A useful approximation for the vibrational potential is to assume a Morse function,

$$V(r) = D_e\{1 - \exp[-\beta(r - r_e)]\}^2 \tag{7.5}$$

where r_e is the value of r at potential minimum and D_e is the classical height of the potential barrier. It is determined from ΔH_0, the enthalpy of reaction at 0 K, which is assumed to be known from thermochemical data (cf. Fig. 7.1). The constant β which determines the steepness of the potential can be estimated from (Appendix 4)

$$\beta = 2\pi\omega_{rc}[\mu/(2D_e)]^{1/2} \tag{7.6}$$

where ω_{rc} is the "frequency" of the reaction coordinate (i.e. of the bond undergoing rupture), and μ is the reduced mass.

As noted in Chapter 9, the interpretation of μ as the reduced mass of the two *atoms* on either side of the breaking bond makes it consistent with the definition of the reaction coordinate as the bond undergoing rupture.

Alternatively, the potential can be obtained from the so-called extended Rydberg potential (ERP) which correctly renders the quadratic, cubic, and quartic force constants. These can be obtained from tabulated values for diatomic bond-stretching potentials (Appendix 4, Eq. (A4.17)), a reasonable approximation to the potential in the reaction coordinate as defined above. The ERP is useful in cases in which uncertainty regarding the reaction-coordinate frequency ω_{rc} would make the Morse β calculated from Eq. (7.6), and hence the entire potential, doubtful [7].

7.2 Implementation

7.2.1 Interpolation

Since the S-D routine for the calculation of the sum of states is based on the logarithm of the inversion integrand (Eq. (4.72)), it is advantageous to interpolate the logarithms of partition functions. For convenience the S-D routine is used in the logarithmic formulation (Eq. (4.75) and following equations); thus $\phi(z)$, the logarithm of the inversion integrand at energy E^* is, in the present instance,

$$\phi(z) = \ln Q^*(z) - k\ln(\ln z^{-1}) - E^*\ln z \tag{7.7}$$

where $\ln z^{-1} = 1/(kT)$, the transform parameter and $k = 1$ since the sum of states is desired.

Let Q_m be the total partition function of the reactant molecule, *exclusive* of the reaction coordinate and the two rotational degrees of freedom (J-rotor) involved in the effective potential (Eq. (3.30)), and let Q_p be the total partition function of the products. The evolution with distance r of the (natural) logarithm of the transition-state partition function Q^* is obtained by interpolation of the logarithms of Q_m and Q_p:

$$\ln Q^*(r) = (\ln Q_m)\,S(r) + (\ln Q_p)\,[1 - S(r)] \tag{7.8}$$

The switching function $S(r)$ is for the moment arbitrary. The logarithm of the r-dependent transition-state partition function $\ln Q^*(r)$ is then substituted directly into Eq. (7.7) and, *at fixed r*, the rest of the inversion routine follows as usual. To obtain the *r*-dependent sum of states the S-D routine is then run repeatedly at several *r*.

Equation (7.8) assumes that all transitional modes convert into free rotations via the same switching function $S(r)$. This approximation is necessary in the absence of other information. It should be clear from the above that interpolation has little effect on "conserved" modes if their frequencies in reactant and products are similar; if they are not (e.g. in C_2H_6, where two CH_3 deformations around $1380\,\text{cm}^{-1}$ connect with $580\,\text{cm}^{-1}$ deformations in the CH_3 radicals – cf. Table 3.1), the interpolation takes this into account.

While Eq. (7.8) is applicable in principle to any system for which one can write down a partition function, the problem is simplified by assuming that vibrational modes are harmonic and rotors are free, although this is not essential, at the price of complicating somewhat the S-D routine (cf. Sections 4.3 and 4.5). In terms of the variable z, the quantum harmonic vibrational and classical rotational partition functions Q_v and Q_r are, respectively,

$$Q_v = \prod_i (1 - z^{\omega_i})^{-1}, \qquad Q_r = \frac{q_r}{(\ell n\ z^{-1})^{r/2}} \qquad (7.9)$$

where ω_i is the "frequency" of the *i*th mode and r is the number of free rotors; in this representation an *r*-dimensional rotation counts for r rotors. As usual, $q_r = Q_r/(\ell T)^{r/2}$.

We now have to distinguish between r_a (the number of "active" rotors in the reactant) and r_t (the number of product rotors that arise from transitional modes in the reactant). The "active" rotors contribute to the density or sum of states in the reactant molecule, e.g. the active *K*-rotor in the symmetric top (which correlates with one fragment rotation), plus any internal rotors present; their total number r_a remains constant throughout the decomposition, and only q_r changes with distance along the reaction coordinate from $q_{r,m}$ (in the molecule, at r_e) to $q_{r,p}$ (in the products, at $r \to \infty$). This change is accomplished by interpolation of the appropriate logarithms and is implicit in Eq. (7.8). The second kind are the r_t rotors (with their own q_t, which does not change with distance) that appear only in the products.

In this perspective the explicit forms of the logarithms of the reactant (subscript m) and product (subscript p) partition functions are

$$\ell n\ Q_m(z) = \ell n \left[\prod_i (1 - z^{\omega_i})^{-1} q_{rm}(\ell n\ z^{-1})^{-r_a/2} \right] \qquad (7.10)$$

$$\ell n\ Q_p(z) = \ell n \left[\prod_j (1 - z^{\omega_j})^{-1} q_{rp}(\ell n\ z^{-1})^{-r_a/2} q_{rt}(\ell n\ z^{-1})^{-r_t/2} \right] \qquad (7.11)$$

where the vibrational "frequencies" of the reactant are indexed i and those of the products are indexed j. After substitution into Eq. (7.8) the function $\phi(z)$ is then obtained from Eq. (7.7).

The rest of the procedure follows exactly as described in Section 4.4: in order to find the saddle point of the inversion integrand at specified E^* and r, the transcendental equation $z\phi'(z) = 0$ has to be solved. On differentiating the explicit form of Eq. (7.7), using Eqs. (7.8), (7.10), and (7.11), we have

$$z\,\phi'(z) = S(r) \sum_i \frac{\omega_i z^{\omega_i}}{1 - z^{\omega_i}} + \frac{1 + \frac{1}{2}r_a}{\ln z^{-1}}$$

$$+ \left(\sum_j \frac{\omega_j z^{\omega_j}}{1 - z^{\omega_j}} + \frac{\frac{1}{2}r_t}{\ln z^{-1}} \right) [1 - S(r)] - E^* = 0 \qquad (7.12)$$

where E^* is the internal distributable energy of the transition state (Fig. 7.1). If the root of this equation is at $z = \theta$, the required sum of states follows from the standard (first-order) formula, Eq. (4.77):

$$G^*(E^*, J, r) = \frac{Q^*(\theta)}{(\ln \theta^{-1})\, \theta^E \, (2\pi \theta^2 \phi''(\theta))^{1/2}} \qquad (7.13)$$

where $Q^*(\theta)$ is the re-assembled interpolated partition function (Problem 7.2) and $\phi''(\theta)$ is the second derivative of $\phi(z)$, both evaluated at $z = \theta$ (Problem 7.3). The r-dependence of $G^*(E^*, J, r)$ arises from θ, which is implicitly a function of r through the switching function, and the J-dependence arises parametrically through E_0^*, which is a function both of r and of J. This is basically an application of the J-shifting approximation introduced in Section 3.4.5, slightly modified.

7.2.2 The zero-point energy

As transitional vibrational modes are converted into product rotations, their zero-point energy "disappears," i.e. is converted into distributable energy E^*, since the rotors, which are assumed to be free, have no zero-point energy. It is therefore necessary to work with a zero-point-energy-corrected effective potential.

The general form of the partition function for the zero-point energy E_z is

$$Q_z = \exp[-E_z/(kT)] \qquad (7.14)$$

In the present case additional subscripts "m" and "p" will distinguish between reactant and products; thus, for the reactant, $E_{zm} = \frac{1}{2}\sum_i \omega_i$, and for the products $E_{zp} = \frac{1}{2}\sum_j \omega_j$. Consistently with the general interpolation scheme of Eq. (7.8), the logarithm of Q_z^*, the partition function for the zero-point energy of the transition state, is then obtained by interpolation between the logarithms of Q_{zm} and Q_{zp} in the manner of Eq. (7.8), which is equivalent to the r-dependent zero-point energy E_z^* of the transition state being given by interpolation between E_{zm} and E_{zp} according to

$$E_z^*(r) = E_{zm} S(r) + E_{zp}[1 - S(r)] \qquad (7.15)$$

From Fig. 7.1 we have by simple conservation of energy

$$E_{zm} + E_r + E = V(r) + E_r^* + E_z^* + E^* \qquad (7.16)$$
$$E_{zm} + E_r + E_0^* = V(r) + E_r^* + E_z^* \qquad (7.17)$$

where $E_r = J(J+1)h/(2\mu r_e^2)$ and $E_r^* = E_r(r_e/r)^2$. The zero-point energy-corrected effective potential is then

$$V_{\mathrm{eff}}(r) = V(r) + E_r^* + E_z^* - E_{zm} \qquad (7.18)$$

From Eq. (7.17) it follows that

$$E_0^* = V_{\mathrm{eff}}(r) - E_r \qquad (7.19)$$

so that Eq. (7.16) yields $E^* = E - E_0^*$ as required. This makes it clear that the J-dependence of $G^*(E^*, J, r)$ arises via E_0^*.

7.2.3 The switching function

The important part of the interpolation concept is the switching function $S(r)$. Empirical relations that relate bond order or force constant to bond length, which are often referred to as Pauling's or Badger's rules [8, 9] suggest that the form of $S(r)$ appropriate for bond frequencies (Eq. (7.3)) *at short distances* is an exponential function:

$$S(r) = \exp[-a(r - r_e)] \qquad (7.20)$$

where r_e is the equilibrium value of r and a is a parameter that depends on the nature of the bond.

In the present context, however, we wish to interpolate logarithms of partition functions, not frequencies; in addition, the interpolation is intended over *large* distances, since the entire distance separating reactant from products is to be covered. Figure 7.2 compares the r-dependences of the sum of states $G(E, r)$, at constant E, obtained by inversion of two types of interpolated partition functions (PF), both of which refer to the same decomposition, but represent different assumptions regarding the evolution of transitional modes.

Case (a): $G(E, r)$ obtained by inversion of a PF calculated on the assumption that vibrational frequencies of transitional modes decline with distance according to Eq. (7.3) with the exponential switching function of Eq. (7.20).

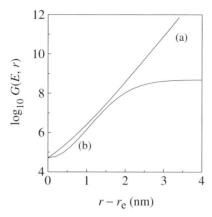

Fig. 7.2. The calculated r-dependence of interpolated $G(E,r)$ at fixed $E = 1000\,\mathrm{cm}^{-1}$ and variable r for the decomposition $CCl_3NO \rightarrow CCl_3 + NO$, assuming the validity of a Morse potential (Eq. (7.5)) with $\beta = 2.76\,\mathrm{nm}^{-1}$. Curve (a) for $G(E,r)$ is obtained by inversion of the partition function for the reactant in which the four transitional mode *frequencies* (C–N–O bend, Cl_3C–NO torsion, and two CCl_3 rockings) are made r-dependent via Eq. (7.3) using the switching function of Eq. (7.20) with $a = \beta/2$. Curve (b) represents $G(E,r)$ by inversion, using $c = 0.45\,\mathrm{nm}^{-2}$ in the half-Gaussian switching function of Eq. (7.21) for the interpolation of logarithms of complete partition functions of reactant *and products* via Eq. (7.8). Here $G(E,r)$ has reached the final value for products at $r - r_e \cong 4\,\mathrm{nm}$.

Case (b): $G(E,r)$ obtained by inversion of a PF interpolated according to Eq. (7.8), using for the switching function the S-shaped half-Gaussian

$$S(r) = \exp[-c(r - r_e)^2] \tag{7.21}$$

Here c is a parameter of dimension (distance)$^{-2}$ and $r \geq r_e$.

Case (a) may be considered to correctly represent the r-dependence of $G(E,r)$ at short distances, in accordance with the roughly exponential decay of frequencies suggested by Pauling's and Badger's rules. The low-r curvature is "concave upward." At large r this $G(E,r)$ goes off to infinity, as expected, which is of course unphysical in the present context. The $G(E,r)$ of case (b) has both the correct curvature at low r and the correct final value at large r, which suggests that a switching function of the type represented by Eq. (7.21) is reasonably well adapted for interpolation of partition functions.

While the calculations in Fig. 7.2 are for a half-Gaussian, any similar S-shaped switching function, of which the half-Gaussian of Eq. (7.21) is merely the simplest example, would produce similar results. One such more-elaborate S-shaped switching function is the two-parameter hyperbolic tangent

$$S(r) = 1 - \tanh[a(r - r_e)^b] \tag{7.22}$$

with parameters a and b. Inasmuch as the predictive value of any theory decreases with increasing number of adjustable parameters, the simplest S-shaped form of $S(r)$ is clearly preferable. Such is the half-Gaussian of Eq. (7.21) with the unique parameter c which has been found adequate for a wide range of reactions.

Insofar as the state count $G^*(E^*, r)$ for an actual transition state is concerned at specified E and J, Fig. 7.2 shows how the "softening" of the transitional modes with increasing r increases the number of states counted at *fixed* $E^* - E - E_0^*$.

However, at fixed total energy ($E = $ constant) and angular momentum J, E^* is not fixed because E_0^* is r-dependent (Eq. (7.19)), so, as r increases, the difference $E^* = E - E_0^*$ decreases at the same time, due to $V_{\text{eff}}(r)$ rising with r between r_e and r_{max}, which tends to decrease the number of states. The result of the two opposing effects is that $G^*(E^*, J, r)$ goes through a minimum at some $r = r^*$. In general, the minimum (and therefore the variational transition state) is located at *smaller* values of r than the centrifugal maximum r_{max}, as suggested in Fig. 7.3.

Specifically, at fixed J, r^* decreases with increasing E^*, as shown in Fig. 7.4, from $r^* \cong r_{\text{max}}$ at threshold, to shorter distances for energies *above* threshold. Thus the transition state starts out "loose" near threshold, i.e. when its internal energy E^* is near zero, and progressively "tightens" as internal energy increases. If only the "loose" and "tight" transition states are taken into account in a simplified calculation of $k(E)$, this is sometimes referred to as "transition-state switching" [10].

The gradual change from "loose" to "tight" as energy increases modifies the temperature dependence of the canonical rate constant k_{uni} (*vide infra*); in fact, the

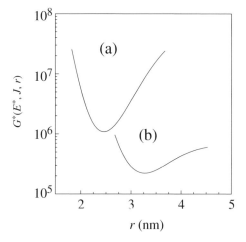

Fig. 7.3. Variation of $G^*(E^*, J, r)$ with r at $J = 20$ and two energies: (a) at $E^* = 15\,434\,\text{cm}^{-1}$ and (b) at $E^* = 10\,211\,\text{cm}^{-1}$. The system is the reaction of Fig. 7.2. The upper curve is shifted 5 log units downward. With rising energy, the minima of the curves shift to distances below the maximum of the effective potential, which, for $J = 20$, is at $r_{\text{max}} = 4.22$ nm.

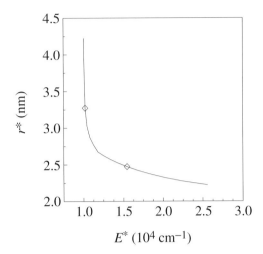

Fig. 7.4. The position of the minimized transition state (r^*) as a function of the excitation energy E^* at $J = 20$, for the same system as in Fig. 7.2. The critical or threshold energy is $E_0^* = 10\,052\,\mathrm{cm}^{-1}$, at which point the variational transition state is at 4.22 nm. Diamonds indicate the positions of the two minima from Fig. 7.3.

difficulty in reconciling with experiment the calculated low- and high-temperature data for dissociation of ethane and the reverse association of methyl radicals was the original motive for the development of variational transition-state theory.

7.2.4 The thermal rate constant

For a selected potential and a given switching function, inversion of $Q^*(r)$ at a given E and J is used to generate $G^*(E^*, J, r)$ at a number of values of r, from which is picked the smallest $G^*(E^*, J, r)$ for substitution into Eq. (7.2); after evaluation of the density $N(E)$ this completes the determination of one particular value of $k^*(E, J)$. The procedure is repeated at a number of values of E and one specified J, from which is obtained the J-dependent strong-collision k_{uni} at the selected J:

$$k_{\mathrm{uni}}^{\mathrm{sc}}(J) = \frac{\omega}{Q_{\mathrm{v}}} \int_0^\infty \frac{k^*(E, J)\, N(E)\, \exp[-E/(kT)]\, \mathrm{d}E}{\omega + k^*(E, J)} \qquad (7.23)$$

Since $k^*(E, J)$ vanishes below threshold, the effective lower limit of the integral for $k_{\mathrm{uni}}^{\mathrm{sc}}(J)$ is the threshold E_0^* which is different for different J.

In a thermal reaction there will be a distribution of J's, so $k_{\mathrm{uni}}^{\mathrm{sc}}(J)$ has to be obtained by repeating the calculation at several J's, followed by averaging over the thermal rotational distribution $P(J)$ of the reactant molecule, given by

$$P(J) = Q_{\mathrm{r}}^{-1}(2J + 1)\exp[-E_{\mathrm{r}}(J)/(kT)] \qquad (7.24)$$

where $E_r(J) = J(J+1)B_e$, with $B_e = \hbar/(2\mu r_e^2)$ the equilibrium rotational constant of the reactant, and $Q_r = \ell T/B_e$ the partition function for the two-dimensional J-rotor. The actual thermal unimolecular rate constant is then the J-averaged $k_{uni}^{sc}(J)$:

$$k_{uni}^{sc} = \left\langle k_{uni}^{sc}(J) \right\rangle_J = \int_0^\infty k_{uni}^{sc}(J) P(J) \, dJ \qquad (7.25)$$

This is the theoretical counterpart of the experimental thermal rate constant. The procedure is the same as that sketched out previously in Section 5.8, and Eq. (7.25) is the same equation as Eq. (5.143) in slightly different notation, expressed in terms of angular momentum rather than rotational energy.

7.2.5 The low-pressure limit

At the low-pressure limit dissociation occurs as soon as the system reaches the critical energy (Section 5.3.6). In the present context this means that dissociation occurs at $E^* = 0$, at which point, at every J, the transition state is located at the corresponding r_{max}. In order to make this clear the critical energy is written $E_0^*(J, r_{max})$. Thus, from Eq. (7.23), we have

$$k_0(J) = \frac{\omega \exp[-E_0^*(J, r_{max})/(\ell T)]}{Q_v} \int_0^\infty N[E + E_0^*(J, r_{max})] \exp[-E/(\ell T)] \, dE$$

$$(7.26)$$

to be followed by averaging over $P(J)$ in order to obtain the thermal rate constant. Thus no minimization is involved, but r_{max} in the effective potential has to be determined at each J.

7.2.6 The high-pressure limit

At high pressures such that $\omega \gg k(E, J)$, the RHS of Eq. (7.23) becomes at specified r

$$k_\infty(J, r) = \frac{\exp[-E_0^*(J, r)/(\ell T)]}{h Q_M} \int_0^\infty G^*(E^*, J, r) \exp[-E^*/(\ell T)] \, dE^* \quad (7.27)$$

where $E_0^*(J, r)$ is written for E_0^* to emphasize its J- and r-dependence. Now

$$\int_0^\infty G^*(E^*, J, r) \exp[-E^*/(\ell T)] \, dE^* = \ell T \, Q_J^*(r) \qquad (7.28)$$

where $Q_J^*(r)$ is the interpolated partition function (Eq. (7.8)) for the transition state at the particular J. Using Eqs. (7.27) and (7.28) $k_\infty(J)$ can be approximated by

$$k_\infty(J) \cong \frac{\mathit{k}T}{h} \frac{(\exp[-E_0^*(J, r)/(\mathit{k}T)] \, Q_J^*(r))_{\min}}{Q_M} \tag{7.29}$$

which requires only the less computationally intensive minimization of $\exp[-E_0^*(J, r)/(\mathit{k}T)] \, Q_J^*(r)$ with respect to r at every J, rather than at every E *and* J. What remains is the averaging of Eq. (7.29) over $P(J)$ (Eq. (7.25)), which is trivial. This short-cut (sometimes referred to as the *canonical* variational result) yields values that are generally about 10%–20% higher than the more accurate Eq. (7.23) with $\omega \to \infty$.

The theoretical rate constant k_r for the *recombination* process $R_1 + R_2 \to M$ is obtained via the equilibrium constant K_c from the rate constant for *dissociation*; in particular, we have at high pressure

$$k_{r,\infty} = \langle k_{d,\infty}(J) \rangle_J \, K_c \tag{7.30}$$

where K_c is given by Eq. (5.49) and $k_{d,\infty}$ is the rate constant for dissociation considered above in Eq. (7.29), with the additional subscript "d" for clarity.

The expressions for the two limiting rate constants may be used to obtain the approximate variational counterparts of the centrifugal correction factors introduced in Section 5.8. Thus, for the low- and high-pressure factors f_0 and f_∞, we have from Eqs. (7.26) and (7.30), respectively,

$$f_0 \cong \frac{\langle k_0(J) \rangle_J}{k_0(J = 0)}, \qquad f_\infty \cong \frac{\langle k_\infty(J) \rangle_J}{k_\infty(J = 0)} \tag{7.31}$$

In general, for the same potential, the above f_0 is quite comparable to the one derived previously, but the above f_∞ usually exhibits a more pronounced temperature coefficient.

7.2.7 The adjustable parameter

Except for the adjustable parameter in the switching function, the entire calculation of the rate constant for thermal dissociation or recombination by Eq. (7.25) or (7.26) requires as its only input the molecular properties of reactant and products, the enthalpy of reaction ΔH_0, and the equilibrium constant in the case of recombination. This input represents static properties of the system that do not contain any dynamical information.

The dynamics of the system is contained in the parameter of the switching function, for the determination of which the above inputs alone are insufficient.

Therefore this parameter cannot be calculated from the present theory but has to be adjusted to fit experiment.

The adjustment involves finding a *unique c* for a given reaction that will produce a fit of the calculated thermal rate constant to its experimental counterparts at several pressures (the fall-off) and temperatures. A useful reference is the value, or at least an estimate, of the experimental limiting high-pressure rate constant at one reference temperature, either for dissociation or for recombination. A preliminary calculation using Eq. (7.29) rapidly narrows down the choice to a value of c that returns roughly the requisite magnitude and temperature coefficient of the thermal rate constant. This preliminary choice of c is then confirmed by calculation of the actual fall-off by the more elaborate Eq. (7.23). In general, fixing the value of c at one reference temperature reproduces fall-off fairly successfully over a range of temperatures.

It should be noted that treating transitional modes as hindered rotors does not do away with the problem of the switching-function parameter for the r-dependence of the hindering potential.

It will readily be appreciated that, given the numerous approximations made in the course of the derivation of the theory, it cannot rival more elaborate and computationally considerably more intensive theories, which, however, always include an adjustable constant. As a trade-off, the present theory offers, rapidly and rather transparently, results that are semi-quantitatively, and in many instances quantitatively, correct, regardless of the complexity of the reactant. In particular, it reproduces the negative temperature coefficient of most thermal rate constants for recombination of radicals [6, 11]. It is known under the acronym MVIPF (microcanonical variational theory by inversion of the partition function) [12].

7.2.8 Applications

Table 7.2 compares experimental data with results obtained by the present theory for the recombination of halogen-substituted methyl radicals at high pressure (the variational routine is actually applied to dissociation, from which the rate of recombination is obtained through the equilibrium constant – cf. Eq. (7.30)). The calculations correctly produce the negative temperature coefficient of $k_{r,\infty}$, which cannot be obtained by standard RRKM theory. The numerical values of this coefficient in Table 7.2 are similar to observed values, within experimental uncertainty.

Table 7.3 lists experimental data for recombinations of the type $CX_3 + O_2$, and Table 7.4 gives calculated results for the same reactions [11]. Since O_2 behaves like a radical, we find the expected negative temperature coefficients, as in other

Table 7.2. *A comparison of calculated and experimental rate constants* $k_{r,\infty}$ *for high-pressure self-combination of radicals* $k_{r,\infty}$ cm^3 $molec^{-1}$ $s^{-1} \cong a(T/298)^b$

	Theory			Experiment		
Radical	c	a	b	a	b	Ref.
CH_3	0.204	6.2	−0.34	4.4	−0.64	(a)
CH_2Cl	0.200	2.9	−0.72	2.8	−0.85	(b)
$CHCl_2$	0.195	1.0	−1.31	0.93	−0.74	(b)
CCl_3	0.1725	0.4	−1.48	0.33	−1.0	(c)

c is the parameter in the switching function $S(r) = \exp[-c(r - r_e)^2]$. a is in units of 10^{-11} cm^3 $molec^{-1}$ s^{-1}. The temperature range considered is 253–363 K.
(a) W. Tsang, *Comb. Flame* **78**, 71 (1989). However, I. R. Slagle, D. Gutman, J. W. Davies and M. J. Pilling, *J. Phys. Chem.* **92**, 4938 (1988) give $a = 6.0$ and $b = -0.4$.
(b) P. B. Roussel, P. D. Lightfoot, F. Caralp, V. Catoire, R. Lesclaux and W. Forst, *J. Chem. Soc. Faraday Trans.* **87**, 2367 (1991).
(c) F. Danis, F. Caralp, B. Veyret, H. Loirat and R. Lesclaux, *Int. J. Chem. Kinet.* **21**, 715 (1989).

Table 7.3. *A summary of experimental rate constants* $k_r = a(T/298)^b$ *for recombinations of the type* $CX_3 + O_2$

		$k_{r,\infty}$ (cm^3 $molec^{-1}$ s^{-1})		$k_{r,0}$ (cm^6 $molec^{-2}$ s^{-1})	
Reactants	Ref.	a	b	a	b
$CH_3 + O_2$	(a)	$(1.2 \pm 0.2) \times 10^{-12}$	$+1.2 \pm 0.04$	$(1 \pm 0.3) \times 10^{-30}$	-3.3 ± 0.4
$CF_3 + O_2$	(b)			$(1.9 \pm 0.2) \times 10^{-29}$	-4.7 ± 0.5
$CFCl_2 + O_2$	(b)	$\approx (9 \pm 3) \times 10^{-12}$	0 ± 1	$(5.5 + 1) \times 10^{-30}$	$-6 + 0.5$
$CCl_3 + O_2$	(c)	$(3.2 \pm 0.7) \times 10^{-12}$	-1.2 ± 0.4	7×10^{-30}	-4.3

(a) M. Keiffer, M. J. Pilling and M. J. C. Smith, *J. Phys. Chem.* **91**, 6028 (1987).
(b) F. Danis, Ph.D. Thesis, Université Bordeaux I (1990).
(c) F. Danis, F. Caralp, M. T. Rayez and R. Lesclaux, *J. Phys. Chem.* **95**, 7300 (1991).

radical–radical reactions, except for $CH_3 + O_2$. This is a rather interesting case, in that the present theory also yields the unexpected result for $CH_3 + O_2$.

On analysis it turns out the reason lies in the zero-point-energy-corrected potential (cf. Eq. (7.15)): even if the "bare" potential (in the absence of angular momentum) for $CH_3O_2 \rightarrow CH_3 + O_2$ is monotonically increasing with distance without a maximum (e.g. $V(r)$ in Fig. 7.1), as in a typical Type-1 reaction, there appears a small barrier in the zero-point-energy-corrected potential owing to interplay

Table 7.4. *Calculated rate constants* $k_{r,\infty}$, $k_{r,0} = a(T/298)^b$ *for recombinations of the type* $CX_3 + O_2$

Reactants	c	$k_{r,\infty}$ (cm^3 molec^{-1} s^{-1})		$k_{r,0}^*$ (cm^6 molec^{-2} s^{-1})	
		a	b	a	b
$CH_3 + O_3$	0.195	1.23×10^{-12}	$+1.19$	3.02×10^{-31}	-2.68
$CF_3 + O_2$	0.23	6.34×10^{-12}	-0.27	1.55×10^{-29}	-4.78
$CFCl_2 + O_2$	0.40	7.01×10^{-12}	-0.77	6.16×10^{-30}	-5.61
$CCl_3 + O_2$	0.535	2.95×10^{-12}	-0.63	1.78×10^{-30}	-6.41

c is the parameter in the switching function $S(r) = \exp[-c(r - r_e)^2]$. The temperature range considered is that shown in Table 7.2.
$k_{r,0}^*$ is the limiting low-pressure *weak-collision* rate constant with β_c factored in. The collision partner is N_2 throughout, except for CH_3O_2, for which it is Ar.

between the high C—H frequencies of the CH_3 moiety in reactant and products, mediated by the switching function.

Some possible refinements of the variational routine are discussed in Section 9.2.

References

[1] D. L. Bunker and M. Pattengill, *J. Chem. Phys.* **48**, 772 (1968).
[2] J. Horiuti, *Bull. Chem. Soc. Japan* **13**, 210 (1938).
[3] W. H. Wong and R. A. Marcus, *J. Chem. Phys.* **55**, 5625 (1971).
[4] S. C. King, J. F. Leblanc and P. D. Pacey, *Chem. Phys.* **123**, 329 (1988).
[5] D. M. Wardlaw and R. A. Marcus, *J. Phys. Chem.* **90**, 5383 (1986); *Adv. Chem. Phys.* **70**, 231 (1987).
[6] W. Forst, *J. Phys. Chem.* **95**, 3612 (1991).
[7] W. Forst and F. Caralp, *J. Phys. Chem.* **96**, 6291 (1992).
[8] L. Pauling, *The Nature of the Chemical Bond*, 3rd Ed. (Cornell University Press, Ithaca, 1960).
[9] D. R. Herschbach and V. W. Laurie, *J. Chem. Phys.* **35**, 458 (1961), Table II.
[10] W. J. Chesnavich, L. Bass, T. Su and M. T. Bowers, *J. Chem. Phys.* **74**, 2228 (1981).
[11] W. Forst and F. Caralp, *J. Chem. Soc. Faraday Trans.* **87**, 2307 (1991).
[12] W. Forst, *QCPE Bull.* **13**, 21 (1993). Program INTERVAR, # QCMP 121.

Problems

Problem 7.1. Justify the use of a two-dimensional hindered rotor in the fragmentation $CH_4 \rightarrow CH_3 + H$ by writing the reactant–product correlation.

Problem 7.2. Show that a convenient way to evaluate the interpolated partition function $Q^*(r)$ at $z = \theta$ is by defining for the molecule

$$L_m = (\ln q_{r,m}) - \sum_i \ln(1 - \theta^{\omega_i})$$

and for the products

$$L_p = \ell_n \, q_{r,p} + \ell_n \left(\frac{q_{r,t}}{(\ell_n \, \theta^{-1})^{r_t/2}} \right) - \sum_j \ell_n (1 - \theta^{\omega_j})$$

Then

$$Q^*(r) = \frac{\exp\{L_m S(r) + L_p[1 - S(r)]\}}{(\ell_n \, \theta^{-1})^{r_a/2}}$$

since $(\ell_n \, \theta^{-1})^{r_a/2}$ is a common factor.

Problem 7.3. Show that the second derivative of $\phi(z)$ is

$$\phi''(z) = S(r) \sum_i \frac{\omega_i^2 z^{\omega_i - 1}}{(1 - z^{\omega_i})^2} + \frac{1 + \frac{1}{2} r_a}{z (\ell_n \, z^{-1})^2}$$

$$+ \left(\sum_j \frac{\omega_j^2 z^{\omega_j - 1}}{(1 - z^{\omega_j})^2} + \frac{\frac{1}{2} r_t}{z (\ell_n \, z^{-1})^2} \right) [1 - S(r)]$$

8

Unimolecular decomposition under a central potential

A central potential, written symbolically as $V(r)$, is an isotropic potential that depends only on the radial coordinate r. Within the context of unimolecular dissociation, $V(r)$ is implicitly the vibrational potential and r is the distance along a one-dimensional reaction coordinate. In a Type-1 bond-fission reaction as represented below in Eq. (8.1) (cf. Section 3.3), $V(r)$ would represent the potential of the R_1–R_2 bond in the molecule M about to break up into fragments $R_1 + R_2$ as r increases from its equilibrium value r_e to infinity:

$$M \rightarrow R_1 + R_2 \qquad (8.1)$$

If M is a polyatomic molecule, at least one of the fragments will be polyatomic; the other may be an atom or a molecule, polyatomic or diatomic.

In the "traditional" definition, the transition state is the configuration at the extremum of the relevant potential. If angular momentum in the simple bond-fission process represented by Eq. (8.1) is ignored, the transition state is located at infinity (and is therefore in effect undefined), since the long-range part of the vibrational potential $V(r)$ is monotonically increasing (Fig. 8.1) and thus has no defined extremum.

8.1 Unimolecular decomposition from microscopic reversibility

8.1.1 A "loose" transition state

If M has angular momentum, which in principle is the case at any temperature above absolute zero, the potential is modified and can be approximated by an *effective* potential $V_{eff}(r)$ (see Fig. 8.1 and Eq. (8.13) below) with an extremum due to a centrifugal barrier, which thus serves to define the position of the transition state. While this potential is discussed in Section 3.4 and more fully in Section 8.2 below, the important aspect in the present context is that the barrier is at relatively

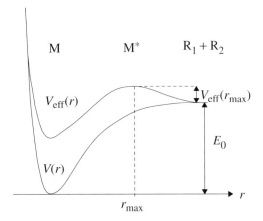

Fig. 8.1. A schematic representation of a section through the potential-energy surface in the simple bond-fission reaction $M \rightarrow R_1 + R_2$ as a function of r, the R_1–R_2 distance, considered to be the reaction coordinate. The lower curve is the purely vibrational potential $V(r)$ which has no definable maximum since it increases monotonically from potential minimum onwards. For finite angular momentum of the reactant the potential is the effective potential $V_{\text{eff}}(r)$ (upper curve) with a definable maximum, $V_{\text{eff}}(r_{\text{max}})$, located at $r = r_{\text{max}}$, which determines the position of the transition state M^*. In this schematic representation zero-point energies are ignored and r_{max} is in reality at a much larger distance except at very large angular momentum.

large interfragment distances (see, for instance, Fig. A4.1 in Appendix 4), so the transition state is "loose," i.e. it resembles the separated fragments more than it does the original bound molecule.

It then becomes more convenient and simpler to invoke microscopic reversibility and consider the reverse process of association of fragments rather than the forward bond fission of the reactant [1]. This opens the way for an alternative approach to the rate constant of the fragmentation (see Section 8.9), and also to the distribution of energy between the fragments (Section 8.10). The difference from the forward process, of course, is that, in the reverse association, there has to be a collision between the fragments, subject to various constraints that will be discussed in due course.

It was shown previously (Section 3.2) that the forward flux (in units of energy^{-1} s^{-1}) through the critical surface at the transition state is equal to $G^*(E - E_0)/h$, referred to the ground state of the *reactant* as the zero of energy. For the present purpose it is more convenient to write the forward flux in the form $G^*(E)/h$ with energy zero at the zero-point energy of the *transition state*. In its simplest form [2, p. 69], assuming that the fragments are point masses, i.e. considering that fragments possess only relative translational energy E_t, the reverse flux through the critical surface from the reaction volume V at specified total energy E can be written in an analogous fashion as a sum over relative translational energies

E_t. Thus, from microscopic reversibility, the forward flux $G^*(E)/h$ is equal to the reverse flux given by

$$\sum_{E_t} \begin{bmatrix} \text{No. of internal} \\ \text{states of separated} \\ \text{fragments at } E - E_t \end{bmatrix} \times \begin{bmatrix} \text{reverse flux from } V \\ \text{per translational} \\ \text{state} \end{bmatrix} \times \begin{bmatrix} \text{density of} \\ \text{translational} \\ \text{states in } V \end{bmatrix} \quad (8.2)$$

On replacing the above summation with integration over E_t, the sum of states $G^*(E)$ in terms of the reverse flux becomes (with square brackets indicating corresponding elements in Eqs. (8.2) and (8.3))

$$G^*(E) = h \int_0^E \left[dE_t\, N_v^f(E - E_t) \right] \left[\sigma(E_t)(v/V) \right] \left[N_t(E_t) \right] \quad (8.3)$$

so the unimolecular rate constant for the forward process remains *formally* the same as before, i.e. $k(E) = G^*(E)/[h N(E + E_0)]$, except that E is now the total energy in the fragments and $E_0 \equiv \Delta H_0$, the enthalpy of the fragmentation reaction (Eq. (8.1)). The actual rate constant based on Eq. (8.3) is discussed more fully in Section 8.9, for the present we need only identify the various factors. $N_v^f(E - E_t)$ is the density of internal (in principle vibrational, plus any internal rotational) states of the fragments (hence the superscript "f") at $E - E_t$, so $N_v^f(E - E_t)\,dE_t$ is the number of such states within dE_t at $E - E_t$; $\sigma(E_t)$ is the cross-section for the association of fragments at E_t (in units of area), v is the relative velocity of the fragments and $N_t(E_t)$ is the density of translational states in the (three-dimensional) volume V (cf. Appendix 1, Section A1.3).

With μ being the reduced mass of the fragments, the explicit form of the last two factors on the RHS of Eq. (8.3) is

$$v = \left(\frac{2E_t}{\mu} \right)^{1/2}, \qquad N_t(E_t) = \frac{2^{5/2}\pi\mu^{3/2}V}{h^3} E_t^{1/2} \quad (8.4)$$

so we can write

$$h\,\sigma(E_t)\,(v/V)\,N_t(E_t) = \frac{\sigma(E_t)}{\pi\lambda^2} \quad (8.5)$$

where λ^2 is the square of the de Broglie wavelength $\lambda = h/\sqrt{2\mu E_t}$. The collision kinematics of the process is thus contained in the dimensionless ratio $\sigma(E_t)/(\pi\lambda^2)$ which refers to what may be called the "external" degrees of freedom of the fragment system (subscript "x"). The internal degrees of freedom of the fragments (vibrational and internal rotational, if any) are taken care of in Eq. (8.3) by the factor $N_v^f(E - E_t)$ (subscript "v"). To emphasize that $G^*(E)$ is now defined by Eq. (8.3) it will henceforth be written $G^*_{vx}(E)$.

8.1.2 The cross-section versus the sum of states

In a statistical theory, like the one discussed here, rates are more conveniently defined in terms of the number or sum of states rather than cross-sections. What is needed, therefore, is the connection between $\sigma(E_t)/(\pi \lambda^2)$ and the sum of states of the "external" degrees of freedom, which can be established by the following elementary argument (adapted from [3]). It takes a particularly simple form since, on the level of the approximation embodied in Eq. (8.5), the $R_1 + R_2$ collision partners are treated as structureless point masses, i.e. atoms.

The thermal rate constant $k(T)$ for a collision between two structureless particles at temperature T can be shown to be the Laplace transform of the product $E_t \sigma(E_t)$ with $s = 1/(kT)$ as the transform parameter [4]:

$$k(T) = \frac{(2s)^{3/2}}{(\pi \mu)^{1/2}} \mathcal{L}\{E_t \sigma(E_t)\} \tag{8.6}$$

From transition-state theory the same rate constant can be represented in the usual way as kT/h (i.e. $1/(hs)$) times the ratio of partition functions for the "transition state" and the "reactant" (with above energy zero):

$$k(T) = \frac{1}{hs} \frac{Q_x(s)}{Q_{tr}(s)} \tag{8.7}$$

where $Q_x(s)$ is the partition function for the "transition state," in this case comprising only the degrees of freedom involved in the kinematics of the $R_1 \cdots R_2$ association, which we have called "external" degrees of freedom (hence the subscript "x"), and $Q_{tr}(s) = [(2\pi \mu)/(sh^2)]^{3/2}$ is the partition function for translation in three dimensions, that is, for the degrees of freedom of the "reactant." By equating the two expressions for $k(T)$ and taking the inverse transform of $Q_x(s)/s$ we obtain

$$E_t \sigma(E_t) = \frac{h^2}{8\pi \mu} \mathcal{L}^{-1}\left\{ \frac{Q_x(s)}{s} \right\} \tag{8.8}$$

We have seen previously (Section 4.1) that the inverse transform of a partition function divided by the transform parameter is the sum of states, so, using the definition of the de Broglie wavelength, Eq. (8.8) can be rearranged to

$$G_x^*(E_t) = \frac{\sigma(E_t)}{\pi \lambda^2} \tag{8.9}$$

from which it is apparent that $\sigma(E_t)/(\pi \lambda^2)$ represents simply the sum of "external" states of the fragments.

In a more realistic case, when one or both particles are molecules with a given symmetry, the "external" states will comprise also degrees of freedom of overall rotation of the fragments, with rotational energy E_r, so E_t in Eq. (8.9) is to be replaced by $y = E_t + E_r$, the total "external" energy of the fragments (see also Eq. (8.22) below). Furthermore, on account of the rotational degrees of freedom, the sum of states for the transition state will acquire a dependence on the total angular momentum J (more on this in Section 8.3), so that Eq. (8.3) has to be rewritten to show the y- and J-dependences explicitly:

$$G^*_{vx}(E, J) = \int_{y_m(J)}^{E} N^f_v(E - y)\, G_x(y, J)\, dy \qquad (8.10)$$

where the lower integration limit anticipates the future result (Eq. (8.53)) that there will be a J-dependent minimum y_m necessary to generate the specified angular momentum J. A trivial change of variables converts the RHS into an integral with limits 0 and $E - y_m(J)$, in which form it will be recognized as standard convolution, at effective energy $E - y_m(J)$, between the density of internal states of the fragments and their J-dependent sum of "external" (rotational–translational) states. The implication is that the unimolecular rate constant also becomes J-dependent, which suggests that proper treatment of the fragmentation of the type shown in Eq. (8.1) should include the effect of angular momentum.

We can also define a *density* of "external" states $N_x(y, J)$ in the usual way, i.e. as the derivative of $G_x(y, J)$ with respect to y at constant J:

$$N_x(y, J) = \frac{\partial}{\partial y} G_x(y, J) \qquad (8.11)$$

Given the rules of convolution, the integral in Eq. (8.10) can then be written equivalently as

$$G^*_{vx}(E, J) = \int_{y_m(J)}^{E} G^f_v(E - y)\, N_x(y, J)\, dy \qquad (8.12)$$

where $G^f_v(E - y)$ is now the *sum* of internal states of the fragments. This alternative version is sometimes useful for practical computations, e.g. exact counting of vibrational states of fragments.

Equations (8.10) and (8.12) are valid for a system with a "loose" transition state, i.e. one that obeys what might be termed "statistics after dissociation," with a sum of states given entirely in terms of the properties of separated fragments. In principle the integral explores the entire phase space of the fragments, i.e. weights equally every possible combination of fragment states, subject to constraints to be specified.

While Eqs. (8.10) and (8.12) look deceptively simple, the important (and difficult) part is the actual evaluation of the properly constrained sum of "external" states $G_x(y, J)$ or the density $N_x(y, J)$. If the only constraints imposed are the constants of motion, i.e. conservation of energy and angular momentum, this is usually referred to as the "phase-space theory." In the following we shall impose an additional constraint that arises from the nature of the interaction potential between fragments.

Inasmuch as the sum of states $G_x(y, J)$ is easier to deal with conceptually than the density $N_x(y, J)$, it will be the principal subject of this chapter.

8.2 The potential

8.2.1 The effective potential

We will consider at the outset the general case when at least one of the fragments is polyatomic so that the association of fragments will involve both translational and rotational motions. A simple way to take account of the effect of fragment rotations under a central potential is to define an *effective* potential $V_{eff}(r)$ in terms of the rotational potential $L(L + 1)\hbar^2/(2\mu r^2)$:

$$V_{eff}(r) = V(r) + \frac{L(L + 1)\hbar^2}{2\mu r^2} \tag{8.13}$$

where L is the orbital angular momentum of the fragments about their common center of gravity, μ is their reduced mass, and r is the radial coordinate representing the R_1–R_2 distance. This is the same effective potential as that introduced previously in Eq. (3.30) to which Eq. (8.13) reduces for $L \cong J$ (see Section 8.3 below). Unlike $V(r)$, the effective potential is no longer a monotonically increasing function of r but, depending on the value of L, exhibits a centrifugal barrier, i.e. a more or less pronounced maximum (Fig. 8.1).

A more general effective potential is introduced in Chapter 9, which also serves to examine the limits of validity of Eq. (8.13).

8.2.2 Orbital angular momentum

Since the centrifugal barrier in the effective potential is far out along r (at least when the orbital angular momentum is not too large), only the attractive part of the interfragment potential intervenes and needs to be considered. The ensuing calculations are considerably simplified if this attractive part is approximated in the form $-C_n/r^n$, where C_n is a constant of dimension energy $\times$ (distance)^n.

As a further but minor simplification, the rotational-potential part of $V_{eff}(r)$ in Eq. (8.13) will be written semi-classically in terms of L^2 rather than the quantum-mechanical $L(L + 1)$:

$$V_{eff}(r) = -\frac{C_n}{r^n} + \frac{L^2\hbar^2}{2\mu r^2} \tag{8.14}$$

The zero of this potential is at $r \to \infty$. It will have a maximum at r_{max} given by $\partial V_{eff}(r)/\partial r = 0$, i.e.

$$r_{max} = \left(\frac{n\mu C_n}{L^2 \hbar^2}\right)^{1/(n-2)} \tag{8.15}$$

Thus r_{max} shifts to lower values of r with increasing L; however, clearly there will be no maximum in r unless $n > 2$. The potential at r_{max} is

$$V_{eff}(r_{max}) = \frac{n-2}{2}\left(\frac{L^2 \hbar^2}{n\mu C_n^{2/n}}\right)^{n/(n-2)} \tag{8.16}$$

so the potential at r_{max} (i.e. the centrifugal barrier) *increases* with L. (These formulas can be "re-quantized" by replacing L^2 with $L(L+1)$.)

For the association of fragments to take place, the relative translational energy of fragments E_t must be sufficient to clear the centrifugal barrier. We must therefore have $E_t \geq V_{eff}(r_{max})$, and this in turn determines the limit on the corresponding orbital-angular-momentum quantum number L (see also Problem 8.1). Thus, for specified E_t, we have from Eq. (8.16)

$$L \leq b_n E_t^{(n-2)/(2n)} \tag{8.17}$$

where b_n is a constant (of dimension energy$^{(2-n)/(2n)}$) that depends only on the potential and fragment masses:

$$b_n = \frac{1}{\hbar}\left(\frac{2n\mu}{n-2}\right)^{1/2}\left(\frac{(n-2)C_n}{2}\right)^{1/n} \tag{8.18}$$

The implication of Eq. (8.17) is that the association of fragments with specified E_t can take place only if the orbital angular momentum L does not exceed L given by the equality sign in Eq. (8.17). Consequently, in terms of b_n, we can also write $V_{eff}(r_{max}) = E_t = (L/b_n)^{2n/(n-2)}$.

The requirement that associating fragments must clear the centrifugal barrier is tantamount to locating the "transition state" at the extremum of the effective potential. For this reason this version of phase-space theory is sometimes referred to as the orbiting-transition-state (OTS) phase-space theory [5, Section 7.4].

8.2.3 Potential parameters

Both the value of C_n and the power n will depend on the nature of R_1 and R_2. If both R_1 and R_2 are non-polar closed-shell structures (i.e. stable molecules), the long-range attraction will be governed mainly by the energy of dispersion, for which $n = 6$. If one is a polarizable neutral and the other an ion, the long-range attraction is governed principally by the energy of induction, for which $n = 4$ [2, p. 133ff; 6,

Table 8.1. *Numerical values of the potential constant C_n*

System ($n = 6$)	ΔH_0 (cm^{-1})	r_e (nm)	C_6 (10^{-59} erg cm^6)
$C_2H_5 + H$	34 280	1.094	2.34
$CH_3 + H$	36 072	1.094	2.46
$NO + O$	25 131	1.197	2.94
$CH_3 + CH_3$	30 642	1.53	15.62
$Br_2 + O$	15 890	2.28	88.6

10^{-59} erg cm^6 = 50 340 cm^{-1} nm^6.

System ($n = 4$)	α (nm^3)	C_4 (10^{-43} erg cm^4)
$C_4H_7^+ + H$	0.667	0.77
$NH_4^+ + OH$	1.24	1.43
$C_3H_5^+ + CH_3$	2.2	2.54
$C_2H_4^+ + C_2H_4$	4.22	4.87

10^{-43} erg cm^4 = 50 340 cm^{-1} nm^4.

Chapter 13]. For $n = 6$ we then have from Eq. (8.18) $b_6 = (54\mu^3 C_6)^{1/6}/h$ and $E_t = (L/b_6)^3$; similarly, for $n = 4$, we have $b_4 = (16\mu^2 C_4)^{1/4}/h$ and $E_t = (L/b_4)^4$.

The constant C_6 can be taken as the long-range part of the Lennard-Jones (L-J) potential (Appendix 4), in which case $C_6 \approx 4\epsilon\sigma^6$, where ϵ is the depth of the L-J potential well and σ the collision diameter, obtained from viscosity or second-virial-coefficient data for the reactant. It may be preferable to refer C_6 to the actual reaction threshold ($\epsilon \approx E_0$) and to the equilibrium interfragment distance $r_e = 2^{1/6}\sigma$ (cf. Appendix 4), so that $C_6 \approx 2E_0 r_e^6$, where $E_0 \equiv \Delta H_0$ is the 0-K enthalpy of the fragmentation reaction. While the different methods give somewhat different numerical results, fortunately the ultimate central-potential results are only very weak functions of C_6 since the latter enters mostly to the power $\frac{1}{6}$.

For ion–molecule reactions the potential constant is generally taken as $C_4 = \frac{1}{2}\alpha q^2$, where α is the polarizability of the neutral molecule and q is the ionic charge in multiples of the elementary electronic charge (4.803×10^{-10} e.s.u.). In most formulas C_4 enters to the power $\frac{1}{4}$, so again ultimate results are weak functions of the numerical value of C_4. Numerical values of C_6 and C_4 for a few typical cases are given in Table 8.1.

In a true fragmentation reaction, as represented by Eq. (8.1), the collision partners R_1 and R_2 are in principle open-shell structures, i.e. radicals or atoms, for which the $1/r^6$ potential may be a poor representation of the actual interfragment potential at large distances. In such cases appropriate C_n and n can be obtained from a fit of the attractive part of a more representative potential, e.g. a Morse function, in which case it generally turns out that $n < 6$ (e.g. Table 1 in [7]). Such n is likely to be non-integral, so some results discussed below are no longer obtainable analytically.

8.3 Angular momentum

Since the fragmentation process of Eq. (8.1) will be dealt with in terms of the reverse association of fragments, i.e. the R_1–R_2 collision, this has an implication for the association of fragments: in addition to the conservation of the constants of motion, the relative translational energy E_t of the fragments must be sufficiently high for them to clear the centrifugal barrier. These two constraints will now be considered.

8.3.1 Conservation of angular momentum

There are two constants of motion: total energy and total angular momentum. If J_1 and J_2 are, respectively, the intrinsic angular momenta of the fragments R_1 and R_2 about their axis of symmetry, the condition of conservation of angular momentum is given by the *vector* sum

$$J = L + J_1 + J_2 \qquad (8.19)$$

where J is the angular momentum of the reactant M, and L is the orbital angular momentum of the fragments about their common center of gravity. (We assume implicitly that angular momenta are expressed in units of $\hbar$, so that the symbols J, L, J_1, and J_2 represent also angular-momentum quantum *numbers*.)

8.3.2 Quasi-diatomic reactant

The simplest case arises if (a polyatomic) M with angular momentum J is approximated as a quasi-diatomic molecule $M \equiv R_1$–R_2; then the fragments R_1 and R_2 are considered to be "atoms," i.e. structureless particles with $J_1 = J_2 = 0$, so that Eq. (8.19) reduces to $J = L$, i.e. there is only one allowed value of L for every J. Since the fragments R_1 and R_2 are not actual atoms, in principle with $J_1 \neq J_2 \neq 0$, consideration of the quasi-diatomic case implies that $L \gg J_1 + J_2$, i.e. the rotational excitation of the fragments is assumed to be small, or, in other words, angular momentum in the products is predominantly orbital. This is a fairly good approximation in many, but not all, cases.

8.3.3 Compounding of angular momenta

In general, however, rotational excitation of polyatomic fragments need not be small, so the actual compounding of angular momenta indicated in Eq. (8.19) has

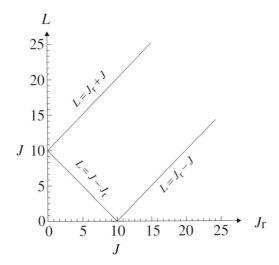

Fig. 8.2. A graphical representation of the angular-momentum sum $L + J_r = J$ for $J = 10$. The three-sided rectangle delimits the values of L and J_r that add vectorially to produce the specified J. Quantum mechanically these should be seen as discrete lattice points inside the rectangle.

to be performed. This may be executed as a two-step process. First J_1 and J_2 are added vectorially to form J_r which may be thought of as an intermediate rotational quantum number:

$$J_r = J_1 + J_2 \tag{8.20}$$

The prescribed J then follows by vector addition of J_r and L:

$$J_r + L = J \tag{8.21}$$

To illustrate the general principles, consider Fig. 8.2, which represents graphically the vector sum $L + J_r = J$. The conservation condition of Eq. (8.19) will reduce to this relatively simple case when one particle, say particle 2, is an atom: then $J_2 = 0$ and $J_r \equiv J_1$. The allowed values of L and J_r that add to specified J are contained inside the three-sided rectangle. Since we are implicitly dealing with discrete angular-momentum quantum numbers, the allowed values can be thought of as represented by discrete lattice points within the rectangle.

8.3.4 Conservation of "external" energy

If fragments have intrinsic angular momenta, i.e. rotational energies, there is an additional conservation condition that has to be satisfied, namely the conservation of the total "external" (translational–rotational) energy of the fragments. This is ensured by y defined as the sum of energies in translation (E_t) and rotation (E_r) of the fragments:

$$y = E_t + E_r \tag{8.22}$$

In terms of J_r, $E_r \geq B_r J_r^2$, where B_r is the effective rotational constant of the two fragments defined as $B_r = B_1 B_2/(B_1 + B_2)$ (see also Section 8.6 and Chapter 9). For specified y, the maximum possible value of J_r, called J_m, arises when all "external" energy is rotational: $y = B_r J_m^2$. For arbitrary $J_r \leq J_m$, writing $E_t = y - E_r$, Eq. (8.17) can therefore be written

$$hL = b_n \left(B_r J_m^2 - B_r J_r^2 \right)^{(n-2)/(2n)} \tag{8.23}$$

When all energy is translational, $y = E_t$, then from Eq. (8.17) the maximum of the orbital angular momentum is $L_m = b_n y^{(n-2)/(2n)}$, so from Eq. (8.23) we have [2, p. 142; 8]

$$\frac{L}{L_m} = \left[1 - \left(\frac{J_r}{J_m} \right)^2 \right]^{(n-2)/(2n)} \tag{8.24}$$

Thus the curve of L versus J_r further delimits the allowed L–J_r combinations that yield the specified J, as shown in Fig. 8.3, which represents $L_m > J_m$, the most common case.

Suppose that the collision is of the type $M + A$, where M is a molecule and A is an atom. The maximum L_m corresponds to the case when all energy y is in the relative translation of M and A, and J_m occurs when all energy y is in the rotation of M. From

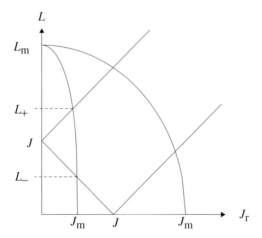

Fig. 8.3. Examples of the curve of L versus J_r (Eq. (8.24)) when $L_m > J_m$. Shown are the cases when $J_m < J$ and $J_m > J$. L_+ and L_- are the integration limits with respect to L in Eqs. (8.42) and (8.43), given by the intersection of the curves of L_m versus J_m with the $J_r \pm L$ lines which define the J-delimited rectangle. Depicted are the limits for the case when $J_m < J$; similar limits (not shown) would apply in the case $J_m > J$ (rightmost curve).

Eqs. (8.17) and (8.18) we see that L_m is proportional to $(\mu_{M-A})^{1/2}C_n^{1/n}y^{(n-2)/(2n)}$, and J_m is proportional to $(\mu_M y)^{1/2}$, so that

$$\frac{L_m}{J_m} \approx \left(\frac{\mu_{M-A}}{\mu_M}\right)^{1/2}\left(\frac{C_n}{y}\right)^{1/n} \tag{8.25}$$

Since $1/n$ is in general small (e.g. $n = 6$ for collisions of two neutral species), the important factor in Eq. (8.25) will be the ratio of the reduced masses. If the atom is very light, e.g. the hydrogen atom, then $\mu_{M-A} \approx 1$, and we will have $L_m < J_m$ only in this special case (e.g. KCl + H, [8, Fig. 11]), particularly when the energy y is high; in virtually all other instances $L_m > J_m$ (e.g. HCl + K, [8, Fig. 11]). A useful approximation is to represent these two opposite cases by the horizontal and vertical cut-offs, respectively, as shown in Fig. 8.4 later and used in Section 8.4 below.

The rotational energy E_r, defined here as the rotational energy of *fragments*, must be carefully distinguished from and not confused with E_r in Section 3.4, which was defined as the "external" rotational energy of the *reactant*. It is also necessary to distinguish between r as a variable signifying interfragment separation, and subscript "r" signifying that the variable in question refers to rotations.

8.3.5 Summations over J_r and L

Vector addition of specified J_r to L yields J satisfying

$$|J_r - L| \le J \le J_r + L \tag{8.26}$$

To each value of J there correspond $2J + 1$ values of m, the components of J about a laboratory (space-fixed) axis in field-free space. The total number of J-states arising from given J_r and L is then obtained by summing $2J + 1$ between the limits given by Eq. (8.26), which, from the standard formula for the sum of an arithmetic progression [9], is

$$\sum_{|J_r - L|}^{J_r + L} (2J + 1) = (2J_r + 1)(2L + 1) \tag{8.27}$$

The probability of specified J, properly normalized, is therefore

$$P(J) = \frac{2J + 1}{(2J_r + 1)(2L + 1)} \tag{8.28}$$

If particle 1 has structure, i.e. symmetry, there will be additional degeneracy $D(J_r)$ due to additional states arising from the components of J_r ($\equiv J_1$, for we are still assuming that particle 2 is an atom with $J_2 = 0$), e.g. $D(J_r) = 2J_r + 1$ for a linear/symmetric top and $D(J_r) = (2J_r + 1)^2$ for spherical-top particles. In addition, owing to the rotational symmetry of the particles, there will appear a symmetry number σ that was encountered previously in connection with rotational partition functions or the rotational sum or density of states (cf. Appendix 1 and Chapter 4).

If particle 2 is *not* an atom, then J_2 is not zero, so that J_r will be the result of the addition of J_1 and J_2 according to Eq. (8.20), with the consequence that $D(J_r)$ will have a more complicated form (see below Section 8.6). Hence for *specified* J_r we have, including now the symmetry numbers of both particles,

$$\sum_L (2L+1)P(J)D(J_r) = \frac{2J+1}{\sigma_1\sigma_2} \sum_L \frac{D(J_r)}{2J_r+1} \tag{8.29}$$

where the sum represents all L-states arising from the vector addition of *specified* J_r to L to produce the prescribed J. However, what we need for the determination of the sum of "external" states is the sum that includes *all* J_r that add to L to yield the specified J. If we denote this sum by $G_x(J_r, L \to J)$, we have quite generally

$$G_x(J_r, L \to J) = \frac{2J+1}{\sigma_1\sigma_2} \sum_{J_r} \sum_L \frac{D(J_r)}{2J_r+1} \tag{8.30}$$

Now the sum of "external" states, and therefore also the factor $2J+1$, appear in the expression for the transition state, Eq. (8.10). As shown in Section 4.5.7, the same factor appears in the density of states for the reactant which is in the denominator of the expression for the microcanonical rate constant, so that the factor $2J+1$ cancels out, except in the very special case that the reactant is a spherical top.

We shall therefore find it convenient to work with an "external" sum of states $\Gamma(y, J)$ that excludes the symmetry numbers and the degeneracy with respect to the space-fixed axis, that is, $\Gamma(y, J) = G_x(J_r, L \to J)$ multiplied by $\sigma_1\sigma_2/(2J+1)$. Let

$$\Gamma(J_r) = \frac{D(J_r)}{2J_r+1} \tag{8.31}$$

then

$$\Gamma(y, J) = \sum_{J_r=0}^{J_m} \sum_{L=0}^{L_m} \Gamma(J_r) \tag{8.32}$$

where J_m and L_m are the maximum values of J_r and L, respectively, and the summand now refers exclusively to J_r.

The double sum in Eq. (8.32) is denoted as a function of (y, J) since it is clear that the final result of the summation over L and J_r is a function both of the total angular momentum J and of the total translational–rotational energy y which determines both J_m and L_m (see the text following Eq. (8.22)).

With respect to the sum of "external" states $G_x(y, J)$ that appears in the convolution in Eq. (8.10) we have

$$G_x(y, J) = \frac{1}{\sigma_1 \sigma_2} \Gamma(y, J) \tag{8.33}$$

which is thus identified as the angular-momentum-conserved sum of "external" states, for fragments with symmetry numbers σ_1 and σ_2, at specified translational–rotational energy y and total angular momentum J, exclusive of the factor $2J + 1$, in the expectation that it will cancel out in the final expression for the microcanonical rate constant. The corresponding density of states $N_x(y, J)$ is then

$$N_x(y, J) = \frac{1}{\sigma_1 \sigma_2} \frac{\partial}{\partial y} \Gamma(y, J) \tag{8.34}$$

The principal subject of the following sections will be $\Gamma(y, J)$, i.e. the evaluation of the double sum over $\Gamma(J_r)$ in Eq. (8.32), or its semi-classical equivalent.

8.4 $\Gamma(y, J)$ for simple cases

Particularly simple cases arise when $L_m \gg J_m$, or $L_m \ll J_m$ (mentioned previously in connection with Eq. (8.25)), for then the curve of L_m versus J_m (Eq. (8.24)) reduces to virtually straight lines parallel to the L or J_r axes, respectively (Fig. 8.4), so the number of states can be determined exactly by analytical enumeration.

8.4.1 Linear + atom

In the case of fragments consisting of a linear particle + atom, $D(J_r) = 2J_r + 1$, so $\Gamma(J_r) = 1$ in Eq. (8.32), which means that each lattice point in the J_r–L space corresponds to just one state. The sum of states $\Gamma(y, J)$ is then given (quantum-mechanically) by the enumeration of individual lattice points (i.e. states) inside the area enclosed by the straight lines $L = |J \pm J_r|$ and the curve of L_m versus J_m of Eq. (8.24).

Consider first the special case $L_m \gg J_m$ which corresponds to a vertical cut-off on J_r as shown in Fig. 8.4(a). The number of L-states for every J_r is $2J_r + 1$, so for $J_m \leq J$ the sum of states from Eq. (8.32) is [9]

$$\sum_{J_r=0}^{J_m} (2J_r + 1) = 1 + 3 + 5 + \cdots + (2J_m + 1) = (J_m + 1)^2 \tag{8.35}$$

Semi-classically, this summation can be approximated by the integral from 0 to J_m of $\int 2J_r \, dJ_r = J_m^2$, which is the "area" inside the triangle in Fig. 8.4(a). If $J_m \geq J$, the sum of states is given up to J by Eq. (8.35), in which J_m is replaced by J, plus the "area" or number of L-states from J to J_m, which is $(2J + 1)(J_m - J)$ (or,

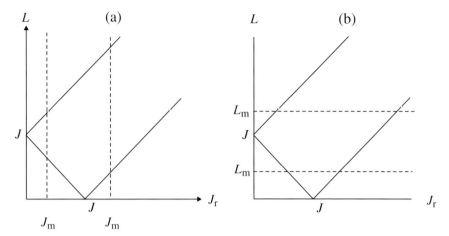

Fig. 8.4. (a) A schematic representation of the condition when $L_{\mathrm{m}} \gg J_{\mathrm{m}}$ ("vertical cut-off"). Allowed values of L are comprised between $|J - J_{\mathrm{r}}|$ and $J + J_{\mathrm{r}}$, for a total of $2J_{\mathrm{r}} + 1$ values of L at every $J_{\mathrm{r}} \leq J$ (leftmost dashed line), and $2J + 1$ values of L at every $J_{\mathrm{r}} \geq J$ (rightmost dashed line). (b) A schematic representation of the condition where $J_{\mathrm{m}} \gg L_{\mathrm{m}}$ ("horizontal cut-off"). The number of values of J_{r} is $2L + 1$ for every $L_{\mathrm{m}} < J$ (bottom dashed line), and $2J + 1$ for every $L_{\mathrm{m}} > J$ (top dashed line).

semi-classically, $2J \int \mathrm{d}J$, integrated from J to J_{m}), so the final result for $L_{\mathrm{m}} \gg J_{\mathrm{m}}$ (vertical cut-off) is

$$\Gamma(y, J) = \begin{cases} (J_{\mathrm{m}} + 1)^2, & J_{\mathrm{m}} \leq J \\ (J + 1)^2 + (2J + 1)(J_{\mathrm{m}} - J), & J_{\mathrm{m}} \geq J \end{cases} \tag{8.36}$$

Consider now the opposite case $L_{\mathrm{m}} \ll J_{\mathrm{m}}$ which represents a horizontal cut-off on L, as shown in Fig. 8.4(b). This merely interchanges L_{m} and J_{m} in Eq. (8.36), so we have directly for $L_{\mathrm{m}} \ll J_{\mathrm{m}}$

$$\Gamma(y, J) = \begin{cases} (L_{\mathrm{m}} + 1)^2, & L_{\mathrm{m}} \leq J \\ (J + 1)^2 + (2J + 1)(L_{\mathrm{m}} - J), & L_{\mathrm{m}} \geq J \end{cases} \tag{8.37}$$

8.4.2 Sphere + atom

A slightly more complicated case arises when $D(J_{\mathrm{r}}) = (2J_{\mathrm{r}} + 1)^2$, which is the case of the spherical top + atom, so that $\Gamma(J_{\mathrm{r}}) = 2J_{\mathrm{r}} + 1$. Since the number of L-states for every J_{r} is again $2J_{\mathrm{r}} + 1$, but every lattice point now represents $2J_{\mathrm{r}} + 1$ states, we have for $J_{\mathrm{m}} \leq J$ and $L_{\mathrm{m}} \gg J_{\mathrm{m}}$ (vertical cut-off) [9]

$$\Gamma(y, J) = \sum_{J_{\mathrm{r}}=0}^{J_{\mathrm{m}}} (2J_{\mathrm{r}} + 1)^2 = 1 + 3^2 + 5^2 + \cdots + (2J_{\mathrm{m}} + 1)^2$$

$$= \frac{1}{3}(J_{\mathrm{m}} + 1)[4(J_{\mathrm{m}} + 1)^2 - 1] \quad (J_{\mathrm{m}} \leq J) \tag{8.38}$$

The integral equivalent to the summation is $\int 4J_r^2 \, dJ_r = \frac{4}{3}J_m^3$, using the same limits, which is the same result if unity is neglected with respect to J_m.

If $J_m > J$, $L_m \gg J_m$, and considering only states at $J_r \geq J$, there are now $2J + 1$ states of L for every $J_r \geq J$, with every lattice point again worth $2J_r + 1$ states, so the number of J_r states *above* J is

$$(2J + 1) \sum_{J_r=J}^{J_m}(2J_r + 1) = (2J + 1)[(J_m + 1)^2 - (J + 1)^2] \tag{8.39}$$

using the result of Eq. (8.35) twice (as sum from 0 to J_m, minus the sum from 0 to J). The total number of states at $J_m > J$ (i.e. below as well as above J) is then the sum of Eq. (8.38) evaluated at $J_m = J$, plus Eq. (8.39):

$$\Gamma(y, J) = \frac{1}{3}(J + 1)[4(J + 1)^2 - 1] + (2J + 1)[(J_m + 1)^2 - (J + 1)^2] \tag{8.40}$$

Integration yields $\Gamma(y, J) = 2J J_m^2 - \frac{2}{3}J^3$, which is the same result if unity is neglected with respect to J and J_m.

The results for sphere + atom with horizontal cut-off ($J_m \gg L_m$) are different but obtainable by a similar argument (Problem 8.2).

8.5 $\Gamma(y, J)$ for the general case

The problem is now to calculate the sum of states $\Gamma(y, J)$ with a more general type of constraint represented by Eq. (8.24), i.e. when there is not a simple vertical or horizontal cut-off. A typical example of a cut-off for fragments other than hydrogen is shown in Fig. 8.3. In such cases, treating angular momenta as continuous variables, semi-classical evaluation of J_1, J_2 and J_r, L combinations in terms of the actual "area" is less onerous than the exact procedure requiring enumeration of individual quantum states as discrete lattice points. A comparison between a quantum and a semi-classical result has shown [10] that the modest increase in accuracy is not worth the complication of summations over discrete quantum states. The development given here is inspired by the work in [11–15], using largely their notation.

Semi-classically, the $\Gamma(y, J)$ of Eq. (8.32) is given by

$$\Gamma(y, J) = \int dL \int dJ_r \, \Gamma(J_r) \tag{8.41}$$

It will be assumed in the following that $\Gamma(J_r)$ is available as a (continuous) function of J_r and L. The actual calculation of this integral for specific more complicated cases is deferred until the next section (Eq. (8.55)). The immediate interest here lies in the integration limits with respect to J_r and L.

8.5.1 Integration in the J_r–L space

Semi-classically, the double sum in Eq. (8.32) becomes the J_r–L area, which will be evaluated as a double integral of $\Gamma(J_r)$, first with respect to J_r, at constant L, and then with respect to L. The integration limits with respect to J_r are obvious: the lower limit is the value of J_r on one of the straight lines $|J-L|$, and the upper limit is on the $J + L$ line or on the curve of L_m versus J_m of Eq. (8.24) (denoted J_r^*), whichever is closer. However, the integration limits with respect to L, which we shall call L_+ and L_- (Fig. 8.3) have to be calculated for each pair (y, J), as described further below.

The double integral takes different forms depending on the relative magnitudes of J and J_m, as can be readily appreciated by examining Fig. 8.3. If $J_m \leq J$, we have

$$\Gamma(y, J) = \int_{L_-}^{L_+} dL \int_{|J-L|}^{J^*} dJ_r \, \Gamma(J_r) \tag{8.42}$$

and, if $J_m \geq J$,

$$\Gamma(y, J) = \int_0^{L_-} dL \int_{|J-L|}^{J+L} dJ_r \, \Gamma(J_r) + \int_{L_-}^{L_+} dL \int_{|J-L|}^{J_r^*} dJ_r \, \Gamma(J_r) \tag{8.43}$$

where J_r^* is the value of J_r on the curve of L_m versus J_m at a given L. From Eq. (8.24), after substitution for L_m and J_m, we have

$$J_r^* = \left(\frac{y - (L/b_n)^{2n/(n-2)}}{B_r} \right)^{1/2} \tag{8.44}$$

Note that the numerator under the square root is in fact $y - E_t$, i.e. the rotational energy E_r (cf. Eq. (8.17)).

It was shown previously in Section 8.4 that, when one fragment is an atom, we have $J_1 \equiv J_r$, so for the case linear fragment + atom the semi-classical result is simply $\Gamma(J_r) = 1$, whereas for spherical fragment + atom it is $\Gamma(J_r) = 2J_r$. In these two cases the above double integrals in Eqs. (8.42) and (8.43) can be evaluated analytically for given L_- and L_+ (Problems 8.3 and 8.4), but for more complicated cases – to be discussed next in Section 8.6 – integration with respect to L has to be done numerically. Analytical expressions (sometimes quite complicated) for $\Gamma(J_r)$ can be obtained for most cases of interest [14, 15], and are shown in Table 8.2.

Table 8.2. *The sum of states* $\Gamma(J_r)$ *(B are rotational constants)*

Linear + atom	1
Sphere + atom	$2J_r$

Sphere + linear
(B_s) (B_1)

$$\begin{cases} \dfrac{4}{3\omega^{1/2}B_s}\left(E_r - B_r J_r^2\right)^{3/2}, & \begin{array}{c} B_r J_r^2 \le E_r \le B_s J_r^2 \\ (\omega = B_s + B_1) \end{array} \\[4mm] \dfrac{2J_r}{\omega}\left[\omega E_r - \left(B_s B_1 + \dfrac{1}{3}B_s^2\right)J_r^2\right], & E_r \ge B_s J_r^2 \end{cases}$$

Sphere + sphere
(B_1) (B_2)

$$\dfrac{8}{3}J_r\left[\dfrac{B_r}{\omega}\left(\dfrac{E_r}{B_r} - J_r^2\right)\right]^{3/2}, \qquad \begin{array}{c} E_r \ge B_r J_r^2 \\ (\omega = B_1 + B_2) \end{array}$$

8.5.2 Determination of L_+ and L_-

Since L_+ and L_- are given by the intersection of J_r^* and one of the $L \pm J$ lines, Eq. (8.44) can be used to determine the integration limits on L by solving for L the two equations

$$y - (L/b_n)^{2n/(n-2)} = B_r(L \pm J)^2 \tag{8.45}$$

The solutions depend on the relative magnitudes of J_m and J. If $J_m < J$, the minus sign inside the square yields two real positive roots as a solution both for L_+ and for L_-; if $J_m > J$, the minus sign yields one real positive root as the solution for L_+, and the plus sign similarly yields the solution for L_-. For non-integral n numerical solution is necessary, but, for $n = 6$ (interaction of two neutral species), Eq. (8.45) amounts to a cubic equation in L, and for $n = 4$ (ion–neutral-species interaction), Eq. (8.45) becomes a quartic in L, both of which can be solved analytically [16].

Note that, for a given L_m, J_m, namely the value of J_r at the intersection of L given by Eq. (8.24) and the straight lines $L = J_r \pm J$ and $L = J - J_r$, is given by the solutions for J_r of

$$L_m\left[1 - \left(\frac{J_r}{J_m}\right)^2\right]^{(n-2)/(2n)} = \begin{cases} J_r \pm J \\ J - J_r \end{cases} \tag{8.46}$$

If these solutions are called J_+ and J_-, substitution into Eq. (8.24) then yields the corresponding L_+ and L_-, which provides an alternative route to the integration limits in L. For $n = 6$ Eq. (8.46) is a cubic equation in J_r, in which case it is necessary to respect the sign of the difference $J - J_r$ (Problem 8.5); for $n = 4$, Eq. (8.46) becomes a quartic in J_r (Problem 8.6).

Procedures based on Eqs. (8.45) and (8.46) give of course the same result for L_+ and L_-. An example of the J-dependence of L_+ and L_- is given in Fig. 8.5 as

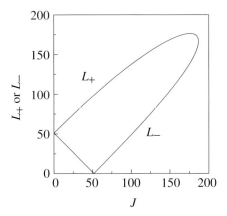

Fig. 8.5. J-dependences of the integration limits L_+ and L_- calculated from Eq. (8.45) or (8.46) at $y = 10^4 \, \text{cm}^{-1}$. The system is two neutral spherical tops ($n = 6$) represented by $CH_3 + CH_3$, with each methyl radical "spherically averaged," i.e. having for rotational constant the geometric average of rotational constants for the methyl radical (($9.45^2 \times 4.73)^{1/3} = 7.503 \, \text{cm}^{-1}$); thus $B_r = 3.7515 \, \text{cm}^{-1}$. The potential constant C_6 in Eq. (8.14) is given in Table 8.1; $L_m = 176$, $J_m = 51$ and $b = 8.177$. The two curves start at $J = 0$ (L_+ increases monotonically with J, but L_- drops to zero at first), and both end at $J = 187$, which is the maximum total angular momentum that can be generated at the specified total translational + rotational energy.

a function of J for the case $n = 6$. The two integration limits are equal at $J = 0$, and again at high J when the available energy becomes insufficient to generate the requisite rotational energy. At this point the $L = J - J_r$ line becomes a tangent (of slope -1) to the curve of L versus J_r (Problem 8.7).

8.5.3 Limiting case 1: "low J"

Before dealing with more complex forms of $\Gamma(J_r)$ at arbitrary J, J_m, and L_m, it is worthwhile to consider first two limiting cases that are of particular interest, namely the sum of states $\Gamma(y, J)$ for an arbitrary $\Gamma(J_r)$ at "low J" and "high J."

When $J = 0$, we have $L = J_r$, so the J-delimited rectangle in Fig. 8.2 collapses into a diagonal ($45°$) line; the "area" in the $L-J_r$ space is then merely the number of lattice points on the diagonal until it crosses the line L_m versus J_m at J_r^*. Thus semi-classically:

$$\Gamma(y, J = 0) = \int_0^{J_r^*} \Gamma(J_r) \, dJ_r \tag{8.47}$$

At constant y, $\Gamma(y, J = 0)$ is thus a constant. If J_r^* is given by Eq. (8.44), the above equation is the exact result (Problem 8.8); however, provided that y is not too large, the contribution of $(L/b_n)^{2n/(n-2)}$ under the square root of Eq. (8.44) can be neglected, so we can take $J_r^* \approx (y/B_r)^{1/2} = J_m$, which represents the vertical cut-off (Fig. 8.4(a)).

When J is not zero but merely "low," the 45° diagonal may be thought of as a heavy line of "thickness" $2J$ (semi-clasically), or more exactly $2J + 1$ (cf. Fig. 8.8 below), so Eq. (8.47) becomes

$$\Gamma(y, \text{"low } J\text{"}) = (2J + 1) \int_0^{J_r^*} \Gamma(J_r) \, dJ_r \qquad (8.48)$$

Since, at a given y, the integral, i.e. $\Gamma(y, J = 0)$, is a constant, we have the result that $\Gamma(y, J)$ increases linearly with J at low J: $\Gamma(y, \text{"low } J\text{"}) = (2J + 1)\Gamma(y, J = 0)$.

Note that, if we had used the semi-classical factor $2J$ instead of $2J + 1$, Eq. (8.48) would yield $\Gamma(y, \text{"low } J\text{"}) = 0$ for $J = 0$, regardless of the value of $\Gamma(J_r)$. As Eq. (8.47) shows, this is not correct, which is a common defect of all semi-classical formulations at the origin of a variable, as encountered previously in Section 4.2 (Eqs. (4.20) and (4.21)). If $\Gamma(J_r)$ is finite, then $J = 0$ cannot mean that the sum of states is actually zero (which would be the case only if J_r and L were all zero), but merely that the J_r and L vectors are equal and anti-parallel. Thus $\Gamma(y, J)$ versus J has a small intercept at the origin.

8.5.4 Limiting case 2: "high J"

By the "high-J" case is meant merely that $J > J_m$ (if J is too "high" for a given y, the sum of states is zero, cf. Fig. 8.7). From Fig. 8.4(a) (the dashed line on the left) we see that, for a vertical cut-off, L goes from $J - J_r$ to $J + J_r$ for a total of $2J_r$ L-states. On inverting the order of integrations, Eq. (8.42) thus becomes

$$\Gamma(y, \text{"high } J\text{"}) \rightarrow \int_0^{J_r^*} dJ_r \int_{J-J_r}^{J+J_r} dL \, \Gamma(J_r) = 2 \int_0^{J_r^*} dJ_r \, J_r \, \Gamma(J_r) \qquad (8.49)$$

The result of the integration of the final equation is a function of $J_r^* \approx J_m = (y/B_r)^{1/2}$ *only* and thus contains no reference to J. The "high J" case therefore means that the sum of states is energy-limited rather than angular-momentum-limited, so the formulas of Chapter 4 for the free-rotor sum of states, which were derived by ignoring angular-momentum constraints, apply as such (Problem 8.9).

The vertical cut-off provides a useful approximation to Eqs. (8.47) and (8.49) for the fairly common case when the line L_m versus J_m is rather steep (cf. Fig. 8.3). The case of the less common horizontal cut-off is the subject of Problem 8.10 (cf. also Problem 8.2).

It can be verified, using the previously determined $\Gamma(J_r) = 1$ for linear + atom and the semi-classical $\Gamma(J_r) = 2J_r$ for sphere + atom, that Eq. (8.47) with $J_r^* \approx J_m$ yields for $J = 0$ the same results as those that can be obtained from Eqs. (8.36) and (8.40), respectively, for the condition that $J_m > J$, if we let $J \to 0$, and neglect 1 with respect to J_m. Similarly, Eq. (8.36) for $J_m < J$ and Eq. (8.38) yield the same result as Eq. (8.49).

8.5.5 Determination of y_m

With $\Gamma(y, J)$ determined from Eqs. (8.42) and (8.43), $G_x(y, J)$ follows from Eq. (8.33), which is then available for the calculation of the convolution integral in Eq. (8.10) at a specified J. The lower integration limit y_m is conditioned by the requirement that the energy y be sufficient for the recombining fragments to clear the centrifugal barrier, i.e. sufficient to generate the combined translational–rotational energy corresponding to J.

Specifically, we shall first determine the minimum L for a given J, a problem we have already considered above from a different angle. The condition, as before, is that the $L = J - J_r$ line (Fig. 8.2) be a tangent to the curve of L_m versus J_m (Fig. 8.6). From Eqs. (8.17) and (8.22) we have

$$\left(\frac{L}{b_n}\right)^{2n/(n-2)} + B_r J_r^2 = y \tag{8.50}$$

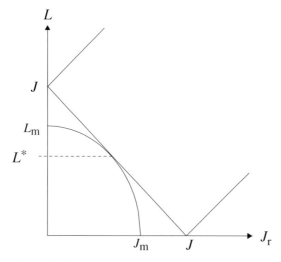

Fig. 8.6. An illustration of L^*, defined as the value of L at which the J-rectangle is at a tangent to the curve of L_m versus J_m (cf. Eq. (8.54)). This condition defines the maximum J for specified y, or minimum y for specified J.

Taking the differential at constant y gives

$$\frac{2n}{n-2} \frac{L^{(n+2)/(n-2)}}{b_n^{2n/(n-2)}} \frac{\mathrm{d}L}{\mathrm{d}J_r} + 2B_r J_r = 0 \tag{8.51}$$

Now the slope $\mathrm{d}L/\mathrm{d}J_r = -1$, and $J_r = J - L$, so the value of L that satisfies the above conditions $(=L^*)$ is given by the solution of

$$2B_r(J - L) - \frac{2n}{n-2} \frac{L^{(n+2)/(n-2)}}{b_n^{2n/(n-2)}} = 0 \tag{8.52}$$

In a given system, L^* obtained from this equation is numerically the same as that obtained in Problem 8.7 (although it is given in terms of different variables). Then y_m follows from Eq. (8.45):

$$y_m = \left(\frac{L^*}{b_n}\right)^{2n/(n-2)} + B_r(L^* - J)^2 \tag{8.53}$$

When $n = 6$ the solution of Eq. (8.52) for L^* is particularly simple:

$$L^* = \frac{B_r b_6^3}{3}\left(\sqrt{1 + \frac{6J}{B_r b_6^3}} - 1\right) \tag{8.54}$$

If J is not too large, the ratio $6J/(B_r b_6^3)$ is small, so the square root in Eq. (8.54) can be expanded: $[1 + 6J/(B_r b_6^3)]^{1/2} \cong 1 + 3J/(B_r b_6^3)$; then $L^* \cong J$ and consequently Eq. (8.53) yields $y_m \cong (J/b_6)^3$, which is in fact E_t (Eq. (8.17)): small J means small rotational energy, so the minimum energy necessary to surmount the centrifugal barrier depends principally on the relative translational energy of the fragments.

At large J the above approximation to y_m tends to be an overestimate; however, at large J the sum of "external" states approaches zero (cf. Fig. 8.7 below) so, insofar as the convolution in Eq. (8.10) is concerned, $y_m \cong (J/b_6)^3$ will be in general a reasonable approximation. Analytical solution of Eq. (8.52) is also possible when $n = 4$ (Problem 8.11). In this case, for $J < 100$, a useful approximation is $y_m \cong (J/b_4)^4$, except when one fragment is a hydrogen atom.

8.5.6 Convolution for $G_{vx}^*(E, J)$

The results from Eqs. (8.42) and (8.43) for $\Gamma(y, J)$, after conversion into $G_x(y, J)$ (Eq. (8.33)), can now be used in principle in the convolution of Eq. (8.10) to obtain the angular-momentum-conserved vibrational–rotational sum of states $G_{vx}^*(E, J)$. Equation (8.10) presents the inconvenience that the vibrational density of states for the fragment $N_v^f(E - y)$ is in principle zero when $E - y$ is less than the the lowest fragment frequency (cf. Section 4.6).

Such numerical problems at low energies can be avoided by making use of the alternative convolution in Eq. (8.12), where a smooth-function approximation to the quantum vibrational *sum* of states $G_v^f(E - y)$, which is unity for energies below the lowest fragment frequency, is convoluted with the essentially classical density $N_x(y, J)$. This density follows from Eq. (8.34) on taking the derivative of $G_x(y, J)$, and the smooth-function approximation for the sum of states $G_v^f(E - y)$ is readily obtained by the S-D method (cf. Section 4.4).

8.6 $\Gamma(y, J)$ for more complex cases

Within the semi-classical approximation, $\Gamma(J_r)$ for two symmetric polyatomic fragments R_1 and R_2 can be represented quite generally by the multiple integral

$$\Gamma(J_r) = \int \int \int \int dK_1 \, dK_2 \, dJ_1 \, dJ_2 \qquad (8.55)$$

where the quantum numbers K_1 and K_2 are the projections of J_1 and J_2, respectively, along the symmetry axes of fragments R_1 and R_2.

Since now one of the fragments is *not* an atom, J_2 is not zero, and the compounding of angular momenta is therefore done most conveniently in two steps, starting with the vector addition $J_1 + J_2 = J_r$ (Eq. (8.20)), which means that $|J_1 - J_2| \leq J_r \leq J_1 + J_2$.

For simplicity we will assume that fragments R_1 and R_2 are each described by a unique rotational constant, B_1 and B_2, respectively, and that their rotational energies are (semi-classically) $B_1 J_1^2$ and $B_2 J_2^2$. This means that the fragments are assumed to be of linear or spherical-top symmetry, the latter not necessarily meant literally but merely as an approximation of the more general class of non-linear fragments (with three different rotational constants B_a, B_b, and B_c), in which case the rotational constant B is replaced in all spherical-top formulas below (and in Tables 8.2–8.7) by the geometric average $(B_a B_b B_c)^{1/3}$.

The total rotational energy E_r of two fragments of linear or spherical-top symmetry is thus

$$B_1 J_1^2 + B_2 J_2^2 = E_r \qquad (8.56)$$

which is the equation of an ellipse with semi-axes $\sqrt{E_r/B_1}$ and $\sqrt{E_r/B_2}$. This equation can be solved for J_1 by eliminating J_2 via $J_2 = |J_r \pm J_1|$, leading to a quadratic equation in J_1 with two real roots. If the two solutions for J_1 are called $J_\pm$, we have

$$J_\pm = \frac{\omega^{1/2}\left(E_r - B_r J_r^2\right)^{1/2} \pm B_2 J_r}{\omega} \qquad (8.57)$$

where $B_r = B_1 B_2 / (B_1 + B_2)$ is the effective (or reduced) rotational constant (cf. also Chapter 9), and $\omega = B_1 + B_2$ (Problem 8.12). Since the expression under the square root must be positive, it is clear that, *in terms of J_r, E_r cannot be less than* $B_r J_r^2$; thus $E_r \geq B_r J_r^2$. From Eq. (8.56) we have

$$\frac{dE_r}{dJ_2} = 2B_2 J_2 \tag{8.58}$$

We can then write Eq. (8.55) in the form

$$\Gamma(J_r) = \int \int \int \int dK_1 \, dK_2 \, dJ_1 \, \frac{dE_r}{2B_2 J_2} \tag{8.59}$$

The explicit form of $\Gamma(J_r)$ depends on the symmetry of the fragment, as is shown below in two examples.

8.6.1 Example 1: two spherical tops

In the case of fragments consisting of two spherical tops with rotational constants B_1 and B_2, the integration limits with respect to K_1 and K_2 are $-J_1$ to $+J_1$ and $-J_2$ to $+J_2$, respectively, which yield upon integration $4J_1 J_2$; Eq. (8.59) thus becomes

$$\Gamma(J_r) = \frac{2}{B_2} \int dE_r \int_{J_-}^{J_+} J_1 \, dJ_1 = \frac{4J_r}{\omega^{3/2}} \int dE_r \left(E_r - B_r J_r^2 \right)^{1/2} \tag{8.60}$$

With respect to E_r, the lower limit is $B_r J_r^2$, as seen above; the upper limit is the available rotational energy at specified (y, L), which, from Eq. (8.44), is $B_r J_r^{*2}$. Thus the result of the integration with respect to E_r in Eq. (8.60) is

$$\Gamma(J_r) = \frac{8}{3} J_r \left(\frac{B_r}{\omega} \left(J_r^{*2} - J_r^2 \right) \right)^{3/2} \tag{8.61}$$

Observe that, if integration with respect to E_r is omitted,

$$\frac{2}{B_2} \int_{J_-}^{J_+} J_1 \, dJ_1 = \frac{4J_r}{\omega^{3/2}} \left(E_r - B_r J_r^2 \right)^{1/2} = N(J_r) \tag{8.62}$$

The result has the dimension (energy)$^{-1}$ and thus represents the *density* of states $N(J_r)$ at specified E_r and J_r; it will be discussed more fully in Section 8.10. Conversely, of course, $N(J_r) = \partial \Gamma(J_r) / \partial E_r$.

For the purpose of evaluating $\Gamma(y, J)$, the $\Gamma(J_r)$ of Eq. (8.61) has to be integrated over the J_r–L space, as indicated in Eqs. (8.42) and (8.43), first with respect to J_r, and then with respect to L. Integration with respect to J_r yields

$$\int \Gamma(J_r) \, dJ_r = -\frac{8}{15} \left(\frac{B_r}{\omega}\right)^{3/2} \left(J_r^{*2} - J_r^2\right)^{5/2} \tag{8.63}$$

This remains to be eveluted within limits on J_r that introduce an L-dependence that depends on the relative magnitudes of J_m and J (cf. Eqs. (8.42) and (8.43)); there remains J_r^*, which is likewise L-dependent (Eq. (8.44)). All of these cause the final expressions to depend on the potential parameter n.

Subsequent integration with respect to L (Eqs. (8.64)–(8.67) below) is compli-cated enough in most cases to require numerical evaluation. On taking the derivative with respect to y of the final result we obtain the corresponding density of states $N(y, J) = \partial \Gamma(y, J)/\partial y$ (note the distinction between $N(J_r)$ and $N(y, J)$). The fi-nal expressions in compact form for the sum and density of states at (y, J) are shown to be

$$\left.\begin{array}{c} \Gamma(y, J) \\ N(y, J) \end{array}\right\} = \left\{\begin{array}{c} C_G \\ C_N \end{array}\right\} \times \text{INT} \tag{8.64}$$

where

$$C_G = \frac{8}{15} \left(\frac{B_r}{B_1 + B_2}\right)^{3/2}, \qquad C_N = \frac{4}{3} \frac{B_r^{1/2}}{(B_1 + B_2)^{3/2}} \tag{8.65}$$

and

$$\text{INT} = \begin{cases} \displaystyle\int_0^{L_+} I_-^m \, dL - \int_0^{L_-} I_+^m \, dL & (J_m > J) \\[4mm] \displaystyle\int_{L_-}^{L_+} I_-^m \, dL & (J_m < J) \end{cases} \tag{8.66}$$

with

$$I_\pm = \frac{y - (L/b_n)^{2n/(n-2)}}{B_r} - (J \pm L)^2 \tag{8.67}$$

where $m = \frac{5}{2}$ for $\Gamma(y, J)$ and $m = \frac{3}{2}$ for $N(y, J)$. The potential parameter takes the values $n = 6$ or 4 for interactions between two neutral species and ion–neutral-species interactions, respectively; similarly the integration limits L_+ and L_- depend on n (cf. Eq. (8.45)).

Figure 8.7 shows an example of the sum of states $\Gamma(y, J)$ calculated from Eq. (8.64). Note that $\Gamma(y, J)$ goes to zero at high J when the total energy available is insufficient to generate the corresponding rotational energy; the same is true of the density of states $N(y, J)$.

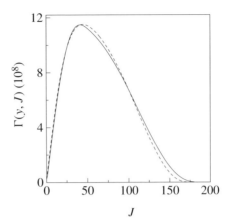

Fig. 8.7. The J-dependence of $\Gamma(y, J)$ at $y = 10^4\ \mathrm{cm}^{-1}$ for two neutral spherical tops ($n = 6$). The solid line was calculated from Eq. (8.64); the integration limits L_+ and L_- for this case are shown in Fig. 8.5. The dashed line is the approximation from Eqs. (8.78) and (8.79). The system in question is described in the caption of Fig. 8.5.

The "low-J" result follows directly from Eqs. (8.48) and (8.63), with the approximation $J_r^* \approx J_m$ (cf. Problem 8.8),

$$\Gamma(y, \text{"low } J\text{"}) = (2J + 1)\frac{8y^{5/2}}{15 B_1 B_2 (B_1 + B_2)^{1/2}} \tag{8.68}$$

Similarly, we have from Eqs. (8.49) and (8.61) for the "high-J" result

$$\Gamma(y, \text{"high } J\text{"}) = \frac{\pi y^3}{6(B_1 B_2)^{3/2}} \tag{8.69}$$

which is the same result as that obtainable from the general formula (4.18) in Chapter 4 for two independent three-dimensional free rotors with rotational constants B_1 and B_2.

8.6.2 Example 2: spherical top + linear

Let the rotational constant of the linear fragment be B_1 and that of the sphere be B_s, with corresponding angular momenta J_1 and J_s, such that $E_r = B_1 J_1^2 + B_s J_s^2$. Since the linear fragment has no K-dependence, Eq. (8.59) becomes

$$\Gamma(J_r) = \int \int \int dK_s\, dJ_1 \frac{dE_r}{2 B_s J_s} = \frac{1}{B_s} \int dE_r \int_{J_-}^{J_+} dJ_1 \tag{8.70}$$

where $J_\pm$ is given by Eq. (8.57) with $B_2 \equiv B_s$, $B_r = B_s B_1/\omega$, $\omega = B_s + B_1$. The result of the integration with respect to J_1 is $(J_+ - J_-)$, which has two forms depending on the relative values of E_r and $B_s J_r^2$; thus the *density* of states $N(J_r)$ at specified

E_r and J_r is (Problem 8.13)

$$N(J_r) = \frac{1}{B_s} \int_{J_-}^{J_+} dJ_1 = \begin{cases} \frac{2}{\omega^{1/2} B_s} \left(E_r - B_r J_r^2\right)^{1/2}, & B_r J_r^2 \leq E_r \leq B_s J_r^2 \\ \frac{2 J_r}{\omega}, & E_r \geq B_s J_r^2 \end{cases} \quad (8.71)$$

On integrating the top expression with respect to E_r from $B_r J_r^2$ to $B_s J_r^2$, we have for the *sum* of states at E_r and J_r

$$\Gamma_1(J_r) = \frac{4}{3\omega^{1/2} B_s} \left(E_r - B_r J_r^2\right)^{3/2}, \quad B_r J_r^2 \leq E_r \leq B_s J_r^2 \quad (8.72)$$

At the upper limit, i.e. at $E_r = B_s J_r^2$, this gives $\Gamma_1(J_r) = 4 B_s^2 J_r^3/(3\omega^2)$, which is the starting point for the integration of the bottom expression in Eq. (8.71) from $B_s^2 J_r$ to E_r; thus, after some manipulation,

$$\Gamma_2(J_r) = \frac{4 B_s^2 J_r^3}{3\omega^{1/2}} + \frac{2 J_r}{\omega} \int_{B_s J_r^2}^{E_r} dE_r$$

$$= \frac{2 J_r}{\omega} \left[\omega E_r - \left(B_s B_1 + \frac{1}{3} B_s^2\right) J_r^2\right], \quad E_r \geq B_s J_r^2 \quad (8.73)$$

Owing to the two forms of $\Gamma(J_r)$, integration in the L–J_r plane is considerably more complicated than in the sphere–sphere case. To begin with, there are now two L/L_m curves, and consequently two J^* values, one of which we may call J_r^*, which is defined by Eq. (8.44), and another called J_s^*, defined similarly but with B_s replacing B_r in the denominator of Eq. (8.44). Since $B_r < B_s$, we have $J_s^* < J_r^*$ at the same rotational energy. If the maximum values corresponding to J_s^* and J_r^* are called J_{ms} and J_{mr}, respectively, there are three cases to be considered, which represent the various relative values of J, J_{mr}, and J_{ms} when the total energy y is large enough to excite both rotational energies $B_s J^2$ and $B_1 J^2$. Two additional cases arise when y is just sufficient to excite $B_r J^2$ but not enough to excite $B_s J^2$ [17]. $\Gamma(J_r)$ and $N(J_r)$ for various symmetries are summarized in Tables 8.2 and 8.3.

We shall consider here only the case $J_{mr}, J_{ms} > J$ (Fig. 8.8), which is useful for calculating the limit of "low J" shown in Figure 8.8 as the dashed rectangle. Equation (8.48) becomes

$$\Gamma(y, \text{``low } J\text{''}) = (2J + 1) \left(\int_0^{J_s^*} \Gamma_2(J_r) \, dJ_r + \int_{J_s^*}^{J_r^*} \Gamma_1(J_r) \, dJ_r \right)$$

$$= (2J + 1) \frac{y^2}{2 B_s^{3/2} B_1^{1/2}} \sin^{-1} \left[\left(\frac{B_1}{B_s + B_1} \right)^{1/2} \right] \quad (8.74)$$

Table 8.3. *The density of states* $N(J_r)$ *(B are rotational constants)*

Linear + atom	$\delta(E_r - B_r J_r^2),$	$E_r \geq B_r J_r^2$
Sphere + atom	$2 J_r \delta(E_r - B_r J_r^2),$	$E_r \geq B_r J_r^2$
Sphere + linear $(B_s) \quad (B_l)$	$\begin{cases} \dfrac{2}{\omega^{1/2} B_s} (E_r - B_r J_r^2)^{1/2}, & B_r J_r^2 \leq E_r \leq B_s J_r^2 \\ & (\omega = B_s + B_l) \\[2mm] \dfrac{2 J_r}{\omega} & E_r > B_s J_r^2 \end{cases}$	
Sphere + sphere $(B_l) \quad (B_2)$	$\dfrac{4 J_r}{\omega^{3/2}} (E_r - B_r J_r^2)^{1/2}$	$\begin{array}{l} E_r \geq B_r J_r^2 \\ (\omega = B_1 + B_2) \end{array}$

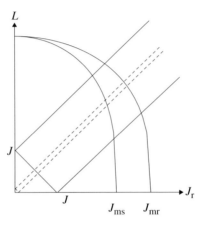

Fig. 8.8. The region of integration in the L–J_r space for the sphere–linear case when J_{ms}, $J_{mr} > J$ (cf. text below Eq. (8.73)). For $J \to 0$ ("low J"), the J-delimited rectangle indicated by the solid line collapses to the one indicated by the dashed lines.

Similarly, the "high-J" result follows from Eq. (8.49) after some algebra:

$$\Gamma(y, \text{"high } J\text{"}) = 2 \int_0^{J_s^*} J_r \, \Gamma_2(J_r) \, dJ_r + 2 \int_{J_s^*}^{J_r^*} J_r \, \Gamma_1(J_r) \, dJ_r$$

$$= \frac{8 y^{5/2}}{15 B_s^{3/2} B_l} \tag{8.75}$$

which again is the same result as that obtainable from the general formula in Chapter 4 for one free three-dimensional rotor combined with one free two-dimensional rotor.

8.6.3 Spherical top + symmetric top

This is a generalization of the two spherical-tops cases; as such it is more complicated and requires considerable algebra, which will not be given here. Analytical

Table 8.4. The sum of states $\Gamma(y, J)$

Fragment symmetry (B are rotational constants)	$J = 0$	"High J"	r
Linear + atom	$(y/B)^{1/2}$	y/B	2
Sphere + atom	y/B	$\dfrac{4}{3}(y/B)^{3/2}$	3
Linear + linear (B_1) (B_2)	$\dfrac{2y^{3/2}\left(\sqrt{B_1}+\sqrt{B_2}-\sqrt{B_1+B_2}\right)}{3B_1 B_2}$	$\dfrac{y^2}{2B_1 B_2}$	4
Sphere + linear (B_s) (B_1)	$\dfrac{y^2 \sin^{-1}\left[\left(\dfrac{B_s}{B_s+B_1}\right)^{1/2}\right]}{2B_s^{3/2}B_1^{1/2}}$	$\dfrac{8y^{5/2}}{15B_s^{3/2}B_1}$	5
Sphere + sphere (B_1) (B_2)	$\dfrac{8y^{5/2}}{15B_1 B_2(B_1+B_2)^{1/2}}$	$\dfrac{\pi y^3}{6(B_1 B_2)^{3/2}}$	6

forms of $\Gamma(J_r)$ both for prolate and for oblate symmetric tops have been published [12, 17] and may be used to show that the error in "spherical averaging," i.e. in substituting for the actual rotational constants of a symmetric top their geometric average, is not large.

Of more interest are the two limiting cases. For "low J" we have in the case sphere + prolate top [12, 17]

$$\Gamma(y, \text{"low } J\text{"}) = (2J+1)\frac{8y^{5/2}\sin^{-1}\left[\left(\dfrac{B_s(A-B)}{A(B_s+B)}\right)^{1/2}\right]}{15B_s^{3/2}[B(A-B)]^{1/2}} \tag{8.76}$$

where B_s is the rotational constant of the spherical top and A and B $(A > B)$ are the rotational constants of the prolate top. At the other extreme we have

$$\Gamma(y, \text{"high } J\text{"}) = \frac{\pi y^3}{6B_s^{3/2}BA^{1/2}} \tag{8.77}$$

It can be shown that as the prolate top becomes more spherical, i.e. as $A \to B$, the above "low-J" and "high-J" forms go over smoothly into the two spherical-top expressions, Eqs. (8.74) and (8.75) (Problem 8.14).

By continuing in this fashion we can build up a table for a number of different fragment symmetries both in the "low-J" limit and in the "high-J" limit of $\Gamma(y, J)$ (Problem 8.15 and Table 8.4) and of $N(y, J)$ (Table 8.5), the latter simply as the derivative $\partial\Gamma(y, J)/\partial y$. Since it is obvious from Eqs. (8.47) and (8.48) that $\Gamma(y, \text{"low } J\text{"}) = (2J+1)\Gamma(y, J = 0)$ (and similarly for the density), it is sufficient to list in the tables only the $J = 0$ result.

Table 8.5. *The density of states* $N(y, J)$

Fragment symmetry (B are rotational constants)	$J = 0$	"High J"	r
Linear + atom	$\frac{1}{2}(yB)^{-1/2}$	$1/B$	2
Sphere + atom	$1/B$	$\dfrac{2y^{1/2}}{B^{3/2}}$	3
Linear + linear (B_1) (B_2)	$\dfrac{y^{1/2}\left(\sqrt{B_1} + \sqrt{B_2} - \sqrt{B_1 + B_2}\right)}{B_1 B_2}$	$\dfrac{y}{B_1 B_2}$	4
Sphere + linear (B_s) (B_1)	$\dfrac{y \sin^{-1}\left\{[B_s/(B_s + B_1)]^{1/2}\right\}}{B_s^{3/2} B_1^{1/2}}$	$\dfrac{4y^{3/2}}{3B_s^{3/2} B_1}$	5
Sphere + sphere (B_1) (B_2)	$\dfrac{4y^{3/2}}{3B_1 B_2(B_1 + B_2)^{1/2}}$	$\dfrac{\pi y^2}{2(B_1 B_2)^{3/2}}$	6

The last column in each table gives r, the number of rotational degrees of freedom involved in each specified fragment combination. Observe that, in Table 8.4, each "high-J" entry for $\Gamma(y, J)$ has y to the power $\frac{1}{2}r$, whereas the corresponding $J = 0$ entry always contains y to the power $\frac{1}{2}(r - 1)$. Thus, insofar as the sum of states $\Gamma(y, J)$ is concerned, the $J = 0$ result looks as if one degree of freedom of rotation were "missing" with respect to the respective "high-J" case. A similar relation exists with respect to the $J = 0$ and "high-J" entries for $N(y, J)$ in Table 8.5.

8.6.4 An approximation for $\Gamma(y, J)$ at arbitrary J

When fragments are of different symmetries, the expressions for $\Gamma(y, J)$ and $N(y, J)$ become sufficiently complicated to discourage their use in routine calculations: witness the above sphere + linear and sphere + symmetric top cases. Note that, even for a single value of y and J, the evaluation of $\Gamma(y, J)$ or $N(y, J)$ requires numerical integration with respect to L; hence the need for a practical simplification, as was recognized early [18]. For the present purpose the sum of states $\Gamma(y, J)$ given by the full numerical integration of Eqs. (8.42) and (8.43) will be considered "exact."

Given the "low-J" and "high-J" forms of $\Gamma(y, J)$, it has been shown [19] that it is possible to approximate the "exact" $\Gamma(y, J)$ and $N(y, J)$ at all J (the solid line in Fig. 8.7) by interpolation and thus avoid the repeated numerical integrations in Eqs. (8.42) and (8.43). To this end a "reduced" representation [20] in terms of the "low-J" and "high-J" forms is quite successful, using the error function erf [21] for interpolation [22].

The interpolated $\Gamma(y, J)$ at arbitrary J is given by

$$\Gamma(y, J) = \Gamma(y^{\#}, \text{"high } J\text{"}) \, \text{erf}\left\{\Gamma\left(\frac{3}{2}\right) x\right\} \tag{8.78}$$

where $y^{\#} = y - y_m$, with y_m approximated by $y_m \cong (J/b_n)^{2n/(n-2)}$, or obtained more accurately from Eq. (8.53). The variable x is defined by

$$x = \frac{\Gamma(y^{\#}, \text{"low } J\text{"})}{\Gamma(y^{\#}, \text{"high } J\text{"})} \tag{8.79}$$

in which $\Gamma(y^{\#}, \text{"low } J\text{"}) = (2J + 1)\Gamma(y^{\#}, J = 0)$, the latter taken directly from Table 8.4. The gamma function in the argument of the error function ensures that the interpolated $\Gamma(y, J)$ has the correct value at $J = 0$ since $\text{erf } x \cong x/\Gamma(\frac{3}{2})$ for small x.

Quite analogous interpolations are defined for the density of states $N(y, J)$ by merely substituting into Eqs. (8.78) and (8.79) the "high-J" and $J = 0$ forms of $N(y^{\#}, J)$ from Table 8.5. The dashed line in Fig. 8.7 shows the result of the above approximation for $\Gamma(y, J)$ as a function of J at constant y. Strictly speaking interpolated $N(y, J)$ is not quite the derivative of interpolated $\Gamma(y, J)$ with respect to y, but the difference is not large (Problem 8.16).

With the help of Tables 8.4 and 8.5, interpolated results for any of the listed fragment symmetries and at any (y, J) can be obtained for $\Gamma(y, J)$ and $N(y, J)$, and, at the price of actually executing the convolution integral, for $G^*_{vx}(E, J)$ as well.

8.7 Reduction of convolution integrals

The numerical evaluation of a convolution integral typified by Eqs. (8.10) and (8.12) requires the calculation of the integrand over a range of y's for each single value of J. Since these integrals appear frequently in the present context, notably later in Section 8.10 in connection with the distribution of energy between fragments, it is of practical interest to show that the S-D method provides a direct route that considerably reduces the numerical computation [19].

8.7.1 Convolution for the sum of states

For "low J," in view of Eq. (8.48), it is sufficient to consider only the $J = 0$ case. Thus Eq. (8.10) becomes, using Eq. (8.33),

$$G^*_{vx}(E, J = 0) = \frac{1}{\sigma_1 \sigma_2} \int_0^E N^f_v(E - y) \Gamma(y, J = 0) \, dy \tag{8.80}$$

(Note that $y_m = 0$ for $J = 0$.) It follows by examining the entries in Table 8.4 that, at $J = 0$, we can write quite generally $\Gamma(y, J = 0) = \Gamma_0 y^{(r-1)/2}$, where Γ_0 is the factor in front of $y^{(r-1)/2}$ in the $J = 0$ column of Table 8.4. With this expression for $\Gamma(y, J = 0)$ the above equation becomes

$$G_{vx}^*(E, J = 0) = \frac{\Gamma_0}{\sigma_1 \sigma_2} \int_0^E N_v^f(E - y) \, y^{(r-1)/2} \, dy \qquad (8.81)$$

Since it can be easily verified that Γ_0 and $q_r(J = 0)$ defined in the next section are related by

$$q_r(J = 0) = \Gamma_0 \left(\frac{r-1}{2}! \right) \qquad (8.82)$$

it is obvious that the integral in Eq. (8.81) is, to within a constant, merely a convolution between the density of vibrational states of the fragments and the *sum* of rotational states for $r - 1$ free rotors (with constant $q_r(J = 0)$), the evaluation of which was considered previously in Section 4.5.6.

The "high-J" form of $G_{vx}^*(E, J)$ is similarly

$$G_{vx}^*(E, \text{"high } J\text{"}) = \frac{\Gamma_h}{\sigma_1 \sigma_2} \int_0^E N_v^f(E - y) \, y^{r/2} \, dy \qquad (8.83)$$

where Γ_h is the coefficient of $y^{r/2}$ in the "high-J" column of Table 8.4. The lower integration limit is again zero, since, as we have seen in Section 8.5, the "high-J" case contains no reference to J. In this case we have

$$q_r(\text{"high } J\text{"}) = \Gamma_h \left(\frac{r}{2}! \right) \qquad (8.84)$$

Hence Eq. (8.83) is a convolution similar to the $J = 0$ case, this time (to within a constant) for the sum of vibrational states plus r free rotors with constant $q_r(\text{"high } J\text{"})$.

In a similar way it is possible to reduce analogous convolutions for the density of internal–"external" states.

8.7.2 Interpolation for arbitrary J

Using the "low-J" and "high-J" forms of $G_{vx}^*(E, J)$, the convolution integral can now be approximated for arbitrary J by an interpolation similar to that used in Eqs. (8.78) and (8.79) [19]. Thus, for example,

$$G_{vx}^*(E, J) \approx G_{vx}^*(y^\#, \text{"high } J\text{"}) \operatorname{erf} \left\{ \Gamma \left(\frac{3}{2} \right) x \right\} \qquad (8.85)$$

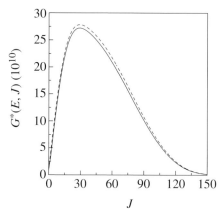

Fig. 8.9. The J-dependence of the vibrational–rotational sum of states $[\sigma_1\sigma_2/(2J + 1)]G^*_{vx}(E, J)$ at $10^4 \, cm^{-1}$ for spherically averaged neutral $CH_3 + CH_3$ [7]. The solid line represents the convolution integral in Eq. (8.12) calculated by 20-point Gaussian quadrature, which may be considered to represent the "exact" result since it makes use of $N_x(y, J)$ obtained from Eq. (8.64) by performing the integrations with respect to L_+ and L_- at every J. The dashed line represents the direct interpolation from Eqs. (8.85) and (8.86) which does away with all integrations. $G^*(E, J)$ in the legend on the vertical axis is an abbreviation for $[\sigma_1\sigma_2/(2J + 1)]G^*_{vx}(E, J)$. Reproduced by permission of The Royal Society of Chemistry on behalf of the PCCP Owner Societies.

where

$$x = \frac{(2J + 1) \, G^*_{vx}(y^\#, J = 0)}{G^*_{vx}(y^\#, \text{"high } J\text{"})} \tag{8.86}$$

with $y^\#$ defined as $y^\# = E - y_m$, the latter determined, as before, from Eq. (8.53). Entirely analogous relations may be defined for $N_{vx}(E, J)$.

Observe that the sum of states in Eq. (8.85) could also be designated $G^*_{vx}(E - y_m, J)$, to signify that it is assigned to $E - y_m$, the energy actually available for redistribution, rather than to E, the nominal total energy.

With $G^*_{vx}(E, J)$ calculated from Eq. (8.85), numerical integrations are dispensed with altogether, so the S-D routine has to be run only twice, each time at a *single* energy E, once for "high J" and once for $J = 0$, instead of at a number of energies $y_m \le y \le E$ as would be required in the numerical integration for the convolution in Eqs. (8.10) and (8.12). An example of this approximation is shown in Fig. 8.9 as the dashed line, which compares quite well with the "exact" evaluation represented by the solid line.

8.8 The *J*-conserved partition function

The conceptually and numerically least-involved way to obtain the sum of internal and "external" states of the fragments $G^*_{vx}(E, J)$, which is the ultimate objective here, is by inversion of the product of the corresponding partition functions for internal and "external" states (on the assumption that there is no coupling between the two). The "internal" (mostly vibrational) states pose no difficulty (see Chapter 4), but the partition function $Q_x(J)$ for "external" states is another matter.

8.8.1 The partition function for "external" states

The calculation of the "exact" J-conserved sum of states $\Gamma(y, J)$ for a given fragment symmetry, as described in Section 8.5 (see also Problems 8.3 and 8.4), yields readily the corresponding "exact" density of states $N(y, J)$ since it is merely the derivative of the "exact" $\Gamma(y, J)$ with respect to y, as shown in Eq. (8.64) for an explicit example.

In principle, therefore, we could write immediately the corresponding "exact" J-conserved partition function $Q_x(J)$ for "external" states (neglecting symmetry numbers):

$$Q_x(J) = \int_{y_m}^{\infty} N_x(y, J)\exp[-y/(k T)]\, dy$$

$$= \exp[-y_m/(k T)] \int_0^{\infty} N_x(y + y_m, J)\exp[-y/(k T)]\, dy \qquad (8.87)$$

which represents the usual definition of a partition function except that it recognizes explicitly that there is a minimum energy required in order to generate the specified angular momentum ($=y_m$), (cf. Eq. (8.10)), which is J-dependent (Eq. (8.53)). Although the actual numerical evaluation of $Q_x(J)$ in Eq. (8.87) poses no problem in principle, it will be readily appreciated that, in the general case, it is computationally intensive.

In the linear-fragment + atom case it is possible to simplify somewhat by using for the density $N_x(y + y_m, J)$ the J-interpolated analog of Eqs. (8.78) and (8.79), and then evaluate the integral in Eq. (8.87), which in this case can be done analytically [23]. With the help of Table 8.5 we have by interpolation

$$N_x(y + y_m, J) = \frac{1}{B}\,\mathrm{erf}\left\{\left(J + \frac{1}{2}\right)\frac{1}{2}\left(\frac{\pi B}{y}\right)^{1/2}\right\} \qquad (8.88)$$

and then, by integration as in Eq. (8.87),

$$Q_x(J) = \exp[-y_m(J)/(k T)]\,\frac{k T}{B}\left\{1 - \exp\left[-\left(J + \frac{1}{2}\right)\left(\frac{\pi B}{k T}\right)^{1/2}\right]\right\} \qquad (8.89)$$

Table 8.6. *The partition function* $\sigma_1\sigma_2 Q_x(J) = q_r(\ell T)^n$

Fragment symmetry (B are rotational constants)	$q_r(J = 0)$, $n = (r - 1)/2$	q_r ("high J"), $n = r/2$	r
Linear + atom	$\dfrac{1}{2}(\pi/B)^{1/2}$	$1/B$	2
Sphere + atom	$1/B$	$\dfrac{\sqrt{\pi}}{B^{3/2}}$	3
Linear + linear $\quad(B_1)\qquad(B_2)$	$\dfrac{\pi^{1/2}\left(\sqrt{B_1} + \sqrt{B_2} - \sqrt{B_1 + B_2}\right)}{2B_1 B_2}$	$\dfrac{1}{B_1 B_2}$	4
Sphere + linear $\quad(B_s)\qquad(B_1)$	$\dfrac{\sin^{-1}\left\{[B_s/(B_s + B_1)]^{1/2}\right\}}{B_s^{3/2} B_1^{1/2}}$	$\dfrac{\sqrt{\pi}}{B_s^{3/2} B_1}$	5
Sphere + sphere $\quad(B_1)\qquad(B_2)$	$\dfrac{\pi^{1/2}}{B_1 B_2 (B_1 + B_2)^{1/2}}$	$\dfrac{\pi}{(B_1 B_2)^{3/2}}$	6

8.8.2 General $Q_x(J)$ by interpolation

For the purpose of ultimate inversion by the S-D method, it is best to work, for any fragment symmetry, with a partition function for "external" states in the form $Q_x = q_r(\ell T)^n$. This is the standard form of the partition function for rotors, for which the inversion routine in combination with oscillators has been worked out in Section 4.5. Since we need Q_x at arbitrary J, the procedure is to interpolate $q_r(\ell T)^n$ between its "high-J" and "low-J" forms.

The requisite forms of $q_r(\ell T)^n$ for various fragment symmetries are obtained directly by taking the Laplace transforms of the $J = 0$ and "high-J" forms of the densities of states in Table 8.5 (or Laplace transforms of sums of states in Table 8.4, divided by ℓT – cf. Eq. (4.11)). The transforms, which involve only simple powers, are dealt with directly (Appendix 2). The results are given in Table 8.6 for $Q_x(J = 0) = q_r(J = 0)(\ell T)^{(r-1)/2}$ and $Q_x("high J") = q_r("high J")(\ell T)^{r/2}$.

By interpolation between "low-J" and "high-J" forms of $Q_x(J)$, which is similar to the interpolations used above in Eqs. (8.85) and (8.86), the resulting interpolated J-conserved partition function $Q_{xi}(J)$ for "external" states at arbitrary J is then

$$Q_{xi}(J) = \exp[-y_m(J)/(\ell T)] \, Q_x("high J") \operatorname{erf}\left\{(2J + 1)\, \Gamma\left(\frac{3}{2}\right) \frac{Q_x(J = 0)}{Q_x("high J")}\right\}$$

(8.90)

Using the explicit forms of $Q_x(J)$ from Table 8.6 for the various fragment symmetries, including symmetry numbers, Eq. (8.90) can be put in the more compact

Table 8.7. *Constants in the J-interpolated partition function of Eq. (8.91)*

$$\sigma_1\sigma_2 Q_{xi}\exp[y_m(J)/(\ell T)] = b(\ell T)^{r/2}\,\mathrm{erf}\left\{\left(J+\frac{1}{2}\right)\frac{a}{(\ell T)^{1/2}}\right\}$$

Fragment symmetry (B are rotational constants)	a	b	r
Linear + atom	$\dfrac{\pi}{2}B^{1/2}$	$\dfrac{1}{B}$	2
Sphere + atom	$B^{1/2}$	$\dfrac{\sqrt{\pi}}{B^{3/2}}$	3
Linear + linear $(B_1)\quad(B_2)$	$\dfrac{\pi}{2}\left(\sqrt{B_1}+\sqrt{B_2}-\sqrt{B_1+B_2}\right)$	$\dfrac{1}{B_1 B_2}$	4
Sphere + linear $(B_s)\quad(B_1)$	$B_1^{1/2}\sin^{-1}\left[\left(\dfrac{B_s}{B_s+B_1}\right)^{1/2}\right]$	$\dfrac{\sqrt{\pi}}{B_s^{3/2}B_1}$	5
Sphere + sphere $(B_1)\quad(B_2)$	$\left(\dfrac{B_1 B_2}{B_1+B_2}\right)^{1/2}$	$\dfrac{\pi}{(B_1 B_2)^{3/2}}$	6

form

$$Q_{xi}(J) = \exp[-y_m(J)/(\ell T)]\,b\frac{(\ell T)^{r/2}}{\sigma_1\sigma_2}\,\mathrm{erf}\left\{\left(J+\frac{1}{2}\right)\frac{a}{(\ell T)^{1/2}}\right\} \qquad (8.91)$$

where the constants a, b, and r are given in Table 8.7 for each fragment combination. Since the $J=0$ and "high-J" forms both refer to the same fragments, the constant a does not contain their symmetry numbers.

The accuracy of the approximation to $Q_{xi}(J)$ represented by Eq. (8.91) is compared with the "exact" partition function in [7, Fig. 1], which shows that the agreement is excellent. Another confirmation is that, in the case linear fragment + atom, the interpolated partition function yields results identical to the analytical result in Eq. (8.89).

8.8.3 Internal–"external" states by inversion

With the above expression for $Q_{xi}(J)$, and the standard partition function Q_v^f for the internal states of the fragments, the sum of states $G_{vx}^*(E, J)$ is obtainable directly at any J from the S-D method by inverting the product $Q_v^f Q_{xi}(J)$.

A general procedure to accomplish this inversion is given in Section 4.5.10. It makes use of the general function $Q_{xx}(J)$ (Eq. (4.109)), which for convenience is displayed again:

$$Q_{xx}(J) = \exp[-c/(\ell T)]\,\frac{b}{\sigma}(\ell T)^{r/2}\,\mathrm{Fn}\left\{\frac{a}{(\ell T)^{\frac{1}{2}}}\right\}$$

This function accommodates both $Q_{xi}(J)$ and the so-called K-rotor partition function $Q_{rK}(J)$ (Appendix 1, Section A1.4) merely by appropriate definition of the constants and the function Fn. In the present instance this means Fn = erf, $c = y_m(J)$, $\sigma = \sigma_1\sigma_2$, and constants a, b, and r from Table 8.7 according to the fragment symmetry. The inversion routine then proceeds as in Eqs. (4.110)–(4.115), with the trivial replacement of Q_v by Q_v^f. The result is that (with $k = 1$) $G_{vx}^*(E, J)$ is obtained directly at any E, J without the necessity to perform any of the intermediate convolutions and interpolations.

The logarithmic derivatives of the quantum form of $Q_v^f(x)$ are available in Chapter 4 in the separable-harmonic-oscillator (Problem 4.14) and Morse-oscillator approximations, including hindered internal rotors if necessary. The inversion of $Q_v^f Q_{xi}(J)$ is invoked further in Chapter 9 in connection with the variational routine of Chapter 7.

8.9 Rate constant by phase-space theory

The preceding sections have provided the wherewithal to obtain the sum of states and partition function for various fragment symmetries under conservation of angular momentum. This provides the ancillary information necessary for an alternative formulation of the unimolecular rate constant, which will now be taken up, both in the microcanonical (specific-energy) form and in the canonical (thermal) form. Let us make it clear that the rate constant detailed below refers to a version of phase-space theory (PST) that contains the constraint that associating fragments must clear the centrifugal barrier, the maximum of which defines the location of the transition state.

8.9.1 The microcanonical rate constant

It was mentioned briefly in connection with Eq. (8.3) in Section 8.1 that, in terms of the reverse flux, the "forward" microcanonical rate constant retains *formally* the usual expression, i.e. the sum of states for the "transition state" divided by the density of states for the reactant multiplied by Planck's constant. The essence of the alternative formulation is the sum of states as defined by Eq. (8.10), in the actual application of which detailed attention has to be paid to zero-point energies and degeneracies.

The various energy parameters are defined in Fig. 8.10. E is now the internal energy of the *products*, ΔH_0 is the 0-K enthalpy of the fragmentation reaction (i.e. the R_1–R_2 bond energy), and, if $E_r(J)$ is the rotational energy of reactant M at a given J, the energy of internal states of M is $E_M = E + \Delta H_0 - E_r(J)$. With the zero of E at the ground state of the *products*, the rate constant

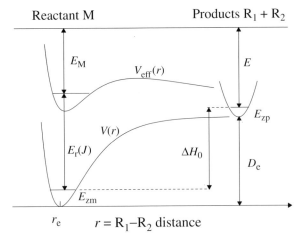

Fig. 8.10. Definition of energy parameters in the PST microcanonical rate constant $k(E, J)$ of Eq. (8.92). $V(r)$ is the purely vibrational potential for the process $M \rightarrow R_1 + R_2$; r_e is the equilibrium R_1–R_2 distance. E is the total (internal + "external") energy of the fragments, E_M is the internal energy of the reactant, $E_r(J)$ is the (overall) rotational energy of the reactant with angular momentum J and $V_{eff}(r)$ is the corresponding effective potential. ΔH_0 is the 0-K enthalpy of the reaction $M \rightarrow R_1 + R_2$, D_e is its classical counterpart, E_{zm} and E_{zp} are the zero-point energies of the reactant and product, respectively. For the purpose of the diagram the position of the products on the R_1–R_2 axis is shown unrealistically close to r_e.

is [1, 13]

$$k(E, J)_{PST} = \frac{(\alpha_f/\alpha_b)G_{vx}^*(E, J)}{h(2J + 1)^m N(E + \Delta H_0 - E_r(J))} \qquad (8.92)$$

where α_f is the (forward) reaction-path degeneracy and α_b is the reaction-path degeneracy for the back reaction $R_1 + R_2 \rightarrow M$.

The $G_x(y, J)$ part of the convolution for $G_{vx}^*(E, J)$ (Eq. (8.10)) involves the degeneracy factor $(2J + 1)/(\sigma_1\sigma_2)$ (cf. Eq. (8.33)). The density of states of the reactant at a given J in the denominator of Eq. (8.92) will also involve the factor $2J + 1$ if the reactant is a symmetric top, or the factor $(2J + 1)^2$ if the reactant is actually a spherical top (cf. Section 4.57). Thus $2J + 1$ cancels out for the symmetric-top reactant ($m = 0$) but not for the spherical-top reactant ($m = 1$). The factor α_b represents the overall symmetry number of the fragments, so $\alpha_b = \sigma_1\sigma_2$.

Insofar as the actual numerical evaluation, at arbitrary J, of $G_{vx}^*(E, J)$ in Eq. (8.92) is concerned, we simply have to invert the partition-function equivalent of $G_{vx}^*(E, J)$ as discussed in the previous section. The density of states in the denominator of Eq. (8.92) is obtained with the same S-D routine by substituting $Q_v(x)$ for $Q_v^f(x)$, deleting all terms related to $Q_{xi}(J)$, setting $k = 0$, and re-defining the energy E as $E + \Delta H_0 - E_r(J)$.

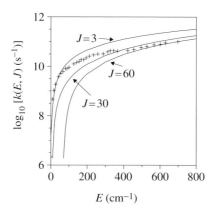

Fig. 8.11. The energy dependence of the microcanonical rate constant $k(E, J)_{\text{PST}}$ for the decomposition $NCNO \rightarrow NC + NO$ [7]. Solid lines: calculated from Eq. (8.92) and the S-D routine (Eqs. (4.110)–(4.115)) at $J = 3, 30$, and 60. Crosses: experimental data at $J = 3$ [36]. Reproduced by permission of the Royal Society of Chemistry on behalf of the PCCP Owner Societies.

If the reactant is a prolate top and contains an active K-rotor, the S-D routine proceeds as above but with $Q_{rK}(J)$ replacing $Q_{xi}(J)$, which amounts merely to re-defining parameters in all terms related to $Q_{xi}(J)$ as indicated in Section 4.5.10. If the reactant is an oblate top with an active K-rotor, the procedure is similar except that the function daw and its derivatives replace the terms related to the error function.

Experimental values for $k(E, J)$, with which the PST results could be compared, are available only for dissociation of three small molecules ($NO_2 \rightarrow NO + O$, $NCNO \rightarrow NC + NO$, and $CH_2CO \rightarrow CH_2 + CO$), and only at low J, because of difficulties of interpretation at large J for large molecules. Consequently only low-J experimental data are available for comparison with the results of the application of Eq. (8.92), an example of which is shown in Fig. 8.11 for the reaction $NCNO \rightarrow NC + NO$ (see [7, Figs. 3 and 7] for other examples). Now, since $G_{vx}^*(E, J)$ in Eq. (8.92) is calculated for the reverse association of fragments, the three reactions cited refer in fact to associations in the category linear + atom, linear + linear, and sphere + linear, respectively, if the fragment CH_2 (an asymmetric rotor) is approximated as a sphere. At low J both $NCNO$ and CH_2CO were treated as prolate tops with properly constrained active K-rotors using the same S-D routine [7].

The other symmetry cases, sphere + atom, represented by $CH_4 \rightarrow CH_3 + H$, and sphere + sphere, represented by $C_2H_6 \rightarrow CH_3 + CH_3$ (with CH_3 "spherically averaged" in all instances), involve large molecules for which PST calculations can be compared only with variational calculations [7, Figs. 5 and 8], and ultimately with experimental thermal results.

The figures show that, in every instance, the PST rate constant is larger than the value found by experiment, though perhaps less so in the case of NO_2. This is an indication that the reactions are not quite at their statistical limit.

8.9.2 The canonical rate constant

With Eq. (8.91) for the partition function of the fragments it is possible to write directly the J-resolved canonical, i.e. thermal, high-pressure PST rate constant $k_\infty(J, T)_{PST}$ for the *dissociation* $M \rightarrow R_1 + R_2$ at specified angular momentum J and specified temperature T as the standard transition-state-theory expression, with particular attention to reaction-path and electronic degeneracies; thus

$$k_\infty(J, T)_{PST} = \frac{\hbar T}{h} \frac{\alpha_f}{\alpha_b} g_e \frac{Q_v^f Q_{xi}(J)}{Q_M} \exp[-\Delta H_0/(\hbar T)] \tag{8.93}$$

where, as before, Q_v^f is the partition function for the internal degrees of freedom of the fragments and $Q_M = Q_v Q_r$ is the total partition function for the reactant molecule factorized into vibrational (Q_v) and rotational (Q_r) parts. The factor g_e is the ratio of effective electronic degeneracies of the fragments and the reactant (i.e. essentially the ratio of electronic partition functions), and α_f and α_b were defined previously. Averaging of $Q_{xi}(J)$ over all J produces the experimentally observable T-dependent thermal *dissociation* rate constant $k_{d,\infty}$:

$$k_{d,\infty} \equiv k_\infty(T)_{PST} = \frac{\hbar T}{h} \frac{\alpha_f}{\alpha_b} g_e \frac{Q_v^f \langle Q_{xi}(J) \rangle_J}{Q_M} \exp[-\Delta H_0/(\hbar T)] \tag{8.94}$$

Here $\langle Q_{xi}(J) \rangle_J$ is the average of $Q_{xi}(J)$ over the thermal distribution of angular momenta $P(J) = (2J + 1)^n \exp[-J(J + 1)B_e/(\hbar T)]$, with B_e the doubly or triply degenerate equilibrium rotational constant of a symmetric-top ($n = 1$) or spherical-top ($n = 2$) reactant, respectively. The normalization constant of $P(J)$ is implicitly included in Q_M as the rotational partition function Q_r.

The corresponding high-pressure *recombination* rate constant can be obtained by application of the equilibrium constant K_c (Section 5.4.5). For the process $R_1 + R_2 \rightleftharpoons M$ the equilibrium constant is

$$K_c = \frac{Q_t Q_M}{g_e Q_t^f Q_v^f Q_r^f} \exp[\Delta H_0/(\hbar T)] = \frac{k_{r,\infty}}{k_{d,\infty}} \tag{8.95}$$

where Q_t/Q_t^f is the ratio of translational partition functions per unit volume (cf. Appendix 1, Eq. (A1.16)) and Q_r^f is the total rotational partition function of the fragments. The high-pressure *recombination* rate constant is then $k_{r,\infty} = k_{d,\infty} K_c$:

$$k_{r,\infty} = \frac{\hbar T}{h} \frac{\alpha_f}{\alpha_b} \frac{Q_t}{Q_t^f} \frac{\langle Q_{xi}(J) \rangle_J}{Q_r^f} \equiv k_{r,\infty}(T)_{PST} \tag{8.96}$$

The power n in the temperature dependence T^n of this recombination rate constant is of some interest. Table 8.8 compares the temperature dependence of $k_{r,\infty}(T)_{PST}$ with experiment for the three examples cited above, plus the two cases $H + CH_3$ and $CH_3 + CH_3$. The comparison shows that, in every case, with the possible exception of $O + NO$, PST produces too much (positive) temperature dependence, which is a reflection of the overestimate of the energy dependence of the microcanonical rate constant, as shown in Fig. 8.11. The reason for the discrepancy must be sought in the fundamental assumption inherent in PST which is not satisfied, namely that the transition state corresponds to essentially separated fragments.

The rate constant $k_{r,\infty}(T)_{PST}$ merits additional scrutiny. The ratio Q_t/Q_t^f is proportional to $(\mu \ell T)^{-3/2}$, where μ is the reduced mass of the fragments. Now $\langle Q_{xi}(J)\rangle_J \approx b(\ell T)^{r/2}\langle \exp[-y_m(J)/(\ell T)]\rangle_J$, but $b(\ell T)^{r/2} \equiv Q_r^f$. Thus, to within better than 10% in the range 300–500 K:

$$k_{r,\infty}(T)_{PST} \approx \text{constant} \times \frac{\langle \exp[-y_m(J)/(\ell T)]\rangle_J}{(\ell T)^{1/2}\mu^{3/2}} \tag{8.97}$$

where the constant includes, apart from numerical factors, the ratio α_f/α_b. The average $\langle \exp[-y_m(J)/(\ell T)]\rangle_J$ is over the rotational distribution $P(J)$ specified above in connection with Eq. (8.94).

For a symmetric-top reactant it turns out that $\langle \exp[-y_m(J)/(\ell T)]\rangle_J \approx \ell T/B_e$, while for a spherical top (methane), $\langle \exp[-y_m(J)/(\ell T)]\rangle_J \approx (\ell T/B_e)^{3/2}$. Thus $k_{r,\infty}(T)_{PST}$ depends principally on μ, and, through $y_m(J)$ and the rotational distribution $P(J)$, on the potential and the rotational constants of the reactant and products. The result is that $k_{r,\infty}(T)_{PST}$ is approximately proportional to $T^{1/2}$ for the symmetric top, and proportional to T for a spherical-top reactant, which is roughly what is shown in Table 8.8. PST tends to overestimate the rate constant because there is no tightening of the "transition state" at high energies as in a variational routine. This has been linked to anisotropy of the potential [24], which PST does not take into account.

8.9.3 Ion–molecule reactions

PST is sometimes better suited to the treatment of ionic systems because the potential may be more attractive than it is in neutral-species–neutral-species systems, with the result that the transition state is "looser," i.e. located at larger intermolecular distances compared with the fragmentation of neutral species. Thus the simplification of taking only the attractive part of the potential, as in Section 8.2 and following sections, is somewhat better justified in ionic systems.

Of particular interest is that ion–molecule reactions are implicated in the chemistry not only of the upper atmosphere [25] but also of interstellar clouds [26], e.g.

Table 8.8. *PST high-pressure recombination rate constants versus experiment,*
$$k_{r,\infty}(T)_{PST} = constant \times T^n$$

Reaction	n (PST)	n (experiment)	References for experimental data
$NO + O \rightarrow NO_2$	0.09	~ 0	(a)
$CH_3 + H \rightarrow CH_4$	0.88	~ 0	(b)
$NC + NO \rightarrow NCNO$	0.28	0 or -0.3	(c) or (d)
$CH_2 + CO \rightarrow CH_2CO$	0.47	Not available	
$CH_3 + CH_3 \rightarrow C_2H_6$	0.49	-0.64	(e)

(a) R. Atkinson, D. L. Baulch, R. A. Cox, R. F. Hampson, J. A. Kerr and J. Troe, *J. Phys. Chem. Ref. Data* **21**, 1125 (1992).
(b) D. L. Baulch, C. J. Cobos, R. A. Cox, C. Esser, P. Frank, Th. Just, J. A. Kerr, M. J. Pilling, J. Troe, R. W. Walker and J. Warnatz, *J. Phys. Chem. Ref. Data* **21**, 411 (1992).
(c) W. Tsang, *J. Phys. Chem. Ref. Data* **21**, 753 (1992).
(d) I. R. Sims and I. W. M. Smith, *J. Chem. Soc. Faraday Trans.* **89**, 1 (1993).
(e) W. Tsang, *Comb. Flame* **78**, 71 (1989). Cf. ref. (a) in Table 7.2.

$C^+ + H_2 \rightarrow CH_2^+$, followed by $CH_2^+ + H_2 \rightarrow CH_3^+ + H$, which starts the synthesis of larger species in intergalactic space, for example $CH_3^+ + HCN \rightarrow CH_3 - HCN^+$ [27].

Closer to earth, a canonical version of PST, with conservation of angular momentum, has been applied [28] with fairly good success to the series of reactions

$$A_2H^+ \leftrightarrow A + AH^+$$

where A may be ammonia or mono-, di-, or tri-methyl amine. A_2H^+ was treated as a spherical top, and the association $A + AH^+$ represented as the case of sphere + sphere. The rate constant for the recombination $A + AH^+$ was calculated as a function of pressure, and the rate constant for dissociation of A_2H^+ as a function of temperature. Except for ammonia, the PST results agreed reasonably well with experiment for all the methyl amines.

Particularly interesting and much investigated reactions are

$$C_2H_4 + C_2H_4^+ \rightarrow C_4H_8^+ \rightarrow \begin{cases} \xrightarrow{k_b} C_4H_7^+ + H & \text{(b)} \\ \xrightarrow{k_c} C_3H_5^+ + CH_3 & \text{(c)} \end{cases}$$

which is something of a workhorse system in ion–molecule reactions, like NO_2 for neutral species. This is actually a chemical-activation process forming excited $C_4H_8^+$ that has two decomposition channels. Channels (b) and (c) were mentioned previously in Section 3.5.2 in connection with $C_4H_8^+$ obtained directly by the PEPICO version of photoionization.

If one now compares the energy- and angular-momentum-averaged branching ratio $R = k_b/k_c$, it is found experimentally [29, 30] that $R = 0.38$ for $C_4H_8^+$ produced by photoionization, and $R = 0.11$ for $C_4H_8^+$ produced in the above ethylene-ion-association reaction. Application of PST, treating $C_2H_4 + C_2H_4^+$ as sphere + sphere, suggests [30] that the difference in branching ratios is due to the different rotational populations of the butene ion produced by the two excitation techniques.

In photoionization only thermal angular momentum is imparted to the butene ion ($J \sim 30$), whereas the association $C_2H_4 + C_2H_4^+$ produces $C_4H_8^+$ with large angular momentum (a distribution with its maximum at $J \sim 80$ [29]) that creates a higher centrifugal barrier, which inhibits the formation of products.

The preponderance of channel (c) over channel (b) in each excitation mode can be understood with respect to the heights of the centrifugal barriers in the two channels. From Eq. (8.16), with $n = 4$ and $C_4 = \frac{1}{2}\alpha q^2$ (Section 8.2.3), we have (assuming that $J \sim L$) $V_{eff}(r_{max}) = (L\hbar)^4/[8\alpha(\mu q)^2]$, so at the same L the barrier will be highest (and, at a given energy, the yield lowest) for the channel with lowest polarizability α and lowest reduced mass μ. This is supported by the data; thus, for channel (b), $\alpha = 0.667$ and $\mu = 0.98$ a.m.u. (it is very low on account of the hydrogen atom), whereas for channel (c) $\alpha = 2.2$ and $\mu = 11$ a.m.u. [29]. More on this later in connection with Fig. 8.16.

In general, therefore, it can be expected that fragmentation channels in which one fragment is a hydrogen atom will be inhibited by a centrifugal barrier. In principle this holds for reactions of neutral species as well, e.g. $CH_4 \rightarrow CH_3 + H$.

8.10 Distribution of energy between fragments

The routines described in the previous sections can be used to calculate, at a given total angular momentum J, the statistical distribution of energy between fragments R_1 and R_2 in a reaction of the type shown in Eq. (8.1). The approach is again semi-classical, meaning that it is assumed implicitly that at least one of the fragments is a polyatomic molecule with a quasi-continuum of energy levels.

The basic premise is, as before, that, for a given interaction potential (taken in the form $-C_n/r^n$), the distribution is given by the phase space occupied by the products, subject to conservation of energy and angular momentum, provided that the system has enough energy to surmount the centrifugal barrier. The implication is that such a distribution is essentially pre-ordained in the transition state, which is assumed to be located at barrier maximum, and is not modified during the actual separation of fragments.

Since in PST the transition state is assumed to consist of essentially separated fragments, all manipulations for the distribution of energy between fragments involve the integrand for the transition-state sum of states $G_{vx}^*(E, J)$ (Eq. (8.10)) and the various ways in which the "external" part of the sum can be put together. By and large, PST is often better at predicting the distribution of energy between fragments than it is at predicting rate constants for fragmentation or recombination.

The problem of the distribution of energy between the fragments can be considered at several levels of detail, as discussed below in progressively finer detail.

8.10.1 Internal–"external" distribution

If the main interest is merely the fraction of energy that is shared between internal and "external" degrees of freedom of the fragments, the corresponding normalized probability is given simply by the integrand of Eq. (8.10), with $G_{vx}^*(E, J)$ the normalization constant:

$$P(E, z, J) = \frac{N_v^f(z)\, G_x(E - z - y_m, J)\, dz}{G_{vx}^*(E, J)}, \qquad 0 \le z \le E - y_m \qquad (8.98)$$

This represents the probability, at specified E and J, that internal degrees of freedom of the fragments shall contain energy z, while "external" degrees of freedom shall contain total energy (relative translational + rotational) $E - z - y_m$. The density of internal states $N_v^f(z)$ follows by straightforward application of the S-D routine (cf. Chapter 4), while the sum of "external" states $G_x(E - z - y_m, J)$ is readily obtained with the help of Eq. (8.23), Table 8.4, and the interpolation routine analogous to Eqs. (8.78) and (8.79). The reduction of $G_{vx}^*(E, J)$ was discussed in some detail in Section 8.7.

The probability in the above form gives all the information about sharing of energy between internal and "external" degrees of freedom when one fragment is an atom; an example is illustrated in Fig. 8.12 for the fragmentation $CF_3I \rightarrow CF_3 + I$. The purpose of Problem 8.17 is to show that, in the case of diatomic + atom, the probability $P(E, z, J)$ can be obtained analytically with the help of Eq. (8.88).

If there are two fragments, both polyatomic, the density of states $N_v^f(z)$ in Eq. (8.98) refers to the total density of internal states of both fragments, so, if we wish to find the probability of specified internal energy z in a particular fragment, say fragment R_1, it is then necessary to convolute the density of fragment R_2 with the sum of "external" degrees of freedom. Applying the rules of convolution (Section 4.1), Eq. (8.98) becomes

$$P(E, J, z) = \frac{N_{v1}^f(z)\, dz \displaystyle\int_{y_m}^{E-z} dy\, N_{v2}^f(E - z - y)\, G_x(y, J)}{G_{vx}^*(E, J)} \qquad (8.99)$$

where the integral in the numerator represents a sum of states, i.e. the convolution of the sum of "external" states with the density of internal states of fragment R_2,

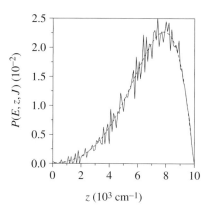

Fig. 8.12. The normalized distribution of energy between the fragments of the reaction $CF_3I \rightarrow CF_3 + I$ at total energy $10^4 \, cm^{-1}$ and $J = 20$. The plot represents $P(E, z, J)$, where z is the vibrational energy of spherically averaged CF_3 with rotational constant $B = 0.283 \, cm^{-1}$, so this is an example of the spherical top + atom case. The discontinuous curve is the result of an exact count of vibrational states of CF_3; the dashed curve is the smooth-function approximation from Eq. (8.98).

at total energy $E - z$ and angular momentum J. Shifting the integration limits and indicating the sum of states explicitly, we can write

$$P(E, J, z) = \frac{N_{v1}^f(z) \, G_{v2,x}^f(E - z - y_m, J) \, dz}{G_{vx}^*(E, J)} \tag{8.100}$$

The sum of states $G_{v2,x}^f(E - y_m - z, J)$ is similar to that encountered in Section 8.7, where it was shown that it can be obtained, directly and at arbitrary J, by interpolation between the $J = 0$ and "high-J" forms (Eqs. (8.85) and (8.86)), by noting that $G_{v2,x}^f(E - z - y_m, J = 0)$ is equivalent, to within a constant, to the sum of states for vibrations plus $r - 1$ free rotors, while $G_{v2,x}^f(E - z - y_m,$ "high J") is the same for r free rotors, r being the number of rotations for a given symmetry category as given in the last column of Table 8.4. The term $G_{vx}^*(E, J)$ in the denominator is still given by Eq. (8.10). An example of the application of Eq. (8.100) is shown in Fig. 8.13 and is the subject of Problem 8.18.

A question of interest in the present context can be put thus: what is the error if conservation of angular momentum is ignored? In practical terms this means that the sum of "external" states in Eq. (8.100) is simply replaced by the sum of internal states of fragment 2: $G_{v2,x}^f(E - z - y_m, J) \Rightarrow G_{v2}^f(E - z)$. The perhaps obvious answer is the dot–dashed curve in Fig. 8.13: the amount of energy carried off in the internal energy of fragment 1 is overestimated since there is no energy going into the "external" degrees of freedom. Note at the same time that ignoring angular momentum is not the same as assuming that there is zero angular momentum (compare the dot–dashed curve with the solid line in Fig. 8.13). This was already

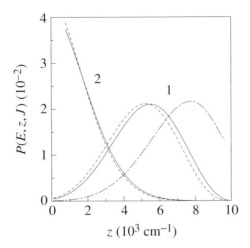

Fig. 8.13. An example of the application of Eq. (8.100): the normalized distribution of energy between the fragments of the reaction $CH_3O_2NO_2 \rightarrow CH_3O_2 + NO_2$ at total energy 10^4 cm^{-1}. Curves labeled 1 refer to energy z in the fragment CH_3O_2, while curves labeled 2 refer to energy z in the fragment NO_2; solid lines are for $J = 0$, and dashed lines for $J = 50$. In each case the remaining energy $E - z$ is contained in internal degrees of freedom of the other fragment, plus "external" degrees of freedom. The dot–dashed curve represents the distribution of energy z in the fragment CH_3O_2 when conservation of angular momentum is ignored.

mentioned in Section 8.5 in connection with Eq. (8.48) and explained by the vectors J_r and L being equal and anti-parallel.

As shown in Chapter 4, a density multiplied by an energy interval, like the quantity $N^f_{v1}(z) \, dz$ in Eqs. (8.98)–(8.100), can be considered as the number of states $W(z)$ at energy z, which is amenable to exact enumeration. Figure 8.12 compares the result of such an exact count with a smooth-function approximation obtained with the S-D routines of Section 8.7. Clearly the approximation provides a very satisfactory average of the discrete count.

8.10.2 *Translational/rotational distribution*

If the main interest is the probability distribution of relative translational and rotational energy individually for a fragment, we need more information than just the sum of "external" states at specified total angular momentum (J) and specified sum $y = E_t + E_r$ of relative translational (E_t) and rotational energy (E_r). To do this, basically we go back to the level of Eq. (8.98) but with a more highly resolved $G_x(y, J)$, keeping the same unresolved total internal fragment states, for which we specify only the total internal (in principle vibrational) energy $E_v = E - y$.

Such a resolved $G_x(y, J)$ is in essence an integral of a density of states but not the previously determined density of states $N(y, J)$ (e.g. Eq. (8.64)), which is not sufficient for this purpose. We require instead the detailed density $N_x(J, E_t, E_r)$, i.e. density (of "external" states) at individually specified J, E_t, and E_r such that $E_t + E_r = y$.

The probability $P_t(E, J, E_t)$ of finding specified relative translational energy E_t, at given E and J is then obtained by the appropriate convolution between the densities of internal states and the detailed density of "external" states of the fragments:

$$P_t(E, J, E_t) = \frac{dE_t}{G^*_{vx}(E, J)} \int_{E_r^{min}}^{E-E_t} N_v^f[(E - E_t) - E_r] N_x[(J, E_t), E_r] dE_r \quad (8.101)$$

In other words, the E_t-distribution is obtained by integrating out the E_r-dependence. For clarity round brackets enclose variables that are kept constant throughout the integration.

This distribution is really a more detailed version of the translational energy distribution $P(E, \epsilon_t)$ obtained on the basis of a very simple argument in Chapter 3 (Eq. (3.40)), where ϵ_t is E_t in the present notation, and the zero of E in $P(E, \epsilon_t)$ is the ground state of the reactant, whereas in Eq. (8.101) the zero of E is the ground state of the products. Since it is difficult to determine E_t for neutral species, calculations and experimental results on the distribution of translational energy (often referred to, rather imprecisely, as the "kinetic energy release") come mostly from measurements on ionic systems [31].

The analogous probability $P_r(E, J, E_r)$ of a given E_r, at specified E and J, is given by a similar convolution between the densities of internal (vibrational) and "external" states of the fragments, such that now the E_t-dependence is integrated out:

$$P_r(E, J, E_r) = \frac{dE_r}{G^*_{vx}(E, J)} \int_{E_t^{min}}^{E-E_r} N_v^f[(E - E_r) - E_t] N_x[(J, E_r), E_t] dE_t \quad (8.102)$$

The lower integration limits E_r^{min} and E_t^{min} are discussed below.

Since Eqs. (8.101) and (8.102) are merely transformations of Eq. (8.98), $G^*_{vx}(E, J)$ is clearly the normalization constant for both P_t and P_r. Symmetry numbers are redundant since they cancel out upon normalization.

8.10.3 The density in the J_r–L space

For the purpose of evaluating the integrals in Eqs. (8.101) or (8.102), integration of the detailed "external" density has to be done at specified J, E_r, or E_t, i.e. at specified J_r^* or L^*, since $J_r^* = (E_r/B_r)^{1/2}$ and $L^* = b_n E_t^{(n-2)/2n}$, cf. Eq. (8.17). Thus the domain of integration is delimited by straight lines parallel to the J_r or

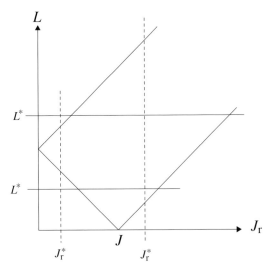

Fig. 8.14. An illustration of integration in the J_r–L plane at specified $J_r^* = (E_r/B_r)^{1/2}$ (dashed vertical line) or $L^* = b_n E_t^{(n-2)/2n}$ (solid horizontal line). The region of integration is represented by the segment of a (vertical or horizontal) line comprised inside the J-delimited rectangle and one solid and one dashed line.

L axes, as shown in Fig. 8.14, where constant L^* represents a horizontal solid line, and constant J_r^* a dashed vertical line. Both the $(L^*, J_r^*) < J$ case and the $(L^*, J_r^*) > J$ case are shown.

$N_x[(J, E_t), E_r]$ in Eq. (8.101) as a function of E_r, at fixed E_t, can be viewed, with the help of Fig. 8.14, in terms of the "area" below the line $L^* = $ constant swept out by the vertical straight line representing J_r^* as it moves along the J_r axis across the J-delimited $L - J_r$ plane from $J_r^* = 0$ to $J_r^* = [(E - E_t)/B_r]^{1/2}$. However, if $L^* < J$, it is clear from Fig. 8.14 that the "area," i.e. the integrand, will be zero unless we have $J_r^* \geq J - L^*$, so the lower limit in Eq. (8.101) is $E_r^{\min} = B_r(J - L^*)^2$, unless $L^* \geq J$, in which case $E_r^{\min}$ is zero.

Similarly, $N_x[(J, E_r), E_t]$ in Eq. (8.102) as a function of E_t, at fixed E_r, can be viewed as the "area" to the left of the vertical line $J_r^* = $ constant swept out by the horizontal straight line representing L^* as it moves along the L-axis from $L^* = 0$ to the maximum given by $L^* = b_n(E - E_r)^{(n-2)/(2n)}$. In this case, if $J_r^* < J$, the "area" will be zero unless $L^* \geq J - J_r^*$, so, under these conditions, the lower limit in Eq. (8.102) is $E_t^{\min} = [(J - J_r^*)/b]^{2n/n-2}$, whereas $E_t^{\min}$ is zero if $J^* \geq J$.

In either case the density that we need is given by integrating $N(J_r)$ in the J-constrained $L - J_r$ space:

$$N_x(J, E_t, E_r) = \int dL \int dJ_r \, N(J_r) \qquad (8.103)$$

This is illustrated by two examples.

8.10.4 Example 1: atom + linear/spherical top

The density $N(J_r)$ for these two systems can be written semi-classically in terms of the δ-function as $N(J_r) = (2J_r)^{r-2}\,\delta(E_r - B_r J_r^2)$ [32, p. 262], where the exponent r takes the value $r = 2$ for linear + atom and $r = 3$ for sphere + atom (cf. Table 8.4). Using the integration properties of the δ-function (see Appendix 3), note that integration with respect to E_r gives $\Gamma(J_r) = (2J_r)^{r-2} \int \delta(E_r - B_r J_r^2)\,dE_r = (2J_r)^{r-2}$, which is the same result as that obtained previously in Section 8.4 if we make the approximation $2J_r + 1 \cong 2J_r$.

In the present context (cf. Eq. (8.103)) we need first the integral of the density with respect to J_r, i.e. $\int N(J_r)\,dJ_r$. We have (Appendix 3)

$$\int (2J_r)^{r-2}\,\delta\!\left(E_r - B_r J_r^2\right) dJ_r = \begin{cases} \dfrac{1}{2(B_r E_r)^{1/2}}, & r = 2 \text{ (linear + atom)} \\[2mm] \dfrac{1}{B_r}, & r = 3 \text{ (sphere + atom)} \end{cases}$$

(8.104)

Evidently, then, $\int N(J_r)\,dJ_r$ is independent of L, and consequently further integration with respect to L (the "L-integral" $L_{int} = \int \ldots dL$) yields results that are the same for sphere + atom and linear + atom. Hence $N_x(J, E_t, E_r)$ for the two cases differs only by the multiplicative factor given by Eq. (8.104).

L_{int} takes somewhat different forms for the E_t- and E_r-distributions. In the case of the E_t-distribution, i.e. for the J_r-dependence at fixed J and L^* (Fig. 8.14, horizontal lines), we have

$$\text{for } J_r \le |L^* - J|: \quad L_{int} = \begin{cases} 0 & \text{if } L^* < J \\ \min(2J_r, 2J) & \text{if } L^* > J \end{cases}$$

$$\text{for } |L^* - J| \le J_r \le L^* + J: \quad L_{int} = L^* - |J - J_r| \tag{8.105}$$

$$\text{for } J_r > L^* + J: \quad L_{int} = 0$$

The E_r-distribution, i.e. the L-dependence at fixed J and J_r^* (Fig. 8.14, dashed vertical lines) is given by

$$\text{for } L < |J - J_r^*|: \quad L_{int} = 0$$

$$\text{for } |J - J_r^*| \le L \le J + J_r^*: \quad L_{int} = L - |J - J_r^*| \tag{8.106}$$

$$\text{for } L > J + J_r^*: \quad L_{int} = \min(2J_r^*, 2J)$$

Note that there is a relation between J_r^* and L^* used here and J_m and L_m defined previously in Eqs. (8.36) and (8.37) in connection with the vertical and horizontal cut-offs in the calculation of the sum of states $\Gamma(y, J)$ (Problem 8.19).

An example of the application of normalized probabilities $P_t(E, J, E_t)\,dE_t$ and $P_r(E, J, E_r)\,dE_r$ to the linear + atom case, using Eqs. (8.104) and (8.106), is shown

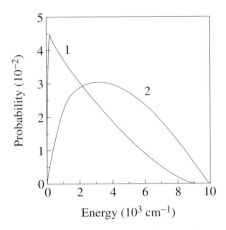

Fig. 8.15. Examples of the normalized distributions of energy between a linear fragment and an atom (Eq. (8.104), with $r = 2$) at total energy $y = 10^4\,\text{cm}^{-1}$ and $J = 10$. The parameters used are applicable to the system $NO + O$ ($C_6 = 1.5 \times 10^{-59}\,\text{erg cm}^6$, $B_r = 1.705\,\text{cm}^{-1}$). NO is assumed to be a harmonic oscillator ($\omega = 1904\,\text{cm}^{-1}$). Curve 1 is the rotational probability $P_r(y, J, E_r)\,dE_r$ as a function of E_r. Curve 2 is the translational probability $P_t(y, J, E_t)\,dE_t$ as a function of E_t.

in Fig. 8.15. The specific model system is $NO_2 \rightarrow NO + O$ but the results are fairly typical of experimental results obtained for the fragmentation of a number of triatomics, e.g. $Ba + O_2$ and $Li + HF$ [33]: a bell-shaped translational probability, and a monotonically decreasing rotational probability.

A good example of the translational probability $P_t(E, J, E_t)\,dE_t$ in the sphere + atom case is the fragmentation $CH_3I^+ \rightarrow CH_3^+ + I$, assuming a spherically averaged methyl ion. Experimental results [34] on energy-selected methyl iodide ions essentially agree with E_t obtainable from Eqs. (8.104)–(8.105).

8.10.5 Example 2: two spherical tops

We start with the density $N(J_r)$ which we have obtained previously (Eq. (8.62)):

$$N(J_r) = \frac{4 J_r}{\omega^{3/2}} \left(E_r - B_r J_r^2 \right)^{1/2} \tag{8.107}$$

Proceeding as in Example 1, consider first the indefinite integral with respect to J_r, the result of which is

$$\int dJ_r\, N(J_r) = -\frac{4}{3} \frac{B_r^{1/2}}{\omega^{3/2}} \left(\frac{E_r}{B_r} - J_r^2 \right)^{3/2} = F(J_r) \tag{8.108}$$

Given the nature of the problem, the lower integration limit with respect to J_r (at fixed L) is always $|J - L|$, while the upper limit is either $J + L$ or J_r^*, depending on the available rotational energy (Fig. 8.14). When $J_r = J_r^*$, $F(J_r) = 0$ since

$J_r^{*2} = E_r/B_r$. Therefore we need consider only $J_r = J \pm L$, in which case the subsequent integral with respect to L becomes (note the minus signs)

$$-R_\pm(L) = \int F(J \pm L)\,dL = -\frac{4\,B_r^{1/2}}{3\,\omega^{3/2}} \int \left(\frac{E_r}{B_r} - (J \pm L)^2\right)^{3/2} dL \quad (8.109)$$

Calculation of this indefinite integral with respect to L yields

$$-R_\pm(L) = \frac{B_r^2(L \pm J)}{3(B_1 B_2)^{3/2}} \left(\frac{E_r}{B_r} - (L \pm J)^2\right)^{1/2} \left(\frac{5}{2}\frac{E_r}{B_r} - (L \pm J)^2\right)$$
$$+ \frac{E_r^2}{2(B_1 B_2)^{3/2}} \sin^{-1}\left(\frac{L \pm J}{(E_r/B_r)^{1/2}}\right) \quad (8.110)$$

$R_\pm(L)$ remains to be evaluated between a number of different upper and lower integration limits with respect to L that depend on the relative values of J, J_r^*, and L^*. The end result is that the density of "external" states for two spherical fragments under all conditions can be expressed in terms of five different combinations of $R_\pm(L)$, denoted below by R_1 through R_5:

$$R_1 = R_-(L^*) + R_4; \qquad R_2 = 2R_+(0) + R_-(L^*) - R_4$$
$$R_3 = 2R_+(0) + R_-(L^*) - R_+(L^*); \qquad R_5 = 2R_+(0) \quad (8.111)$$

These results make use of the following reductions:

$$R_+(J_r^* - J) = \frac{\pi E_r^2}{4(B_1 B_2)^{3/2}} = R_4; \qquad R_-(J \pm J_r^*) = \pm R_4$$
$$R_+(0) - R_-(0) = 2R_+(0)$$

The actual calculation of the explicit expressions for the density of states, while tedious, is fairly straightforward (Problem 8.20).

Figure 8.16 shows the E_t-dependence of the normalized probability $P_t(E, J, E_t)\,dE_t$ for the sphere–sphere system $C_3H_5^+ + CH_3$, calculated from Eqs. (8.101) and (8.110) and using the results of Problem 8.20. For comparison the E_t-dependence for the system $C_4H_7^+ + H$ (sphere–atom) is shown as an illustration of the effect of the increased centrifugal barrier due to the hydrogen atom.

The case of the spherical + linear fragments is considerably more complicated in that the density of "external" states (Eq. (8.103)) yields 17 different results [17] on account of the two forms of the density $N(J_r)$ (Eq. (8.71)).

8.10.6 Non-statistical distribution

The assumption inherent in the results presented here, as stated at the outset, is that the distribution of energy is given by the phase space occupied by the products,

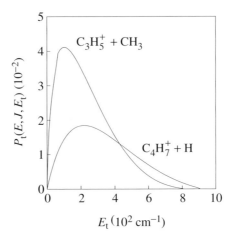

Fig. 8.16. The distribution of translational energy between the fragments of the two product channels in the decomposition of $C_4H_8^+$, calculated from Eq. (8.110) for $C_3H_5^+ + CH_3$ (channel c), and from Eq. (8.105) for $C_4H_7^+ + H$ (channel b), both at $J = 10$ and $E = 1000\,cm^{-1}$. For clarity, the curve for channel b is normalized with respect to $\frac{1}{2}$, while the other is normalized to unity. The effect of the larger centrifugal barrier in channel b is to reduce the probability of low translational energies. The calculation agrees qualitatively with experimental results [30, Figs. 8 and 9].

subject only to conservation requirements. In particular, the fraction of energy that is retained by the "external" degrees of freedom depends on the corresponding sum or density of states. Examination of Table 8.4 or 8.3 shows that the energy dependence of the sum or density decreases in the sequence sphere + sphere → linear + atom, i.e. in the sense of decreasing fragment symmetry.

By assigning lower (fictitious) fragment symmetries than would be appropriate for actual fragments, it is therefore possible to decrease, within limits, the amount of "external" energy carried off by fragments in a given reaction. The effect is the same as if not all phase space were available to the fragments. If such a reduced symmetry assignment is necessary in a given reaction in order to obtain agreement with experiment, there is a strong suggestion of a non-statistical effect. This is a somewhat crude way to achieve the same contraction of phase space similar to that postulated by the separated-statistical-ensembles method [35].

References

[1] C. E. Klots, *J. Phys. Chem.* **75**, 1526 (1971); *Z. Naturforsch.* **27a**, 553 (1972).
[2] W. Forst, *Theory of Unimolecular Reactions* (Academic Press, New York, 1973).
[3] H. Eyring, S. H. Lin and S. M. Lin, *Basic Chemical Kinetics* (Wiley, New York, 1980), pp. 245ff.
[4] M. A. Eliason and J. O. Hirschfelder, *J. Chem. Phys.* **30** 1426 (1959).

[5] T. Baer and W. L. Hase, *Unimolecular Reaction Dynamics* (Oxford University Press, New York, 1996).

[6] J. O. Hirschfelder, C. F. Curtiss and R. B. Bird, *Molecular Theory of Gases and Liquids* (Wiley, New York, 1967).

[7] W. Forst, *Chem. Phys. Phys. Chem.* **1**, 1283 (1999).

[8] P. Pechukas, J. C. Light and C. Rankin, *J. Chem. Phys.* **44**, 794 (1966).

[9] L. B. W. Jolley, *Summation of Series*, 2nd Ed. (Dover, New York, 1961).

[10] W. J. Chesnavich and M. T. Bowers, *Prog. Reaction Kinet.* **11**, 137 (1982).

[11] W. J. Chesnavich and M. T. Bowers, *J. Am. Chem. Soc.* **98**, 8301 (1976). There is a misprint in Eq. (10).

[12] W. J. Chesnavich and M. T. Bowers, *J. Chem. Phys.* **66**, 2306 (1977).

[13] W. J. Chesnavich and M. T. Bowers, *J. Am. Chem. Soc.* **99**,1705 (1977).

[14] W. J. Chesnavich and M. T. Bowers, *J. Chem. Phys.* **68**, 901 (1978). There is a misprint in Eq. (18).

[15] W. J. Chesnavich and M. T. Bowers, in *Gas Phase Ion Chemistry*, Vol. 1, M. T. Bowers, Ed. (Academic Press, New York, 1979), pp. 119ff.

[16] V. Uspensky, *Theory of Equations* (Academic Press, New York, 1948), Ch. 5.

[17] W. J. Chesnavich, Ph.D. Thesis, University of California, Santa Barbara 1976. There are numerous misprints.

[18] M. E. Grice, K. Song and W. J. Chesnavich, *J. Phys. Chem.* **90** 3503 (1986).

[19] W. Forst, *Chem. Phys. Lett.* **262**, 539 (1996).

[20] M. Olzmann and J. Troe, *Ber. Bunsenges. Phys. Chem.* **96**, 1327 (1992). See also J. Troe, *J. Chem. Phys.* **79**, 6017 (1983), Appendix C.

[21] M. Abramowitz and I. A. Stegun, *Handbook of Mathematical Functions* (NBS Applied Mathematics Series 55, Washington, 1972), p. 297.

[22] C. E. Klots and J. Polach, *J. Phys. Chem.* **99**, 15 396 (1995).

[23] I. S. Gradshteyn and I. M. Ryzhik, *Tables of Integrals, Series and Products* (Academic Press, New York, 1965), p. 649, item 6.284.

[24] J. Troe, *Ber. Bunsenges. Phys. Chem.* **101**, 438 (1997).

[25] R. E. Weston, Jr and H. A. Schwarz, *Chemical Kinetics* (Prentice-Hall, Englewood Cliffs, NJ, 1972), pp. 225ff.

[26] E. Herbst and W. Klemperer, *Physics Today*, June 1976, p. 32.

[27] L. M. Bass, P. R. Kemper, V. G. Anicich and M. T. Bowers, *J. Am. Chem. Soc.* **103**, 5283 (1981); S. C. Smith, M. J. McEwan and R. G. Gilbert, *J. Chem. Phys.* **90**, 4265 (1989).

[28] L. Bass, W. J. Chesnavich and M. T. Bowers, *J. Am. Chem. Soc.* **101**, 5493 (1979).

[29] G. G. Meisels, G. M. L. Verboom, M. J. Weiss and T. C. Hsieh, *J. Am. Chem. Soc.* **101**, 7189 (1979).

[30] W. J. Chesnavich, L. Bass, T. Su and M. T. Bowers, *J. Chem. Phys.* **74**, 2228 (1981).

[31] C. E. Klots, *J. Chem. Phys.* **64**, 4269 (1976).

[32] E. E. Nikitin, *Theory of Elementary Atomic and Molecular Processes in Gases* (Clarendon Press, Oxford, 1974).

[33] L. Bonnet and J. C. Rayez, *Chem. Phys.* **201**, 203 (1995); *J. Chem. Phys.* **102**, 9512 (1995).

[34] T. Baer, U. Buchler and C. E. Klots, *J. Chim. Phys.* **77**, 739 (1980).

[35] C. Wittig, I. Nadler, H. Reisler, M. Noble, J. Catanzarite and G. Radhakrishnan, *J. Chem. Phys.* **83**, 5581 (1985).

[36] L. R. Khundkar, J. L. Knee and A. H. Zewail, *J. Chem. Phys.* **87**, 77 (1987).

[37] S. J. Klippenstein, L. R. Khundkar, A. H. Zewail and R. A. Marcus, *J. Chem. Phys.* **89**, 4761 (1988).

Problems

Problem 8.1. Equations (8.15)–(8.17) can be expressed in terms of the impact parameter b defined by $\hbar L = \mu v b = b(2\mu E_t)^{1/2}$ (v is the relative velocity). The critical impact parameter b_c (cf. Section 5.2.4) is determined by the condition $V_{eff}(r_{max}) = E_t$. Show that, for the potential $-C_n/r^n$,

$$b_c^2 = \left(\frac{n}{n-2}\right)\left(\frac{(n-2)C_n}{2E_t}\right)^{2/n}$$

which can then be used to define the corresponding orbital angular momentum L. Show that the re-quantized result is

$$b_c^2 = \frac{\hbar^2}{2\mu E_t}\sum_{L=0}^{L}(2L+1) = \frac{\hbar^2(L+1)^2}{2\mu E_t}$$

which is in effect Eq. (8.17) with the equality sign (note the distinction between b_c and b_n defined in Section 8.2.2).

Problem 8.2. Show that, in the sphere + atom case with horizontal cut-off, the $L_m > J$ and $L_m < J$ results are both

$$\Gamma(y, J) = (2J+1)(L_m+1)^2$$

Problem 8.3. Assuming that L_+ and L_- are known, show that, for arbitrary J and $n = 6$, the J-conserved sum of states for the case linear + atom is given by

$$\Gamma_1(y, J) = I - \frac{1}{2}[(L_+ - J)^2 \pm (L_- - J)^2] \begin{cases} + \text{ sign for } L_- < J \\ - \text{ sign for } L_- > J \end{cases}$$

for $(y/B)^{1/2} \equiv J_m < J$; and

$$\Gamma(y, J) = \Gamma_1(y, J) + \begin{cases} L_-^2, & \text{if } L_- < J \\ J^2 + 2J(L_- - J), & \text{if } L_- > J \end{cases}$$

for $(y/B)^{1/2} \equiv J_m \geq J$, where I is the integral

$$I = \int_{L_-}^{L_+}\left(\frac{y - (L/b_6)^3}{B}\right)^{1/2}dL$$

which has to be obtained numerically.

Problem 8.4. Assuming that L_+ and L_- are known, show that the J-conserved sum of states for the case sphere + atom, at arbitrary J and for $n = 6$, is given by

$$\Gamma(y, J) = \frac{y}{B}(L_+ - L_-) - \frac{L_+^4 - L_-^4}{4Bb_6^3} + \frac{1}{3}[(J + L_-)^3 + (J - L_+)^3 - 2J^3]$$

for $J < (y/B)^{1/2}$; and

$$\Gamma(y, J) = \frac{y}{B}(L_+ - L_-) - \frac{L_+^4 - L_-^4}{4Bb_6^3} + \frac{1}{3}[(J - L_+)^3 + (J - L_-)^3]$$

for $J \geq (y/B)^{1/2}$.

Problem 8.5. Show that, for $n = 6$, the solution for J_r from Eq. (8.46) takes three forms, depending on the magnitude of J with respect to L_m and J_m:

$$(1) \qquad J_r^3 + J_r^2 \left(\frac{L_m^3}{J_m^2} + 3J \right) + 3J_r J^2 - L_m^3 + J^3 = 0$$

$$(2) \qquad J_r^3 + J_r^2 \left(\frac{L_m^3}{J_m^2} - 3J \right) + 3J_r J^2 - L_m^3 - J^3 = 0$$

$$(3) \qquad J_r^3 - J_r^2 \left(\frac{L_m^3}{J_m^2} + 3J \right) + 3J_r J^2 + L_m^3 - J^3 = 0$$

This is because in this case the RHS of Eq. (8.46) is an odd power of $J_r \pm J$ or $J - J_r$, so the relative magnitudes of J_r and J are important. Verify that, if $L_m > J$, L_+ is obtained from (1); similarly, if $J_m > J$, L_- is obtained from (2). If both L_m and J_m are larger than J, L_+ and L_- are both obtained from (3). In general, each equation will have one real positive solution; if there are two real positive solution, the higher one will be J_+, the lower one J_-.

Problem 8.6. Show that, for $n = 4$, Eq. (8.46) is a quartic that takes only two forms:

$$J_r^4 \pm 4J_r^3 J + J^2 \left(6J^2 + \frac{L_m^4}{J_m^2} \right) \pm 4J_r J^3 + J^4 - L_m^4 = 0$$

where the upper $(+)$ sign applies to the solution for L_+, the lower $(-)$ sign for L_-.

Problem 8.7. If L^* is the value of L at which the slope of the curve of L_m versus J_m is -1, show, starting from Eq. (8.24), that L^* is the solution of the equation

$$aL^{z+4/(n-2)} + L^z - L_m^z = 0$$

where $z = 2n/(n-2)$ and $a = [(z/2)J_m/L_m^{z/2}]^2$. For $n = 6$, this gives the quartic $aL^4 + L^3 - L_m^3 = 0$, where $a = 9J_m^2/(4L_m^3)$.

Problem 8.8. Show that the exact upper limit of the integral in Eq. (8.47) is

$$J_r^* = \left(\frac{y - (L^*/b_n)^{2n/(n-2)}}{B_r} \right)^{1/2}$$

where L^* is the solution of $(L/b_n)^{2n/(n-2)} + B_r L^2 - y = 0$. This L^* is of course $L_+ \equiv L_-$ for $J = 0$ (cf. Eq. (8.45)). For $n = 6$ the equation is a cubic in L at specified y. Using the trigonometric solution of the cubic, show that

$$L^* = \frac{1}{3} B_r b_6^3 \left\{ \cos \left[\frac{1}{3} \cos^{-1} \left(\frac{27}{2} \frac{y}{B_r^3 b_6^6} - 1 \right) \right] - 1 \right\}$$

Show that, for $n = 4$, the equation is a quartic with the solution

$$L^* = \sqrt{\frac{1}{2} b_4^2 \left(\sqrt{B_r^2 b_4^4 + 4y} - B_r b_4^2 \right)}$$

Relative to the exact result, the approximation $J_r^* \approx (y/B_r)^{1/2} \equiv J_m$ amounts to only a small overestimate, except at very high energies. For example, in the spherically averaged $CH_3 + CH_3$ case $(n = 6)$, the approximation overestimates the exact J_r^* by 0.1% at $y = 100 \, cm^{-1}$, by 1% at $y = 10^4 \, cm^{-1}$, and by 10% at $y = 10^6 \, cm^{-1}$.

Problem 8.9. Verify that the "high-J" results for the sum and density of states given in Tables 8.4 and 8.5 for the various fragment combinations are indeed obtainable from the general formula for free rotors.

Problem 8.10. Show that for the horizontal cut-off ($L_m \gg J_m$), the general "high-J" result (i.e. for $L_m < J$, see Fig. 8.4(b), bottom line) is

$$\Gamma(y, \text{"high } J\text{"}) = \int_0^{L_m} dL \int_{J-L}^{L+J} dJ_r \, \Gamma(J_r)$$

whereas for "low J" (i.e. for $L_m > J$, see Fig. 8.4(b), top line) it is

$$\Gamma(y, \text{"low } J\text{"}) = \int_0^{J} dL \int_{J-L}^{L+J} dJ_r \, \Gamma(J_r) + \int_{J}^{L_m} dL \int_{L-J}^{L+J} dJ_r \, \Gamma(J_r)$$

where $L_m = b_n y^{(n-2)/(2n)}$.

Problem 8.11. For $n = 4$, Eq. (8.52) reads $L^3 + \frac{1}{2} B_r b_4^4 (L - J) = 0$. Show that the solution is [12]

$$L^* = \left(\tfrac{1}{2} B_r b_4^4 \right)^{1/3} \left[\left(S + \tfrac{1}{2} J \right)^{1/3} - \left(S - \tfrac{1}{2} J \right)^{1/3} \right]$$

where

$$S = \left(\frac{1}{4} J^2 + \frac{1}{54} B_r b_4^4 \right)^{1/2}$$

For J not too large ($J < 100$), the solution is $L^* \cong J$.

Problem 8.12. Show, from Eq. (8.57), that (a) $J_+ = -J_- = B_2 J_r / \omega$ when $E_r = B_r J_r^2$; (b) $J_- = 0$ when $E_r = B_2 J_r^2$; and (c) $J_+ = J_r$ when $E_r = B_1 J_r^2$.

Problem 8.13. Verify Eq. (8.71).

Problem 8.14. Show that Eq. (8.76) for the sphere–prolate symmetry reduces to the sphere–sphere case when $A \to B$, i.e. as the prolate top becomes spherical.

Problem 8.15. (a) Using Eqs. (8.47) and (8.49) verify the $J = 0$ and "high-J" forms of $\Gamma(y, J)$ for linear + atom and sphere + atom in Table 8.4.
(b) Show that, for two linear fragments, the $J = 0$ form is

$$\Gamma(y, J = 0) = \frac{2}{3} \frac{\left(\sqrt{B_1} + \sqrt{B_2} - \sqrt{B_1 + B_2} \right)}{B_1 B_2} y^{3/2}$$

and the "high-J" form is

$$\Gamma(y, \text{"high } J\text{"}) = \frac{y^2}{2 B_1 B_2}$$

Problem 8.16. The interpolated sum of states (Eq. (8.78)) can be written $\Gamma(y, J) = a y^{r/2} \text{erf}(b/y^{1/2})$, where $a = G_h$, $b = (2J + 1)\Gamma(\frac{3}{2}) G_0 / G_h$, the factors G_h and G_0 being constants in front of powers of y in the $J = 0$ and "high-J" columns of Table 8.4, and r the number of rotational degrees of freedom in the last column of Table 8.4. Show that the corresponding density defined as the derivative with respect to y is

$$\frac{\partial \Gamma(y, J)}{\partial y} = \frac{1}{2} ary^{(r-2)/2} \operatorname{erf}(b/y^{1/2}) - \frac{ab}{\sqrt{\pi}} y^{(r-3)/2} \exp(-b^2/y)$$

Except at low J ($J \leq 5$), the second term will be very small because of the exponential; moreover, since $N_h = \frac{1}{2}ar$, $G_0/G_h \approx N_0/N_h$, N_h and N_0 being the analogous terms in Table 8.5. Consequently the first term in the above expression will dominate, so it is a very good approximation to the interpolated density $N(y, J)$.

Problem 8.17. (a) Show that, to within a normalization constant, the probability $P(E, z, J)$ for the case diatomic + atom (Eq. (8.98)) reduces to

$$P(E, z, J) = \frac{E - z - y_m}{h\nu B} \operatorname{erf}\left[\left(J + \frac{1}{2}\right) \left(\frac{\pi B}{E - z - y_m}\right)^{1/2}\right] dz$$

if the diatomic fragment, with rotational constant B, is approximated by a harmonic oscillator of frequency ν, and the sum of "external" states is interpolated as in Eqs. (8.78)–(8.79) using the data in Table 8.4. At fixed E and J this probability decreases monotonically with z (the energy in the diatomic fragment).
(b) Using the approximation $\operatorname{erf} x \cong 2x/\pi^{1/2}$ for small x, show that a good approximation for this normalized probability at $J = 0$ and large E is

$$P(E, z, J = 0) \cong \frac{3}{2} \frac{(E - z)^{1/2}}{E^{3/2}} dz$$

which is a result similar to that of Problem 8.18 with $v_1 = 1$, $v_2 = 0$, and $n = \frac{1}{2}$.

Problem 8.18. If the angular-momentum dependence is approximately represented by the parameter n in the expression below, show that the normalized classical version of Eq. (8.100) is

$$P(E, z) = \frac{(v_1 + v_2 + n)! z^{v_1 - 1} (E - y_m - z)^{v_2 + n}}{(v_1 - 1)! (v_2 + n)! (E - y_m)^{v_1 + v_2 + n}} dz$$

where v_1 is the number of vibrational degrees of freedom of fragment 1, v_2 is that for fragment 2, and n is the number of "external" degrees of freedom of the system. Show that this function has a maximum at

$$z_{max} = \frac{(E - y_m)(v_1 - 1)}{(v_1 + v_2 + n - 1)}$$

and that the maximum of the distribution is

$$P(E, z_{max}) = \frac{(v_1 + v_2 + n)!(v_1 - 1)^{v_1 - 1}(v_2 + n)^{v_2 + n}}{(v_1 - 1)!(v_2 + n)!(v_1 + v_2 + n - 1)^{v_1 + v_2 + n - 1}} \frac{dE}{(E - y_m)}$$

Thus, with increasing energy, the distribution moves to higher energies, while at the same time $P(E, z_{max})$ decreases.

Problem 8.19. (a) Show that $\Gamma(y, J)$ of Eq. (8.37) is obtained by summing the E_t-distribution of Eq. (8.105) over all L^*-states.
(b) Show that $\Gamma(y, J)$ of Eq. (8.36) is obtained by summing the E_r-distribution of Eq. (8.106) over all J_r^*-states.

Problem 8.20. (a) If the density of "external" states is written for brevity as N_x, show that N_x as a function of J_r (i.e. for the determination of the E_t-distribution) takes the following forms for two spherical fragments:

$$\text{for } J_r < |J - L^*|: \quad N_x = \begin{cases} 0, & \text{if } L^* < J \\ 2R_4, & \text{if } L^* > J \end{cases}$$

$$\text{for } |J - L^*| \le J_r < J: \quad N_x = R_1$$
$$\text{for } J \le J_r \le J + L^*: \quad N_x = R_2$$
$$\text{for } J_r > J + L^*: \quad N_x = R_3$$

(b) Show that, for two spherical fragments, N_x as a function of L (i.e. for the determination of the E_r-distribution) takes the following forms.

(1) If $J_r^* < J$,

$$\text{for } L < J - J_r^*: \quad N_x = 0$$
$$\text{for } J - J_r^* \le L \le J + J_r^*: \quad N_x = R_1$$
$$\text{for } L > J + J_r^*: \quad N_x = 2R_4$$

(2) If $J_r^* > J$,

$$\text{for } L \le J_r^* - J: \quad N_x = R_3$$
$$\text{for } J_r^* - J \le L \le J_r^* + J: \quad N_x = R_2$$
$$\text{for } L > J_r^* + J: \quad N_x = R_5$$

Use R_1 through R_5 from Eq. (8.111).

9

Non-central potential and exit-channel effects

This is an elementary introduction to exit-channel effects in Type-1 reactions when, in addition to a simple radial dependence, the effective potential has also an angular dependence. These effects are illustrated by the fragmentation of the triatomic molecule ABC into A + BC, which is examined in some detail. Implications for the fragmentation of larger polyatomic molecules are briefly discussed. This chapter covers in part the material introduced previously in Section 3.4 and in Section 8.2, but raises some details that have not yet been considered.

9.1 Triatomic systems

9.1.1 The central potential

Important information in a unimolecular reaction is the change of effective potential energy associated with the degrees of freedom directly involved in the chemical transformation. To a good approximation this reduces in Type-1 reactions (simple fission reactions, as defined in Section 3.4) just to one degree of freedom, e.g. the C–C stretching vibration in the dissociation of ethane. The C–C separation thus defines a one-dimensional reaction coordinate r and the progress of the reaction then depends on the change of the effective potential as r increases to infinity.

In the present context the center of interest is the exit channel, i.e. the actual separation of fragments, so that implicitly $r > r_{max}$, where r_{max} is the maximum in the effective potential. This is in contrast with r as used in Eq. (3.30), which defines the entrance channel, where implicitly $r_e < r < r_{max}$.

A potential is termed *central* when it has only a radial (r-) dependence. A more complicated case arises when the potential is *non-central*, i.e. has also an angular dependence. In either case the Type-1 fragmenting molecule becomes progressively stretched out along r so that it maintains roughly the symmetry of a prolate top. This means that the decomposing system will have one small moment of inertia (large rotational constant), basically representing rotation about the r-axis, and two large and roughly equal moments of inertia (small rotational constants), representing rotation about the other two axes approximately perpendicular to r.

The usual notation for a symmetric top (cf. Appendix 1, Section A1.4) is A for the large rotation constant, and B for the small, doubly degenerate rotational constant. To avoid conflicts of notation in the present context, these rotation constants will be designated A_x and B_x, since they concern what are called in Chapter 8 the "external" degrees of freedom (subscript "x").

A convenient model on which to demonstrate the principal aspects of a central potential and a non-central potential is the Type-1 fragmentation of a planar tri-atomic molecule ABC into A + BC, for example $NO_2 \rightarrow NO + O$.

9.1.2 Principal moments of inertia

The inertial properties of a body, in particular the moments of inertia, are defined with respect to its center of mass. These are called the *principal* moments of inertia, and are obtained by resolving the moment-of-inertia tensor I_{ik}, the components of which are [1]

$$I_{ik} = \begin{pmatrix} \sum m(y^2 + z^2) & -\sum mxy & -\sum mxz \\ -\sum myx & \sum m(x^2 + z^2) & -\sum myz \\ -\sum mzx & -\sum mzy & \sum m(x^2 + y^2) \end{pmatrix} \quad (9.1)$$

The summations are over masses (m) and coordinates (x, y, z) of all atoms, e.g. $-\sum_i m_i x_i y_i$, etc.; for clarity, the indices are omitted. In a planar molecule situated in the x–y plane, all z-coordinates are zero, so that the standard form of this tensor for a planar ABC system of the type shown in Fig. 9.1 will be a symmetric matrix that has always the form

$$\begin{pmatrix} a & b & 0 \\ b & c & 0 \\ 0 & 0 & a+c \end{pmatrix} \quad (9.2)$$

The principal moments of inertia are the eigenvalues of this matrix, which are easily obtained analytically (Problem 9.1). The result is

$$I_1 = a + c, \qquad I_2 = \frac{1}{2}[(a + c) + d], \qquad I_3 = \frac{1}{2}[(a + c) - d] \quad (9.3)$$

where $d = [(a - c)^2 + 4b^2]^{1/2}$. For most triatomics, especially if the angle β is not too small, $I_1 \approx I_2$, $I_3 < I_1$, I_2 for virtually any A–BC distance r. As r increases, $I_1 \approx I_2$ increase also, while I_3 remains approximately constant, so that the molecule remains roughly a prolate top during the course of the decomposition, as expected from the semi-quantitative argument above. If the system is defined in

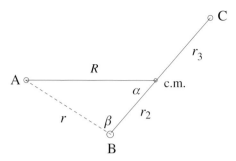

Fig. 9.1. Fragmentation of triatomic ABC into A + BC. R is the distance between atom A and the center of mass (c.m.) of diatomic BC, α is the angle between the direction of R and the axis of BC, r is the A–B separation, which is assumed to be variable (i.e. the reaction coordinate), r_2 and r_3 are the B–c.m. and c.m.–C distances, respectively, both of which are fixed, and β is the angle between r and r_2. If m_B and m_C are the masses of B and C, and the equilibrium B–C distance in the diatomic is $r_e = r_2 + r_3$, then $r_2 = m_C r_e/(m_B + m_C)$, $r_3 = m_B r_e/(m_B + m_C)$. In NO_2, for example, $r = r_e = 1.151$ nm and $\beta = 134°$ at equilibrium.

terms of r, $r_e = r_2 + r_3$, $s = m_A + m_B + m_C$, and angle β (Fig. 9.1), calculation gives (Problem 9.2)

$$I_1 = a + c = \frac{1}{s}\left[m_A(m_B + m_C)r^2 + m_C(m_A + m_B)r_e^2 - 2m_A m_C\, r\, r_e \cos\beta\right]$$

$$= \mu_{A-BC}\, r^2 + \mu_{AB-C}\, r_e^2 - (2m_A m_C\, r\, r_e \cos\beta)/(m_A + m_B + m_C) \quad (9.4)$$

and more complicated analytical results follow for I_2 and I_3.

9.1.3 The effective potential

A simple way to take account of the effect of overall rotations under a central potential is to define an effective potential $V_{\text{eff}}(r)$. This was considered before in Section 3.4 from the point of view of the decomposing molecule, but in the present instance we wish to concentrate on the potential insofar as it affects the separation of the fragments, in somewhat more detail than was done in Chapter 8. We start with Eq. (8.13) for the effective potential:

$$V_{\text{eff}}(r) = V(r) + \frac{L(L+1)\hbar^2}{2\mu r^2} \quad (9.5)$$

where L is the orbital-angular-momentum quantum number for motion of the fragments about the center of gravity of the decomposing system, μ is the reduced mass of the fragment, and r is the radial coordinate representing the distance between atom A and atom B of BC (Fig. 9.1).

The effective potential as defined in terms of the radial coordinate r in Eq. (9.5) will be an exact representation of the fragmentation of a diatomic molecule

$AB \rightarrow A + B$. This can also serve as a model for the fragmentation of a poly-atomic molecule $M \rightarrow R_1 + R_2$ if R_1 and R_2 are viewed as "atoms," with each "atom" located at the center of mass of each fragment; thus M is viewed as the "diatomic" $R_1 R_2$. An example would be the fragmentation $NCNO \rightarrow NC + NO$. $V(r)$ will then be the vibrational potential of the R_1–R_2 bond (the C–N bond in the example) as a function of r, the $R_1 \cdots R_2$ bond extension. Thus r on the RHS of Eq. (9.5) has the same significance in both terms. Another example is the inter-action involving planar or spherically symmetrized methyl with an atom, such as $CH_4 \rightarrow CH_3 + H$ (see below).

9.1.4 The non-central potential

Consider now the fragmentation of a bent planar triatomic molecule ABC into $A + BC$ (Fig. 9.1). The approximately equal moments of inertia of such a system are $I_1 \cong I_2$, so that $\hbar^2/(2I_1)$, with I_1 given by Eq. (9.4), can be considered analogous to the doubly degenerate rotational constant $B_e(r_e/r)^2$ involved in the effective potential of Eq. (3.30). As an "improvement" we could use I_1 to define a more general effective potential, which now becomes both distance (r-) and angle (β-) dependent [2]. This is emphasized by writing $I_1(r, \beta)$ for I_1:

$$V_{\text{eff}}(r, \beta) = V(r) + \frac{L(L + 1)\hbar^2}{2I_1(r, \beta)} \qquad (9.6)$$

This effective potential has lost its isotropy since now it has both a radial (r-) and an angular (β-) dependence, and so has become a non-central potential. In principle the angular dependence of $V_{\text{eff}}(r, \beta)$ can arise either from the poten-tial itself or from the angular dependence of the orbital angular momentum, or, more exactly, from the angular dependence of the orbital moment of inertia, as above.

9.1.5 The orbital moment of inertia

Now the separation of fragments is governed by the orbital angular momentum, for which the relevant distance is the A–c.m. (center of mass) distance R (Fig. 9.1) rather than the reaction coordinate r. To this end the expressions in Eq. (9.4) for the principal moments of inertia have to be modified [3]. With reference to the center-of-mass axis R in Fig. 9.1, we can define the doubly degenerate orbital moment of inertia by

$$I_o = \mu_{\text{A–BC}} R^2 \qquad (9.7)$$

which treats the system as a diatomic with the total mass of B and C concentrated at the c.m. The smallest moment of inertia of the bent ABC system may be defined by the perpendicular distances of B and C from the R-axis:

$$I_x = \mu_{BC} \, r_e^2 \sin^2 \alpha \qquad (9.8)$$

Since the center-of-mass axis R is not necessarily one of the axes of the principal moments of inertia, these moments can be approximated [3] by $I_1 \cong I_2 \cong I_o + I_x$ and $I_3 \cong I_x$. Such approximations perform moderately well at short R but at large R they yield essentially the exact result as obtained by resolution of the moment-of-inertia tensor in Eq. (9.2). An approximation for I_3 that performs somewhat better at short distances is $I_3 \cong (I_o^{-1} + I_x^{-1})^{-1}$; at larger distances it reduces to $I_3 \cong I_x$ as above.

On defining the rotational constant $A_x = \hbar^2/(2I_3)$ for motion about the R-axis, using one of the above approximations for I_3, and rotational constant $B_o = \hbar^2/(2I_o)$ for orbital motion, we have $B_x = \hbar^2/2I_1 \cong (B_o^{-1} + A_x^{-1})^{-1}$ for the doubly degenerate rotational constant of the entire system, according to the approximation $I_1 \cong I_o + I_x$. Observe that A_x is in effect the rotational constant of the K-rotor of the fragmenting system considered as a symmetric top (Section 3.3).

9.1.6 Potential with angular dependence

If the orbital part of the effective potential for triatomic ABC is written, as in principle it should be, in terms of the center-of-mass distance R (Eq. (9.7)), then the potential in Eqs. (9.5) and (9.1) contains two different distances, since the vibrational potential $V(r)$ is defined in terms of the reaction coordinate r (Fig. 9.1).

The situation can be resolved by expressing r in terms of R, or vice-versa, thereby introducing explicit reference to angle α or β. From standard relations in a plane triangle, we have, with reference to Fig. 9.1, $r^2 = R^2 + r_2^2 - 2Rr_2 \cos \alpha$, so, if $V(r)$, the vibrational potential in Eq. (9.5), is written in the simplified form as $V(r) \cong -C_n/r^n$, then, using Eq. (9.7),

$$V_{\text{eff}}(R, \alpha) = -\frac{C_n}{\left(R^2 + r_2^2 - 2Rr_2 \cos \alpha\right)^{n/2}} + \frac{L(L+1)\hbar^2}{2\mu_{A\text{-}BC}R^2} \qquad (9.9)$$

Alternatively, R can be expressed in terms of r and angle β (Fig. 9.1): $R^2 = r^2 + r_2^2 - 2rr_2 \cos \beta$, in which case it is the orbital moment of inertia that becomes angle-dependent:

$$V_{\text{eff}}(r, \beta) = -\frac{C_n}{r^n} + \frac{L(L+1)\hbar^2}{2\mu_{A\text{-}BC}\left(r^2 + r_2^2 - 2rr_2 \cos \beta\right)} \qquad (9.10)$$

This equation is essentially another version of Eq. (9.6).

In principle, there is no reason why the center-of-mass distance R could not be used to define the reaction coordinate [4]. However, on analysis it appears [5] that the reaction coordinate r (the distance between atoms of the breaking bond, as in Fig. 9.1) is a better choice than R, particularly in connection with a variational approach (cf. Chapter 7).

9.1.7 The angle-averaged effective potential

The angle-dependent effective potentials of Eqs. (9.9) and (9.10) could be exploited in a detailed kinematic analysis of the fragmentation processes regarding the dependence of α on R, or of β on r. In the absence of such information, we can assume that the orientation of fragments is random and simplify the angle-dependent effective potentials by averaging over all angles. This stratagem was used for the fragmentation $HNCO \rightarrow H + NCO$ [6].

With reference to Fig. 9.1, if BC assumes all possible orientations with respect to the R-axis at fixed α, the tip B of BC will describe a circle of circumference $2\pi r_2 \sin \alpha$, which is the number of orientations for the specified α. Since α can be in principle anywhere between 0 and π, the probability $P(\alpha) \, d\alpha$ of a given orientation for a given α is obtained by normalizing this expression between 0 and π, with the result $P(\alpha) \, d\alpha = \frac{1}{2} \sin \alpha \, d\alpha$ (Problems 9.3 and 9.4). The average effective potential is then

$$\langle V_{\text{eff}}(R) \rangle_\alpha = \frac{1}{2} \int\limits_0^\pi V_{\text{eff}}(R, \alpha) \sin \alpha \, d\alpha \qquad (9.11)$$

For an alternative average see Problem 9.5.

In the case of the effective potential given by Eq. (9.9), averaging affects only the vibrational potential $V(R, \alpha)$. Taking the most common case of neutral fragments, for which $n = 6$, we have (cf. [6])

$$\langle V(R, \alpha) \rangle_\alpha = -\frac{1}{2} C_6 \int\limits_0^\pi \frac{\sin \alpha \, d\alpha}{\left(R^2 + r_2^2 - 2Rr_2 \cos \alpha \right)^3} = -C_6 \frac{R^2 + r_2^2}{\left(R^2 - r_2^2 \right)^4} \qquad (9.12)$$

At large R $(R \gg r_2)$ this angle-averaged vibrational potential reduces to the simple $-C_6/R^6$ potential, but at intermediate values of R it is more attractive, i.e. more negative; the orbital part of the effective potential remains unaffected.

In a reaction of an ion with a (neutral) polar molecule the vibrational potential is $V(r) \cong -C_4/r^4$, so the angle-averaged vibrational potential in Eq. (9.12) for $n = 4$ is calculated to be $\langle V(R, \alpha) \rangle_\alpha = -C_4/(R^2 - r_2^2)^2$, which reduces to $-C_4/R^4$ at large R; at intermediate R the potential is likewise more attractive.

The result of the averaging in both cases is thus only a weakly modified original $-C_n/R^n$ potential, which can be fitted with a modified constant C_n and/or modified power n. Consequently, to a fairly good approximation, the potential can still be written in the form of Eq. (9.5) if r is interpreted as R. This justifies the use of the $-C_n/r^n$ potential in Chapter 8 even in the case of a triatomic.

For the effective potential given by Eq. (9.10) averaging over β affects only the orbital part. If we write the average in the form

$$\frac{1}{\pi} \int_0^{\pi} \frac{d\beta}{\left(r^2 + r_2^2 - 2rr_2 \cos \beta\right)} = \frac{1}{r^2 - r_2^2} \qquad (9.13)$$

the result is an increase in the angle-averaged rotational constant for orbital motion $B_o = \hbar^2/[2\mu_{\text{A-BC}}(r^2 - r_2^2)]$ at shorter distances, relative to the "normal" rotational constant $B_o = \hbar^2/(2\mu_{\text{A-BC}}r^2)$.

9.1.8 Ion–molecule reactions

In a reaction involving an ion and a (neutral) polar molecule the angle-averaged vibrational potential follows from Eq. (9.11): $\langle V(R, \alpha) \rangle_\alpha = -C_4/(R^2 - r_2^2)^2$. However, if the polar molecule has a permanent dipole moment μ_D, the potential depends on angle α of the dipole with the line of centers:

$$V(R, \alpha) \approx -\frac{\alpha_p q^2}{2R^4} - \frac{q\mu_D}{R^2} \cos \alpha \qquad (9.14)$$

where α_p is the polarizability of the neutral molecule and q is the ionic charge (cf. Section 8.2; here α_p is used for polarizability to distinguish it from the angle α). Figure 9.1 can be used to represent schematically the ion–molecule interaction: let the ion be a point charge at A, and let BC be the neutral molecule with separation $r_2 + r_3$ between the dipoles; R is the line of centers and α the angle of the dipole. The simplest approach is to assume that $\alpha = 0$ (a "locked dipole"), but a better approximation is to determine α as a function of r, and then average α over all r, which is known as the ADO ("average dipole orientation") theory. The averaging is more complicated than that in Eq. (9.12) because the field of the ion influences the rotation of the dipole [7, 8].

9.2 Polyatomic systems

9.2.1 Polyatomic fragments

The above relations for the "external" rotational constants also apply to the more general case of a Type-1 reaction whereby a polyatomic molecule M undergoes fission into two polyatomic fragments, R_1 and R_2, each with its unique rotational constant, B_1 and B_2; this requires that both R_1 and R_2 be of linear or spherical symmetry (or, failing that, be "spherically averaged," see the caption of Fig. 8.5).

The natural reaction coordinate here is the R_1–R_2 bond (e.g. the C–C bond in the fission $H_3C \cdots CH_3$). Therefore in this case the effective potential of Eq. (9.5) expressed in terms of bond extension r is a reasonably valid representation.

If the individual moments of inertia of the fragments R_1 and R_2 for motion about their common axis (i.e. the $R_1 \cdots R_2$ axis) are i_1 and i_2, respectively (and hence $B_1 = \hbar^2/(2i_1)$ and $B_2 = \hbar^2/(2i_2)$), the total moment of inertia about the same axis is $I_x = i_1 + i_2$. We still can define the corresponding rotational constant by $A_x = \hbar^2/(2I_x)$, so that $A_x^{-1} = B_1^{-1} + B_2^{-1}$. This can be written $A_x = B_1 B_2/(B_1 + B_2)$, which will be immediately recognized as the reduced rotational constant B_r of Section 8.6.

The orbital moment of inertia I_o is again given by Eq. (9.7) but with μ now the reduced mass of the $R_1 \cdots R_2$ system, which is indicated by writing μ_R for μ, while the center-of-mass distance R is replaced by r. The doubly degenerate rotational constant of the entire system is then, as before, $B_x = (B_o^{-1} + A_x^{-1})^{-1}$, but with a rotational constant A_x that now has two terms; thus

$$B_x = \left(B_o^{-1} + B_1^{-1} + B_2^{-1} \right)^{-1} \tag{9.15}$$

where $B_o = \hbar^2/(2\mu_R r^2)$. Strictly speaking, therefore, we should then write the effective potential of Eq. (9.5) in the form

$$V_{\text{eff}}(r) = V(r) + L(L+1)B_x \tag{9.16}$$

However, at large r we will have always $B_o^{-1} \gg B_1^{-1} + B_2^{-1}$, so that Eq. (9.5), where the effective potential is written in terms of $B_o = \hbar^2/(2\mu r^2)$ alone, is a good approximation over most of the range. We are therefore back to the diatomic-like representation, with the substitution $\mu \to \mu_R$. This justifies the procedure used in Chapter 8.

9.2.2 A example: the system $CH_3 + H$

To illustrate the above notions, consider the fragmentation of methane, in which a hydrogen atom separates from planar CH_3 along one of the two-fold rotational axes of the methyl group (the r-axis). In this particular case, therefore, $r = R$. From simple geometric consideration the moment of inertia of methyl is $I_x = \frac{3}{2}m_H r_e^2$ (the rotational constant $A_x = \hbar^2/(2I_x)$), where m_H is the mass of the hydrogen atom and r_e the equilibrium C–H distance (if the hydrogen separates along the three-fold rotational axis of CH_3, $I_x = 3m_H r_e^2$). The orbital moment of inertia is $I_o = \mu r^2$ (the rotational constant $B_o = \hbar^2/(2I_o)$), where, with m_C being the mass of a carbon atom, $\mu = (3m_H + m_C)m_H/(4m_H + m_C)$ is the reduced mass of the CH_3–H system and r the distance between the receding hydrogen atom and the center of mass of CH_3. The rotational constant of the entire fragmenting system with respect to the two axes perpendicular to r is $B_x = (B_0^{-1} + A_x^{-1})^{-1}$, doubly degenerate. The overall system is thus a symmetric top, with one small moment of inertia (large rotational constant A_x) and two large r-dependent moments of inertia (small r-dependent rotational constants B_x). At large r we have $A_x^{-1} \ll B_0^{-1}$, so that $B_x \approx B_0$.

9.2.3 Refinement of the variational procedure

The preceding arguments define in a more precise way the evolution of a decomposing system both at small and at large r, which may serve to introduce a better-characterized r-dependence into the simple variational technique set out in Chapter 7.

To begin with, an effective potential more accurate than Eq. (3.30) is the potential of Eq. (9.16), which uses the r-dependent rotational constant B_x (Eq. (9.15)). Consequently, a better expression for the rotational energy E_r^* used in Chapter 7 is $E_r^* = J(J+1)B_x$ instead of $E_r^* = E_r(r_e/r)^2$ (cf. text below Eq. (7.17)).

Of more consequence would be the replacement of the simple partition function Q_p for the products (Eq. (7.8)) in the interpolation scheme by the product $Q_v^f Q_{xi}(J)$ of partition functions for internal–external states of products at specified J, which was introduced in Section 8.8.3 (the actual inversion is discussed in Section 4.5.10). This removes the original approximation $J \approx L$ and introduces into the variational routine of Chapter 7 proper conservation of angular momentum.

The expression for $Q_{xi}(J)$ (Eq. (8.91)) involves $y_m(J)$, the minimum energy necessary to generate the required angular momentum. For a $1/r^6$ potential a good approximation is $y_m(J) \cong (J/b_6)^3$ (Section 8.5.5), which contains no reference to the distance r, that is, the reaction coordinate, since implicitly the fixed reference point for r is r_{max}, the maximum of the effective potential.

In the present instance we require y_m as a function both of J and of r. A good approximation is [10]

$$y_m(J, r) \cong J^2 B_x \qquad (9.17)$$

This expression looks quite different from the original $y_m(J)$ but it can be shown (Problem 9.6) that, in the vicinity of r_{max}, the two are comparable since $(J/b_6)^3 \cong \frac{2}{3} J^2 B_x$.

References

[1] See, for example, L. D. Landau and E. M. Lifshitz, *Mechanics*, 2nd Ed. (Pergamon Press, Oxford, 1969), pp. 98ff.
[2] J. Troe, *J. Chem. Phys.* **75**, 226 (1981).
[3] S. C. Smith, *J. Chem. Phys.* **95**, 3404 (1991).

[4] D. M. Wardlaw and R. A. Marcus, *Adv. Chem. Phys.* **70**, 231 (1988).

[5] S. J. Klippenstein, *Chem. Phys. Lett.* **170**, 71 (1990); *J. Chem. Phys.* **94**, 6469 (1991).

[6] M. Zyrianov, A. Samov, Th. Droz-Georget and H. Reisler, *J. Chem. Phys.* **110**, 10 774 (1999).

[7] T. Su and M. T. Bowers, *J. Chem. Phys.* **58**, 3027 (1973).

[8] R. A. Barker and D. P. Ridge, *J. Chem. Phys.* **64**, 4411 (1976).

[9] I. S. Gradshteyn and I. M. Ryzhik, *Table of Integrals, Series and Products* (Academic Press, New York, 1965), p. 383.

[10] W. Forst, *Chem. Phys. Lett.* **262**, 539 (1996).

Problems

Problem 9.1. Show that eigenvalues of matrix in Eq. (9.2) are given by Eq. (9.3).

Problem 9.2. Show that the elements a, b, and c in Eq. (9.3) are

$$a = \frac{1}{s}\left[(m_A + m_B)m_C\, r_e^2 \sin^2 \beta\right]$$

$$b = -\frac{1}{s}\left[(m_A + m_B)m_C\, r_e^2(\sin \beta)(\cos \beta) - m_A m_C\, r\, r_e \sin \beta\right]$$

$$c = \frac{1}{s}\left[(m_A + m_B)m_C\, r_e^2 \cos^2 \beta - 2m_A m_C\, r\, r_e \cos \beta + m_A(m_B + m_C)r^2\right]$$

where $s = m_A + m_B + m_C$.

Problem 9.3. Let J be given by the vector sum $J = L + J_1$. Assuming that the vectors L and J_1 are randomly oriented in space, with angle φ between the two (Fig. 9.2), show that the average J is

$$\langle J \rangle = L + \frac{1}{3}\frac{J_1^2}{L}$$

Problem 9.4. Let J be given by the vector sum $J = L + J_1$. Assuming, as in Problem 9.3, that the vectors L and J_1 are randomly oriented in space, with angle φ between the two,

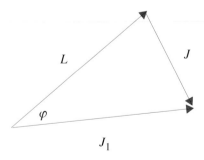

Fig. 9.2. A geometric representation of the vector sum $L + J_1 = J$, with angle φ between L and J_1.

show that the average J can be calculated also as

$$\langle J \rangle = \frac{2}{\pi}(L + J_1) \int_0^{\pi/2} (1 - m \cos^2 \varphi)^{1/2} \, d\varphi$$

where $m^{1/2} = 2(L J_1)^{1/2}/(L + J_1)$. The integral is the elliptic function $\mathscr{E}(m)$ encountered previously in Section 4.3.2. The average J calculated here is within a few percent of the result of Problem 9.3.

Problem 9.5. Alternatively the average of $V(R, \alpha)$ can be written [6]

$$\langle V(R, \alpha) \rangle_\alpha = -\frac{C_6}{\pi} \int_0^{\pi} \frac{d\alpha}{\left(R^2 + r_2^2 - 2R r_2 \cos \alpha\right)^3}$$

Show that the result is $\langle V(R, \alpha) \rangle_\alpha = -(C_6/\pi)I$, where [9]

$$I = \frac{\pi}{4\left(R^2 - r_2^2\right)} \left(\frac{3}{2(R - r_2)^4} + \frac{1}{(R^2 - r_2^2)^2} + \frac{3}{2(R + r_2)^4} \right)$$

The effect is to render this average potential still more attractive than that obtained from Eq. (9.12).

Problem 9.6. Show that, for the $1/r^6$ potential, $(J/b_6)^3 \cong \frac{2}{3}J^2 B_x$ in the vicinity of r_{max}.

Appendix 1
The sum of states, and the partition function as a semi-classical phase-space volume

Outlined below is a semi-classical approach that is useful for calculations involving *independent* degrees of freedom. The quantized energy levels ϵ_i of a particle may be represented by $\epsilon_i = f(n_i)$, where $f(n_i)$ is some function of the quantum numbers n_i. Semi-classically, the total number of states of a collection of N such non-interacting particles at specified total energy E is then the volume integral in phase space given by Eq. (A1.1) below, which uses the quantum expression for the energy levels but considers the quantum numbers n to be continuous variables:

$$G(E) = \int \cdots \int_{\Sigma_i \epsilon_i \leq E} \prod_{i=1}^{N} g_i \, dn_i$$

(A1.1)

Here g_i is the degeneracy of state n_i and integration is over all ϵ_i such that $0 \leq \Sigma_i \epsilon_i \leq E$. In many instances this is less cumbersome than integration over momenta and coordinates as suggested by Eq. (4.7) of Chapter 4. The corresponding density of states $N(E)$ follows on differentiation with respect to E (cf. Eq. (4.6)).

The condition $0 \leq \Sigma_i \epsilon_i \leq E$ implies that the particles are non-interacting, in other words that there are no cross-terms in the energies. This is an important simplification that has to be borne in mind.

A1.1 Vibrations

For N classical oscillators, $\epsilon_i = n_i h \nu_i$, where ν_i is the frequency of the ith oscillator and all $g_i = 1$; thus $g_i \, dn_i = d\epsilon_i/(h\nu_i)$. For N particles the N-dimensional integral of $\prod_i d\epsilon_i$ subject to $\Sigma_i \epsilon_i \leq E$ defines the hypervolume $E^N/\Gamma(1+N)$, so Eq. (A1.1) yields

$$G_v(E) = \frac{1}{\prod_{i=1}^{N} h\nu_i} \int \cdots \int_{\Sigma_i \epsilon_i \leq E} \prod_{i=1}^{N} d\epsilon_i = \frac{E^N}{\Gamma(1+N) \prod_{i=1}^{N} h\nu_i}$$

(A1.2)

which is known as the classical formula for the total number of states of classical oscillators at energy E. The corresponding classical vibrational partition function is

$$Q_v = \prod_{i=1}^{N} \int_0^{\infty} \exp[-n_i h\nu_i/(kT)] \, dn_i = \prod_{i=1}^{N} \frac{kT}{h\nu_i}$$

(A1.3)

Table A1.1. *Rotational symmetry numbers σ [2]*
(spectroscopic point-group designation in parentheses)

σ	Example	Symmetry elements
1	HCN	No symmetry
2	H_2O, H_2CO	One C_2 (C_{2v})
3	CH_3Cl, NH_3	One C_3 (C_{3v})
4	Cyclobutane C_4H_8	One C_4, two C_2 (D_{4h})
6	Rigid C_2H_6	One C_3, three C_2 (D_{3d} or D_{3h})
12	{ Benzene C_6H_6	One C_6, six C_2 (D_{6h})
	{ CH_4	Three C_2, four C_3 (T_d)
24	SF_6	Three C_4, four C_3, six C_2 (O_h)

C_n denotes an n-fold symmetry axis (n identical configurations for one complete revolution).
Non-planarity and/or isotopic substitution reduce σ. Free internal rotation increases σ.

If we define $q_v = Q_v/(\ell T)^N$, Eq. (A1.2) becomes simply $G_v(E) = q_v E^N / \Gamma(1 + N) = q_v E^N / N!$.

A1.2 Rotations

In this case we rewrite Eq. (A1.1) with the explicit expression for the degeneracy in the form $g_i = (2n_i)^z$:

$$G_r(E) = \int \cdots \int \prod_{i=1}^{N} (2n_i)^z \, dn_i, \qquad \text{subject to} \quad \sum_i X_i n_i^2 \leq E \qquad (A1.4)$$

where the summation for E is understood to run over positive n_i only and X_i is a constant that appears in the explicit expression for the energy levels $\epsilon_i = X_i n_i^2$ of the system. Equation (A1.4) is a form of Dirichlet's integral [1], the general solution of which is, with $a = (z + 1)/2$,

$$G_r(E) = \frac{2^{(z-1)N} (\Gamma(a))^N E^{Na}}{\Gamma(1 + aN) \prod_{i=1}^{N} X_i^a} \qquad (A1.5)$$

A1.2.1 The symmetry number

The concept of rotations and rotors cannot be divorced from consideration of the symmetry properties of the molecular species the rotor is intended to represent. In molecules containing identical nuclei not all rotational levels occur, which reduces the available phase-space volume, and therefore also reduces the sum of states and the partition function by a factor σ, called the symmetry number. It is defined as "the number of indistinguishable positions into which the molecule can be turned by simple rigid rotations" [2, p. 508]; it is different for species of different symmetries. Symmetry numbers for some of the more common molecules are listed in Table A1.1.

A1.2.2 The three-dimensional rotor

The simplest example is the spherical top with rotational energy given by $\epsilon = J(J+1)B$ or, classically, $\epsilon \approx J^2 B$, where $B = \hbar^2/(2I)$ is the rotational constant corresponding to the moment of inertia I ($\hbar = h/(2\pi)$) and J is the rotational quantum number, which is considered as a continuous variable in the semi-classical approximation. The particularity of a spherical top is that the moments of inertia (and hence the rotational constants) about each of the three rotational axes are the same on account of the spherical symmetry. The degeneracy of each energy level is $g_i = (2J+1)^2 \approx (2J)^2$ [2, p. 446], so $z = 2$ in Eqs. (A1.4) and (A1.5). For N such rotors this equation yields immediately, with $X_i \equiv B_i$,

$$G_r(E) = \frac{(\pi E^3)^{N/2}}{\Gamma\left(1 + \dfrac{3N}{2}\right) \displaystyle\prod_{i=1}^{N} \sigma_i B_i^{3/2}} \tag{A1.6}$$

The partition function for a collection of N spherical-top rotors is

$$Q_3 = \prod_{i=1}^{N} \frac{1}{\sigma_i} \int_0^\infty (2J)^2 \exp[-J^2 B_i/(kT)]\, dJ = \frac{[\pi(kT)^3]^{N/2}}{\displaystyle\prod_{i=1}^{N} \sigma_i B_i^{3/2}} \tag{A1.7}$$

Both Eqs. (A1.6) and (A1.7) are written to include σ_i, the symmetry number of the ith rotor. If we let $q_3 = Q_3/(kT)^{3N/2}$, Eq. (A1.6) can be written more compactly as $G_r(E) = q_3 E^{3N/2}/\Gamma(1+3N/2)$.

In the case of a classical three-dimensional rotor with three different rotational constants A, B, and C and symmetry number σ, it can be shown [2, 3] that the partition function is, to a very good approximation,

$$Q_3 = \frac{(kT)^{3/2}}{\sigma} \left(\frac{\pi}{ABC}\right)^{1/2} \tag{A1.8}$$

which is a generalization of Eq. (A1.7) for $N = 1$.

A special case of a three-dimensional rotor is the symmetric top, for which two rotational constants are equal and different from the third. It merits separate discussion in Section A1.4 below.

A1.2.3 The two-dimensional rotor

The energy of a two-dimensional rotor is given by $\epsilon = J(J+1)B \approx J^2 B$, where, as before, J is the rotational quantum number and B is the rotational constant. The degeneracy of each energy level is $2J+1 \approx 2J$, so $z = 1$ in Eq. (A1.4), and Eq. (A1.5) yields immediately for N such rotors, again with $X_i \equiv B_i$,

$$G(E) = \frac{E^N}{\Gamma(1+N) \displaystyle\prod_{i=1}^{N} B_i \sigma_i} \tag{A1.9}$$

The additional factor $\prod_i \sigma_i$ appears in the denominator for the same reasons of symmetry as above.

The partition function for a collection of N rotors of this kind is

$$Q_2 = \prod_{i=1}^{N} \frac{1}{\sigma_i} \int_0^{\infty} (2J+1) \exp[-J(J+1)B_i/(\ell T)]\,dJ = \frac{(\ell T)^N}{\prod_{i=1}^{N} B_i \sigma_i} \qquad \text{(A1.10)}$$

(The result is the same if we write the integrand $2J \exp[-J^2 B_i/(\ell T)]$.) If we let $q_2 = Q_2/(\ell T)^N$, then Eq. (A1.9) can be written more compactly as $G(E) = q_2 E^N / \Gamma(1+N)$.

A1.2.4 The one-dimensional rotor

The energy levels of a one-dimensional rotor are given by $\epsilon = m^2 A$, where A is the rotational constant. The quantum number m takes on both positive and negative values, so each energy level is doubly degenerate. Thus Eq. (A1.4) applies with $z=0$ but with the factor 2^n in front to correct for the constant degeneracy (this ignores the fact that the first level with $m=0$ is actually non-degenerate). Hence

$$G(E) = \frac{(\pi E)^{N/2}}{\Gamma(1+N/2) \prod_{i=1}^{N} \sigma_i A_i^{1/2}} \qquad \text{(A1.11)}$$

where, as above in Eq. (A1.9), the additional factor $\prod_i \sigma_i$ in the denominator appears due to the reduction of available phase-space volume.

The partition function for the collection of N such rotors is

$$Q_1 = \prod_{i=1}^{N} \frac{1}{\sigma_i} \int_{-\infty}^{+\infty} \exp\left[-m_i^2 A_i/(\ell T)\right] dm_i = \frac{(\pi \ell T)^{N/2}}{\prod_{i=1}^{N} \sigma_i A_i^{1/2}} \qquad \text{(A1.12)}$$

Now, if we define in this case, as before, $q_1 = Q_1/(\ell T)^{N/2}$, Eq. (A1.11) becomes $G(E) = q_1 E^{N/2}/\Gamma(1+N/2)$.

A1.2.5 The general formula

It is convenient to let r represent the number of rotations such that a single n-dimensional rotor counts for n rotations, so that $N = 3r$ in Eqs. (A1.6) and (A1.7), $N = 2r$ in Eqs. (A1.9) and (A1.10), and $N = r$ in Eqs. (A1.11) and (A1.12). Equations (A1.6), (A1.9), and (A1.11) can then all be written for r rotations in the more compact form

$$G_r(E) = \frac{q_r E^{r/2}}{\Gamma(1+r/2)} = \frac{q_r E^{r/2}}{(r/2)!} \qquad \text{(A1.13)}$$

where q_r stands for the corresponding partition function in which the $(\ell T)^{r/2}$ term has been dropped. The density of rotational states is then obtained by differentiation:

$$N_r(E) = \frac{q_r E^{(r/2)-1}}{\Gamma(r/2)} = \frac{q_r E^{(r/2)-1}}{[(r/2)-1]!} \qquad \text{(A1.14)}$$

These formulas, which are quite general, show that, for the purpose of determining the rotational sum or density of states, it is sufficient to know the corresponding partition function, i.e. rotational constant(s) and symmetry number(s). In particular, Q_3 of Eq. (A1.8) with $B = C$ is equal to the product $Q_1 Q_2$ with $N = 1$ in Eqs. (A1.12) and (A1.10) and $\sigma = \sigma_1 \sigma_2$ (the product of symmetry numbers of Q_1 and Q_2).

On the other hand, Eq. (A1.12) for $N = 3$ shows that the product of partition functions for three independent one-dimensional rotors with rotational constants A, B, and C and symmetry numbers $\sigma = \sigma_1 \sigma_2 \sigma_3$ is equal to $(\pi \Bbbk T)^{3/2}/[\sigma (ABC)^{1/2}]$, which differs by the factor π from Eq. (A1.8). Thus, in this respect, one three-dimensional rotor is not quite equivalent to three one-dimensional rotors, which applies to partition functions and therefore also to the sum and density of states.

Note, however, that these considerations apply only to *independent* rotors. In particular, the symmetric top is strictly speaking a case of *coupled* one- and two-dimensional rotors, as is discussed in more detail in Section A1.4.

A1.3 Translations

Let there be a particle of mass μ in (three-dimensional) volume V. The translational energy levels of a particle confined in a one-dimensional space of length ℓ_i are $\epsilon_i = h^2 n_i^2/(8 \mu \ell_i^2)$, where n_i is the quantum number. Let the total translational energy be $\Sigma_i \epsilon_i = E$, where the index i runs from 1 to 3, since the particle has three degrees of freedom.

Define new variables $\alpha_i = n_i/\ell_i$ so that $dn_i = \ell_i \, d\alpha_i$; thus the three-dimensional integral of $\Pi_i dn_i$ subject to $\Sigma_i \epsilon_i \leq E$ becomes $\Pi_i \ell_i$ times the integral of $\Pi_i \, d\alpha_i$ subject to $\Sigma_i \alpha_i^2 \leq 8 \mu E/h^2$, which represents the volume of a three-dimensional sphere of radius $(8 \mu E/h^2)^{1/2}$, to be divided by 2^3 since only positive values of the α_i are allowed. The volume of the sphere for positive α_i is thus $4\pi (2 \mu E)^{3/2}/(3 h^3)$. Inasmuch as the total available volume is $V = \Pi_i \ell_i$, the final result is

$$G(E) = \frac{4\pi (2 \mu E)^{3/2} V}{3 h^3} \tag{A1.15}$$

The partition function is

$$Q = \prod_{i=1}^{3} \int_0^\infty \exp \left(-\frac{h^2 n_i^2}{8 \mu \ell_i^2 \Bbbk T} \right) dn_i = \frac{(2\pi \mu \Bbbk T)^{3/2} V}{h^3} \tag{A1.16}$$

so that, if, as usual, we define $q_t = Q/(\Bbbk T)^{3/2}$, Eq. (A1.15) becomes, for $t = 3$ translational degrees of freedom (cf. Eq. (A1.13)),

$$G_t(E) = \frac{q_t E^{3/2}}{\Gamma \left(1 + \frac{3}{2} \right)} = \frac{q_t E^{3/2}}{\frac{3}{2}!} \tag{A1.17}$$

In bimolecular gas reactions the required quantity is the density of translational states *per unit volume*, so that $q_t = (2\pi \mu)^{3/2}/h^3 = 3.2424 \times 10^{20} \mu^{3/2} \, (\text{cm}^{-1})^{-3/2}$ for μ in atomic mass units.

The semi-classical formulas for the sum and density of states of vibrations, rotations, or translations in the factorial version are easier to remember in that the factorial in the denominator always matches the power of the energy in the numerator.

A1.4 The symmetric top

The symmetric top [6] is a special case of a three-dimensional rotor in which two rotational constants are equal to each other ($B = C$) and different from the third (A). The rotational energy of such a top is characterized by quantum numbers J and K:

$$E_r(J, K) = J(J + 1)B + (A - B)K^2 \qquad \text{(A1.18)}$$

where B is considered as doubly degenerate, i.e. the "J-rotor" is two-dimensional. K refers to the component of angular momentum along the molecule-fixed axis, in this case the a-axis, and is restricted to integral values between $-J$ and $+J$ (with zero included). This can be written as the sum

$$E_r(J, K) = E_r(J) + E_r(K)$$

in which $E_r(J) = J(J + 1)B$ is the energy of the "J-rotor" and $E_r(K) = (A - B)K^2$ is the energy of the "K-rotor."

A1.4.1 The prolate top

The condition $A > B = C$ (the two *smallest* rotational constants are equal) defines a prolate symmetric top. The J-resolved partition function corresponding to $E_r(J)$ of a prolate top is

$$Q_r(J) = (2J + 1)\exp[-J(J + 1)B/(\ell T)] \int_{-J}^{+J} \exp[-K^2(A - B)/(\ell T)]\, dK \qquad \text{(A1.19)}$$

where the integral is $Q_{rK}(J)$, the (semi-classical) partition function of the prolate K-rotor for specified J. It can be evaluated analytically by means of the error function erf (which is defined below in Eq. (A1.23))

$$Q_{rK}(J) = \left(\frac{\pi \ell T}{A - B}\right)^{1/2} \text{erf}\left\{\left(J + \frac{1}{2}\right)\left(\frac{A - B}{\ell T}\right)^{1/2}\right\} \qquad \text{(A1.20)}$$

For large x we have erf$\,x \to 1$, so that $Q_{rK}(J) \to [\pi \ell T/(A - B)]^{1/2}$ for large J, which is the standard partition function for a one-dimensional rotor. When x is small, erf$\,x \to 2x/\pi^{1/2}$, so, in order to ensure that $Q_{rK}(J) \to 1$ at $J \to 0$, $J + \frac{1}{2}$ is written for J in Eq. (A1.20).

The complete partition function is obtained by integrating $Q_r(J)$ over all J, with the result shown in Problem 4.3.

A1.4.2 The oblate top

An oblate top corresponds to $A < B = C$ (the two *largest* rotational constants are equal), with energy levels again given by Eq. (A1.18) (and B again doubly degenerate), except that now $A - B$ is negative, so each K reduces the energy of the J-rotor, with the result that $E_r(J, K)$ may become negative. Therefore, in addition to the limitation $-J \le K \le +J$, the quantum number K is subject to an additional constraint by virtue of the requirement that $E_r(J, K)$ remain positive, i.e. that K_{max} be the integer nearest

$$K_{max} = \pm\sqrt{\frac{J(J + 1)B}{B - A}} \qquad \text{(A1.21)}$$

If we write $J(J + 1)/(1 - A/B)$ for the term under the square root, then, for an oblate top, we have always $(1 - A/B) < 1$, so Eq. (A1.21) is only a very weak constraint.

The K-rotor partition function for specified J in the case of an oblate top ($B > A$) is similarly (taking $K_{\max} \cong J$)

$$Q_{rK}(J) = \int\limits_{-J}^{+J} \exp[K^2(B - A)/(\ell T)]\, dK$$

$$= 2\left(\frac{\ell T}{B - A}\right)^{1/2} \mathrm{daw}\left\{\left(J + \frac{1}{2}\right)\left(\frac{B - A}{\ell T}\right)^{1/2}\right\} \tag{A1.22}$$

where daw z is the modified Dawson's integral which can be obtained by a trivial modification of the erf z routine, as shown below.

The error function is defined by [4, p. 297, item 7.1.1]

$$\mathrm{erf}\, z = \frac{2}{\sqrt{\pi}} \int\limits_0^z \exp(x^{-2})\, dx = \frac{2}{\sqrt{\pi}} \sum_{n=0}^{\infty} \frac{(-1)^n z^{2n+1}}{n!(2n + 1)} \tag{A1.23}$$

where the normalization constant is chosen so that erf $\infty = 1$. The modified Dawson's integral required here is [4, p. 298, items 7.1.17 and 7.1.18]

$$\mathrm{daw}\, z = \int\limits_0^z \exp(x^2)\, dx = \sum_{n=0}^{\infty} \frac{z^{2n+1}}{n!(2n + 1)} \tag{A1.24}$$

For small z, daw $z \to z$, so that, when $J \to 0$, the oblate-top partition function $Q_{rK}(J)$ of Eq. (A1.22) approaches unity, as required.

The corresponding sum or density of states cannot be obtained analytically by inversion of the prolate or oblate forms of $Q_{rK}(J)$ but can be calculated numerically, as shown in Chapter 4.

A1.4.3 Symmetry considerations

Molecules that do not possess symmetry – a large majority – are asymmetric tops, in which all three rotational constants are more or less different. Such tops have energy levels that are much more complicated to calculate [5]; however, if the asymmetry is not too large, the energy levels of the asymmetric top may be approximated by those of a corresponding symmetric top using an average of the most nearly equal rotational constants [2, p. 48]. Thus, for an asymmetric top that is somewhat prolate ($A > B \approx C$) or oblate ($A < B \approx C$), the energy levels can be approximated by taking the arithmetic or geometric average of B and C. The two averages produce very similar results.

Symmetry considerations for a top depend on the symmetry properties of the molecule the top is meant to represent. If the molecule in question possesses an element of symmetry, not all values of the quantum number K necessarily apply. For example, if the nuclear spin of identical atoms is zero and the molecule belongs to the symmetry group C_{2v} (e.g. formaldehyde and ozone), only one-half of all K values will appear [2, p. 507], which corresponds to symmetry number $\sigma = 2$. Similarly, rigid ethane with frozen internal rotation and eclipsed or staggered methyl groups is of symmetry D_{3h} or D_{3d}, respectively (Table A1.1), for either of which $\sigma = 6$, but, if internal rotations are free, $\sigma = 18$ since there are in addition three indistinguishable positions of the two CH_3 groups with respect to each other.

References

[1] See, e.g. R. C. Tolman, *The Principles of Statistical Mechanics* (Oxford University Press, Oxford, 1962), Appendix II; P. J. Robinson and K. A. Holbrook, *Unimolecular Reactions* (Wiley-Interscience, London, 1972), Appendix 4.

[2] G. Herzberg, *Molecular Spectra and Molecular Structure. II. Infrared and Raman Spectra of Polyatomic Molecules* (D. van Nostrand, Princeton, NJ, 1960).

[3] K. S. Pitzer, *Quantum Chemistry* (Prentice-Hall, Englewood Cliffs, NJ, 1960), Appendix 9.

[4] M. Abramowitz and I. A. Stegun, *Handbook of Mathematical Functions* (NBS Applied Mathematics Series 55, Washington, 1970).

[5] S. D. Augustin and W. H. Miller, *J. Chem. Phys.* **61**, 3155 (1974).

[6] W. Forst, *Phys. Chem. Chem. Phys.* **1**, 1283 (1999), Appendix.

Appendix 2
Summary of the properties of the Laplace transform

A2.1 Definitions

Let $F(s)$ be the Laplace transform [1] (symbol $\mathcal{L}$) of $f(E)$:

$$\mathcal{L}\{f(E)\} \equiv \int_0^\infty f(E)\exp(-sE)\,dE = F(s) \qquad \text{(A2.1)}$$

It is assumed that the transform parameter s is large enough to make the integral converge. This relation is in principle invertible, i.e., for any pair $f(E)$ and $F(s)$, $f(E)$ is the inverse transform (symbol $\mathcal{L}^{-1}$) of $F(s)$:

$$f(E) = \mathcal{L}^{-1}\{F(s)\} \qquad \text{(A2.2)}$$

If E has the dimension of energy, the dimension of s is energy^{-1}. The transform $F(s)$ (Eq. (A2.1)) is then dimensionless, and the inverse transform $f(E)$ (Eq. (A2.2)) has the dimension of energy^{-1}.

$\mathcal{L}$ is a linear operator:

$$\mathcal{L}\{f_1(E) + f_2(E)\} = \mathcal{L}\{f_1(E)\} + \mathcal{L}\{f_2(E)\} \qquad \text{(A2.3)}$$

and similarly for the inverse-transform operator $\mathcal{L}^{-1}$.

A2.2 General properties

Direct transforms

1. Scaling (a is a real, positive constant):

$$\mathcal{L}\{f(aE)\} = \frac{1}{a}F\left(\frac{s}{a}\right)$$

2. Zero shift (a is a real, positive constant):

$$\mathcal{L}\{\exp(-a)\,f(E)\} = F(s+a)$$
$$\mathcal{L}\{f(E-a)\theta(E-a)\} = \exp(-as)\,F(s)$$

where $\theta(E - a)$ is the Heaviside step function:

$$\theta(E - a) = 0, \qquad E \leq a$$
$$\theta(E - a) = 1, \qquad E > a$$

3. Convolution: if $F_1(s) = \mathcal{L}\{f_1(E)\}$ and $F_2(s) = \mathcal{L}\{f_2(E)\}$, then

$$\mathcal{L}\left\{ \int_0^E f_1(x) f_2(E - x)\, dx \right\} = \mathcal{L}\left\{ \int_0^E f_2(x) f_1(E - x)\, dx \right\}$$
$$= F_1(s)\, F_2(s)$$

4. Integration:

$$\mathcal{L}\left\{ \int_0^E f(x)\, dx \right\} = \frac{F(s)}{s}$$

This is a special case of repeated n-fold integrals:

$$\mathcal{L}\left\{ \int_0^E \cdots \int_0^E f(x)(dx)^n \right\} = \frac{F(s)}{s^n}, \qquad n = 1, 2, \ldots$$

5. Differentiation:

$$\mathcal{L}\{(-E)^n f(E)\} = \frac{d^n}{ds^n} F(s)$$

6. A simple power of E:

$$\mathcal{L}\{E^n\} = \frac{\Gamma(n + 1)}{s^{n+1}} \qquad n = 0, 1, 2, \ldots$$

7. A delta function:

$$\mathcal{L}\{\delta(E - a)\} = \exp(-as), \qquad a \geq 0$$

Inverse transforms

1. Scaling (a is a real, positive constant):

$$\mathcal{L}^{-1}\left\{ \frac{1}{a} F\left(\frac{s}{a} \right) \right\} = f(aE)$$

2. Zero shift (a is a real, positive constant):

$$\mathcal{L}^{-1}\{F(s + a)\} = \exp(-a)\, f(E)$$
$$\mathcal{L}^{-1}\{\exp(-as)\, F(s)\} = f(E - a)\theta(E - a)$$

3. Convolution: if $F_1(s) = \mathcal{L}\{f_1(E)\}$ and $F_2(s) = \mathcal{L}\{f_2(E)\}$, then

$$\mathcal{L}^{-1}\{F_1(s) F_2(s)\} = \int_0^E f_1(x) f_2(E - x)\, dx = \int_0^E f_2(x) f_1(E - x)\, dx$$

4. Integration:

$$\mathscr{L}^{-1}\left\{\frac{F(s)}{s}\right\} = \int_0^E f(x)\,\mathrm{d}x$$

Repeated n-fold integrals:

$$\mathscr{L}^{-1}\left\{\frac{F(s)}{s^n}\right\} = \int_0^E \cdots (n\text{-fold}) \cdots \int_0^E f(x)(\mathrm{d}x)^n, \qquad n = 1, 2, \ldots$$

This inverse transform has the dimension energy^{n-1}.

5. Differentiation:

$$\mathscr{L}^{-1}\left\{\frac{\mathrm{d}^n}{\mathrm{d}s^n}F(s)\right\} = (-E)^n f(E)$$

6. A simple power of s:

$$\mathscr{L}^{-1}\left\{\frac{1}{s^{n+1}}\right\} = \frac{E^n}{\Gamma(n+1)}$$

where $\Gamma(n)$ is the gamma function, $\Gamma(n) = (n-1)!$

7. A delta function:

$$\mathscr{L}^{-1}\{\exp(-as)\} = \delta(E - a)$$

Reference

[1] E.g. G. E. Roberts and H. Kaufman, *Table of Laplace Transforms* (W. B. Saunders Co., Philadelphia, 1966).

Appendix 3
Review of properties of the delta function

Basic property [1, 2]:

$$\delta(x - y) = \begin{cases} 1, & \text{if } x = y \\ 0, & \text{if } x \neq y \end{cases}$$

$$x\,\delta(x) = 0$$

$$x\,\delta(x - x_0) = x_0\,\delta(x - x_0)$$

$$g(x)\,\delta(x - x_0) = g(x_0)\,\delta(x - x_0)$$

$$\int_{-\infty}^{+\infty} dx\,\delta(x - x_0) f(x) = f(x_0)$$

$$\delta[g(x)] = \sum_i \frac{1}{|g'(x_i)|}\,\delta(x - x_i)$$

where x_i are the solutions of $g(x) = 0$, $g'(x_i)$ is the first derivative of $g(x)$ evaluated at x_i and the sum is taken over all simple zeros of $g(x)$. For one simple zero we have, using the above results

$$\int dx\,\delta[g(x)] = \frac{1}{|g'(x_i)|}$$

$$\int dx\, f(x)\,\delta[g(x)] = \frac{1}{|g'(x_i)|} \int dx\, f(x)\,\delta(x - x_i) = \frac{f(x_i)}{|g'(x_i)|}$$

In the specific instance considered in Chapter 8, $g(x) \rightarrow E_r - B_r J_r^2$. Here the physically meaningful zero is $J_r = +\sqrt{E_r/B_r}$, so $g'(x_i) \rightarrow 2\sqrt{E_r\,B_r}$. If $f(x) \rightarrow 2J_r$, $f(x_i)/g'(x_i) = 1/B_r$.

References

[1] L. D. Landau and E. M. Lifshitz, *Quantum Mechanics* (Pergamon Press, Oxford, 1965), p. 17.
[2] C. Cohen-Tannoudji, B. Diu and F. Laloë, *Mécanique Quantique* (Hermann, Paris, 1986), Vol. II, p. 1462.

Appendix 4
Potentials

The potentials described below are simple central (spherically symmetric) potentials for interaction between two structureless particles, e.g. interaction between "atoms" of a diatomic-like molecule. They provide a reasonably correct representation and their main virtue is that they allow analytical, or at least fairly simple, solutions in many cases.

A4.1 The Lennard-Jones potential

The so-called Lennard-Jones (L-J) 6–12 potential, as a function of the distance r between two non-polar neutral particles, is given by

$$V(r) = 4\epsilon \left[\left(\frac{\sigma}{r} \right)^{12} - \left(\frac{\sigma}{r} \right)^{6} \right] \tag{A4.1}$$

where σ is the "collision diameter," i.e. the distance of closest mutual approach of the two particles. From $\partial V(r)/\partial r = 0$ the potential minimum is found at $r_e = 2^{1/6}\sigma$. The significance of the parameter ϵ then follows on noting that the potential at minimum is $V(r_e) = -\epsilon$, whereas at large r the potential tends to zero ($V(r \to \infty) = 0$); thus ϵ is the depth of the potential well.

Considering only the attractive part, the potential simplifies to $V(r) \approx -4\epsilon(\sigma/r)^6$, which is useful for transport phenomena (e.g. viscosity) in which the collision diameter is the relevant parameter. For reaction dynamics it is preferable to refer the potential to potential minimum at r_e, so that the attractive part of the potential is then $V(r) \approx -2\epsilon(r_e/r)^6$, in which form it is used in Chapters 3 and 8.

Interaction between a polarizable neutral molecule and an ion is described by a 12–4 Lennard-Jones potential. If both particles are polar, the potential may be approximated by the Stockmayer potential ([1], p. 35) which is angle-dependent.

A4.2 The collision number under a L-J potential

The general relation for the collision number Z is Eq. (5.15), which is given in terms of relative velocity v, but, for the present purpose, it is more convenient to work in terms of relative translation energy $E_t = \frac{1}{2}\mu v^2$. The reason is that we already have an expression for the critical impact parameter b_c^2 in terms of E_t under the attractive part of the L-J potential in the general form $-C_n/r^n$ (cf. Problem 8.1):

$$b_c^2 = \left(\frac{n}{n-2}\right)\left(\frac{(n-2)C_n}{2E_t}\right)^{2/n} \tag{A4.2}$$

The collision number Z then becomes (writing $b_c^2(E_t)$ for emphasis)

$$Z = \pi \int_0^\infty b_c^2(E_t)\left(\frac{2E_t}{\mu}\right)^{1/2} F(E_t)\,dE_t \tag{A4.3}$$

where $F(E_t)$ is the Maxwell–Boltzmann distribution,

$$F(E_t) = \frac{\pi}{(\pi kT)^{3/2}} E_t^{1/2} \exp[-E_t/(kT)] \tag{A4.4}$$

Evaluating the integral in Eq. (A4.3) yields [2]

$$Z = \left(\frac{8\pi}{\pi}\right)^{1/2}\left(\frac{(n-2)C_n}{2}\right)^{2/n}(kT)^{(n-4)/(2n)}\,\Gamma\left(1-\frac{2}{n}\right) \tag{A4.5}$$

The hard-sphere counterpart Z_{hs} follows from Eq. (A4.3) on taking $b_c^2(E_t) = d^2$, where d is the mean diameter of the collision partners, which are assumed to be hard spheres, so that the collision cross-section is πd^2, a constant. The equivalent of the Maxwell–Boltzmann distribution of Eq. (A4.3) for velocity v is

$$F(v) = 4\pi\left(\frac{\mu}{2\pi kT}\right)^{3/2} v^2 \exp[-\mu v^2/(2kT)]\,dv \tag{A4.6}$$

from which the average velocity is $\langle v \rangle = (8kT/\pi\mu)^{1/2}$; hence $Z_{hs} =$ (collision cross-section) $\times$ average velocity; thus, for unlike collision partners, in concentration units

$$Z_{hs} = \pi d^2[8kT/(\pi\mu)]^{1/2} \tag{A4.7}$$

With d in nanometers, μ in amu, $Z_{hs} = 4.571 \times 10^{-12}d^2(T/\mu)^{1/2}$ cm^3 molec^{-1} s^{-1}. Note that, for two neutral species ($n = 6$), the collision number Z has a lower temperature coefficient ($T^{1/6}$) than that of the hard-sphere counterpart $Z_{hs}(T^{1/2})$.

Equation (A4.5) can be written $Z = Z_{hs}\Omega$, where Ω is the collision integral

$$\Omega = \left(\frac{(n-2)C_n}{2kT}\right)^{2/n}\frac{\Gamma(1-2/n)}{d^2} \tag{A4.8}$$

If we use only the attractive part of the Lennard-Jones potential with $C_6 \cong 2\epsilon d^6$, Eq. (A4.8) becomes

$$\Omega = \left(\frac{4\epsilon}{kT}\right)^{1/3}\Gamma\left(\frac{2}{3}\right) \tag{A4.9}$$

The expression for the collision integral under the complete Lennard-Jones 6–12 potential (Eq. (A4.1)) is a good deal more complicated [1, pp. 484ff]. A fairly accurate approximate expression for Ω is [3]

$$\Omega = \frac{1.161\,45}{(T^*)^{0.148\,74}} + \frac{0.524\,87}{\exp(0.7732T^*)} + \frac{2.161\,78}{\exp(2.437\,887T^*)} \tag{A4.10}$$

where $T^* = \ell T / \epsilon$ is the reduced temperature and ϵ / ℓ is interpreted as the geometric average of the L-J well-depth (in kelvins) for the two collision partners. Ω of Eq. (A4.9) gives a somewhat stronger temperature dependence than does Ω of Eq. (A4.10).

It follows from Eq. (A4.5) that, for collisions between an ion and a neutral species $(n = 4)$, $Z = \pi (8C_4/\mu)^{1/2}$, which is temperature-independent.

A4.3 The Morse potential

This potential as a function of distance r is given by [4]

$$V(r) = D_e\{1 - \exp[-\beta(r - r_e)]\}^2 \tag{A4.11}$$

where r_e is the value of r at potential minimum, defined to be the zero of potential ($V(r_e) = 0$), D_e is the classical height of the potential barrier ($V(r \to \infty) = D_e$) and β is a constant in units of distance^{-1} that determines the steepness of the potential. For interaction between two fragments in a bond-fission reaction the constant β can be determined from

$$\beta = 2\pi \omega [\mu / (2D_e)]^{1/2} \tag{A4.12}$$

where ω is the "frequency" of the bond undergoing rupture, and μ is the reduced mass of the fragments. For ω and D_e in cm^{-1} and μ in amu, $\beta = 0.121\,80\,\omega\,(\mu/D_e)^{1/2}$ in nm^{-1}. In the case of polyatomic fragments the parameter μ can be interpreted as the reduced mass of the fragment *moieties* (i.e. in terms of the actual masses of the fragments), or as the masses of the *atoms* on either side of the breaking bond. Alternatively, β can be determined from the quadratic force constant f_2 as

$$\beta = [f_2/(2D_e)]^{1/2} \tag{A4.13}$$

where f_2 is determined as described below.

A4.4 The Extended Rydberg Potential

This is a potential function that correctly renders the quadratic, cubic, and quartic force constants if they are known from other information. With the same zero as the above Morse potential, this potential is given by [5]

$$V(R) = D_e - D_e(1 + a_1 R + a_2 R^2 + a_3 R^3) \exp(-a_1 R) \tag{A4.14}$$

with $R = r - r_e$, where r_e and D_e have the same significance as before; in particular, $V(R = 0) = 0$ and $V(R \to \infty) = D_e$. By taking successive derivatives of $V(R)$ it can readily be shown that the three derivatives $V^{(n)} = \partial^n V/\partial R^n$ ($n = 2, 3, 4$) are related to the potential constants a_i by

$$
\begin{aligned}
V^{(2)} &= D_e(a_1^2 - 2a_2) \\
-V^{(3)} &= D_e(2a_1^3 - 6a_1 a_2 + 6a_3) \\
V^{(4)} &= D_e(3a_1^4 - 12a_1^2 a_2 + 24a_1 a_3)
\end{aligned}
\tag{A4.15}
$$

By successive elimination, a_1 is obtained from Eq. (A4.15) in terms of the derivatives as the positive solution of

$$D_e a_1^4 - 6V^{(2)}a_1^2 - 4V^{(3)}a_1 - V^{(4)} = 0 \tag{A4.16}$$

The derivatives of the potential are related to force constants f_n, which, for a diatomic bond-stretching potential between atoms A and B, may be obtained from

$$(-1)^n f_n = 10^{-[r_e - x_{ij}(n)]/y_{ij}(n)}, \qquad n = 2, 3, 4 \tag{A4.17}$$

where $x_{ij}(n)$ and $y_{ij}(n)$ are parameters tabulated by Herschbach and Laurie (HL) [6] for bonds between atoms of rows i and j of the periodic table. For the HL definition of f_n in Eq. (A4.17), the relation to the derivatives $V^{(n)}$ is $f_n = (n!/2)V^{(n)}$.

A4.5 The effective Lennard-Jones potential

This potential is obtained from the standard L-J potential by the addition of the centrifugal term

$$V_{\text{eff}}(r) = 4\epsilon \left[\left(\frac{\sigma}{r} \right)^{12} - \left(\frac{\sigma}{r} \right)^6 \right] + \frac{J^2 h^2}{2\mu r^2} \tag{A4.18}$$

where J is the angular momentum and μ is the reduced mass of the interacting particles; the other symbols have the same meanings as in Eq. (A4.1). This effective potential, simplified by including only the attractive L-J part, is considered in Sections 3.4 and 8.2.

If we wish to consider the full 6–12 L-J potential, it is convenient to introduce the dimensionless variables [7]

$$R = r/\sigma, \qquad q = J^2/(2\mu\epsilon\sigma^2), \qquad \phi_{\text{eff}}(R) = V_{\text{eff}}(R)/\epsilon \tag{A4.19}$$

in terms of which

$$\phi_{\text{eff}}(R) = 4(R^{-12} - R^{-6}) + q/R^2 \tag{A4.20}$$

Differentiating with respect to R gives

$$\frac{\partial \phi_{\text{eff}}(R)}{\partial R} = -\frac{48}{R^{13}} + \frac{24}{R^7} - \frac{2q}{R^3} \tag{A4.21}$$

The roots of $\partial\phi_{\text{eff}}(R)/\partial R = 0$ define $R_{\min}$ and $R_{\max}$, the locations of the minimum and maximum in $\phi_{\text{eff}}(R)$. The condition for $\partial\phi_{\text{eff}}(R)/\partial R = 0$ is

$$q = \frac{12}{R^4} - \frac{24}{R^{10}} \tag{A4.22}$$

On substituting Eq. (A4.22) into (A4.20) we get $\Phi(R)$, the potential at the extrema,

$$\Phi(R) = \frac{8}{R^6} - \frac{20}{R^{12}} \tag{A4.23}$$

which gives the location of the extrema. On differentiating again with respect to R we have

$$\frac{\partial \Phi(R)}{\partial R} = -\frac{48}{R^7} + \frac{240}{R^{13}} \tag{A4.24}$$

and

$$\frac{\partial q(R)}{\partial R} = -\frac{48}{R^5} + \frac{240}{R^{11}} = R^2 \frac{\partial \Phi(R)}{\partial R} \tag{A4.25}$$

Thus both q and Φ have one maximum determined by $\partial\Phi(R)/\partial R = 0$, from which $R_{\max} = 5^{1/6}$, and, from Eq. (A4.22), $q(R_{\max}) = 5^{1/3}(36/25)$. In other words, the potential has an inflection point at $r = 5^{1/6}\sigma$, and the potential at the inflection point is, from Eq. (A4.23),

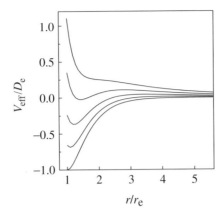

Fig. A4.1. The effective Morse potential of Eq. (A4.27) in dimensionless form, for the C–H bond in methane at several values of the angular momentum J, with equilibrium rotational constant $\hbar/(2\mu r_e^2) = 5.29\,\mathrm{cm}^{-1}$; $\beta = 1.7247$ is calculated from Eq. (A4.12) using masses of atoms C and H in μ. From top to bottom, the curves are for $J = 125, 100, 75, 50,$ and 0. The inflection point in this case is at $J_{\mathrm{infl}} = 119$, so the topmost curve represents the rotationally dissociated bond. The same result is obtained from Eq. (A4.26) for $\sigma \cong 3\,\mathrm{nm}$.

$\Phi(R_{\mathrm{max}}) = \frac{4}{5} = 0.8$, or, in terms of $V_{\mathrm{eff}}(R)$, 0.8ϵ. The maximum angular momentum at the inflection point is then

$$J_{\mathrm{infl}} = \frac{6}{5} 5^{1/6}(2\mu\epsilon\sigma^2)^{1/2} \qquad (A4.26)$$

Effective Morse or Extended-Rydberg Potentials can be defined similarly by the addition of the same centrifugal term; thus, for example, the effective Morse potential becomes (Fig. A4.1)

$$V_{\mathrm{eff}}(r) = D_e\{1 - \exp[-\beta(r - r_e)]\}^2 + J(J + 1)\hbar^2/(2\mu r^2) \qquad (A4.27)$$

but requires a numerical solution for the maximum and minimum.

References

[1] J. O. Hirschfelder, C. F. Curtiss and R. B. Bird, *Molecular Theory of Gases and Liquids* (Wiley, New York, 1967).

[2] See, for example, H. S. Johnston, *Gas Phase Reaction Rate Theory* (Ronald Press, New York, 1966), p. 145, Eq. (9.44).

[3] R. C. Reid and T. K. Sherwood, *The Properties of Gases and Liquids*, 3rd Ed. (McGraw-Hill, New York, 1977).

[4] P. M. Morse, *Phys. Rev.* **34**, 57 (1929).

[5] J. N. Murrell, S. Carter, S. C. Farantos, P. Huxley and A. J. C. Varandas, *Molecular Potential Energy Functions* (Wiley, New York, 1984).

[6] D. R. Herschbach and V. W. Laurie, *J. Chem. Phys.* **35**, 458 (1961), Table II.

[7] S. K. Kim and J. Ross, *J. Chem. Phys.* **42**, 263 (1965); J. Davidsson and G. Nyman, *Chem. Phys.* **125**, 171 (1988).

Appendix 5
Analytical solution at the low-pressure limit

The purpose here is to provide the outline of an analytical solution of the master equation, which is intended as a concrete adjunct to the numerical solution given in Chapter 5. This is done at the price of simplification, which nevertheless preserves the essential features of the solution, and, it is to be hoped, gives some feeling for the way in which the solution is put together.

A5.1 Eigenvalue–eigenfunction expansion

Equation (5.32) represents a set of coupled linear differential equations. It can be shown [1] that the set can be transformed into the equivalent integral equation

$$\int_0^{E_0} S(x, y)\, \Phi_i(y)\, dy = \mu_i \Phi_i(x) \tag{A5.1}$$

where μ_i and Φ_i are the eigenvalues and eigenfunctions, respectively, of the symmetrized kernel $S(x, y)$ (cf. Eq. (5.16))

$$S(x, y) = B(x)^{-1/2} p(x, y) B(y)^{1/2} \tag{A5.2}$$

The transition probability $p(x, y)$ will be assumed to satisfy the detailed-balance condition of Eq. (5.2) in the form

$$p(y, x)\exp[-x/(\ell T)] = p(x, y)\exp[-y/(\ell T)] \tag{A5.3}$$

i.e. with constant density of states, which makes $p(x, y)$ perforce a quasi-diatomic model. The Boltzmann distribution $B(x)$ in Eq. (A5.2) therefore has the form $B(x) = (\ell T)^{-1}\exp[-x/(\ell T)]$, normalized in $(0, \infty)$.

Under these conditions the eigenvalue–eigenfunction expansion for the population density takes the form (with dimensionless μ_i and t)

$$n(x, t) = \exp[-x/(2\ell T)] \sum_{i=0}^{\infty} h_i \Phi_i(x)\exp(\mu_i t) \tag{A5.4}$$

where the h_i depend on the initial population $n(x, 0)$. The similarity with Eq. (5.26) is obvious: the product $\exp[-x/(2\ell T)]\Phi_i(x)$ is analogous to the product $[B(x_i)]^{1/2} f_k(x_i) = \phi_k(x_i)$; similarly the eigenvalues μ_i are in principle negative.

A5.2 The exponential model

The exponential model of the transition probability $p(x, y)$ that allows an analytical solution is [2]

$$p(x, y) = \begin{cases} C(x)\exp[-(y-x)/\alpha], & x < y \text{ (down)} \\ C(y)\exp[-(x-y)/\beta], & x > y \text{ (up)} \end{cases} \tag{A5.5}$$

which is similar to the simple exponential model of Eq. (5.4), except for the normalization. The normalization constant $C(w)$ is

$$C(w) = \frac{1}{\alpha + \beta}\left(1 + \frac{\alpha}{\beta}\exp(-w/C_2)\right) \tag{A5.6}$$

where $w = x$ for $x < y$, $w = y$ for $x > y$, and $C_2 = \alpha\beta/(\alpha + \beta)$. The constant α has the same function as in the simple-exponential model, i.e. it is a measure of the strength of the coupling to the heat bath; likewise $\beta = \alpha \mathcal{k} T/(\alpha + \mathcal{k} T)$.

A5.3 The solution

For this model, the solution of the integral equation (Eq. (A5.1)) in terms of the expansion in Eq. (A5.4) has the form [3] of an exponential (hyperbolic sine) plus a sum of oscillatory terms:

$$n(x, t) = A_0 \exp[-x/(2\mathcal{k} T)] \sinh\left(\frac{Sx}{2\mathcal{k} T} + \frac{a}{2}\right)\exp(\mu_0 t)$$

$$+ \sum_{j=1}^{\infty} A_j \exp[-x/(2\mathcal{k} T)] \sin\left(\frac{R_j x}{2\mathcal{k} T} + \frac{a_j}{2}\right)\exp(\mu_j t) \tag{A5.7}$$

where $a = -2\tanh^{-1} S$ and $a_j = -2\tan^{-1} R_j$ are the phase angles and the constants A_0 and A_j depend on the initial conditions. The exponential term is a function of real S that is the solution of

$$\frac{2S}{2 + [\alpha/(\mathcal{k} T)](1 - S^2)} = \tanh\left(\frac{SE_0}{2\mathcal{k} T}\right) \tag{A5.8}$$

The oscillatory terms arise from imaginary $S = iR$, where $i = \sqrt{-1}$ (R real). The R_j are the solutions of

$$\frac{2R_j}{2 + [\alpha/(\mathcal{k} T)](1 + R_j^2)} = \tan\left(\frac{R_j E_0}{2\mathcal{k} T}\right) \tag{A5.9}$$

If R_1 is the first solution, then, given the periodicity of the tan function, the higher solutions will be roughly at $R_1 + 2\pi n\mathcal{k} T/E_0, n = 1, 2, \dots$.

After some reduction, the smallest (in terms of its absolute value) dimensionless eigenvalue μ_0 is given by [4]

$$-\mu_0 = \left(1 + \frac{4[1 + \alpha/(\mathcal{k} T)]}{(1 - S^2)[\alpha/(\mathcal{k} T)]^2}\right)^{-1} = k_0^{wc} \tag{A5.10}$$

where k_0^{wc} is the (dimensionless) first-order limiting low-pressure weak-collision rate constant. As $E_0 \to \infty$, the solution of Eq. (A5.8) approaches $S \to 1$, and therefore $\mu_0 \to 0$; thus the high-threshold limit of the expansion in Eq. (A5.7) correctly reduces to the non-reactive case. For E_0 that is not too small ($E_0/(\mathcal{k} T) > 10$), the solution of Eq. (A5.8)

Table A5.1. *Low-pressure dissociation of* $SO_2 + Ar \rightarrow SO + O + Ar$

T (K)	$\omega\tau_{exp}$	α (cm^{-1})	k_0^{wc} (calc) (cm^3 molec^{-1} s^{-1})	k_0^{wc} (expt)
2000	630	111	1.57×10^{-22}	2.0×10^{-21}
4000	180	340	8.80×10^{-16}	1.1×10^{-15}
6000	94	473	6.76×10^{-14}	8.9×10^{-14}

$\omega\tau_{exp} = (1.445 \times 10^4/T^{5/6})\exp(40.32/T^{1/3})$ is the experimental dimensionless relaxation time, based on $Z_{hs}\Omega$ (Appendix 4, Eq. (A4.9)). From shock-tube data of V. V. Kishore, S. V. Babu and V. S. Rao, *Chem. Phys.* **46**, 297 (1980).
α is calculated from $\omega\tau_{exp} = |\mu_j|^{-1}$ in Eq. (A5.12) for $j = 1$.
k_0^{wc} (calc) is calculated from Eq. (A5.21); k_0^{wc} (expt) are shock-tube data of K. Saito, T. Yokubo and I. Murakami, *J. Chem. Phys.* **73**, 3017 (1980).

is $1 - S^2 \cong 4\{\exp[-E_0/(\&T)]\}[1 + \alpha/(\&T)]^{-1}$, so, from Eq. (A5.10), the *dimension-bearing* k_0^{wc}(s^{-1}) reduces to

$$k_0^{wc} \cong \frac{\omega\exp[-E_0/(\&T)]}{(1 + \&T/\alpha)^2} \tag{A5.11}$$

The higher dimensionless eigenvalues ($j = 1, 2, \ldots$) are

$$-\mu_j = \left(1 + \frac{4[1 + \alpha/(\&T)]}{(1 + R_j^2)[\alpha/(\&T)]^2}\right)^{-1} \tag{A5.12}$$

With explicit analytical expressions [4] for the initial-value constants A_0 and A_j, all elements are now available for the calculation of each term in the expansion in Eq. (A5.7), from which all properties of the relaxing system can be obtained for specified exponential transition probability parameter α.

A5.4 The population distribution

The expansion in Eq. (A5.7) is in principle an infinite series, but at steady state only the first term effectively contributes, which, assuming that $\exp(\mu_0 t) \cong 1$, can be reduced to [4]

$$n(x) \cong \frac{\exp[-x/(\&T)]}{\&T\{1 - \exp[-E_0/(\&T)]\}}\left(1 - \frac{\exp[-(E_0 - x)/(\&T)]}{1 + \alpha/(\&T)}\right) \tag{A5.13}$$

This equation represents the equilibrium distribution for a diatomic molecule normalized in $(0, E_0)$, multiplied by a correction factor that measures the deviation from equilibrium. As $E_0 \rightarrow \infty$, the steady-state population in Eq. (A5.13) becomes the equilibrium population, as expected.

In these developments, particularly for application to real molecules, it is implicitly assumed that the parameter α and its possible temperature dependence are known from other information. One possible source is high-temperature relaxation-time measurements in a shock tube, which can be put into the general form $\omega\tau_{exp} = aT^{-5/6}\exp(bT^{-1/3})$ (a and b are constants). If the (second-lowest) eigenvalue $|\mu_1|^{-1}$ is identified with $\omega\tau_{exp}$, the parameter α can be obtained by solving Eq. (A5.12) (see Table A5.1 above).

A5.5 The effective threshold

Another consideration regarding application to real molecules is that these are generally polyatomic, so the above quasi-diatomic model with a constant density of states is a poor representation. A remedy is to replace E_0 by an effective threshold E_0^* with a view to improving the rate constant given by Eq. (A5.11), in recognition of the fact that the rate constant is most sensitive to E_0 since it depends on the threshold exponentially. The effective threshold may be defined by [3]

$$E_0^* = -\ell T \ln \left(Q_v^{-1} \int_{E_0}^{\infty} N_v(x) \exp[-x/(\ell T)] \, dx \right) \tag{A5.14}$$

where Q_v and $N_v(x)$ are the partition function and density of states, respectively, of the fragmenting molecule, assuming that only vibrational degrees of freedom are involved (subscripts "v"). This is an improvement on the effective threshold considered previously in Problem 5.14. With this definition of E_0^* the original diatomic-like version of the strong-collision low-pressure rate constant k_0^{sc} becomes

$$k_0^{sc} = \frac{\omega}{\ell T} \int_{E_0^*}^{\infty} \exp[-x/(\ell T)] \, dx = \frac{\omega}{Q_v} \int_{E_0}^{\infty} N_v(x) \exp[-x/(\ell T)] \, dx \tag{A5.15}$$

which is in fact the correct *polyatomic* version of k_0^{sc} (Eq. (5.52)). We can reasonably assume that using the effective threshold in the diatomic-like weak-collision rate constant of Eq. (A5.11) will likewise translate into an approximately correct *polyatomic* version of k_0^{wc}, with the result

$$k_0^{wc} \cong \frac{\omega \exp[-E_0^*/(\ell T)]}{(1 + \ell T/\alpha)^2} = \frac{\omega}{Q_v(1 + \ell T/\alpha)^2} \int_{E_0}^{\infty} N_v(x) \exp[-x/(\ell T)] \, dx \tag{A5.16}$$

Using Eq. (A5.15) for k_0^{sc} and taking the ratio, we have

$$\frac{k_0^{wc}}{k_0^{sc}} \cong \frac{1}{(1 + \ell T/\alpha)^2} \tag{A5.17}$$

so that all reference to the density of states and its energy dependence cancels out. It is in this form that the ratio is used in Section 5.4 (Eq. (5.63)) to define the collision efficiency β_c.

As an example of the application of Eq. (A5.7) to a real molecule with effective threshold E_0^*, consider the incubation time t_{inc} (cf. Section 5.3.7), which is given by

$$\mu_0 t_{inc} = -\ln \left[-2\ell T A_0 \exp[-E_0^*/(2\ell T)] \cosh\left(\frac{a}{2}\right) \sinh\left(\frac{SE_0^*}{2\ell T}\right) \right] \tag{A5.18}$$

Using the parameter α obtained from relaxation times, this expression (after some reduction) was used to calculate t_{inc} in the high-temperature system $O_2 + Ar$ [3, Fig. 9] with satisfactory results.

A5.6 Vibrational–rotational energy transfer

If collisions transfer vibrational as well as rotational energy, we can adopt simplifications similar to those used in Section 5.8, namely that the transition probability factors into vibrational and rotational parts, the latter having the strong-collision form. In addition, assume that the process of rotational dissociation occurs at fixed vibrational energy, and that the reaction is of Type-1 with the barrier separating bound and dissociated states given by [6]

$$E_v + E_r/2 = E_0 \qquad (A5.19)$$

which is essentially a linearized Extended-Rydberg Potential barrier (Fig. 3.3).

The solution for the rate constant in the harmonic quasi-diatomic case, assuming the validity of the exponential model of the vibrational transition probability and the strong-collision model of the rotational transition probability, can be shown to be [7, 8]

$$k_0^{wc} \cong \frac{\omega \exp[-E_0/(\ell T)]}{1 + \ell T/\alpha} \qquad (A5.20)$$

On comparing this with Eq. (A5.11) we see that rotational energy transfer is predicted to increase the low-pressure rate constant, relative to purely vibrational energy transfer, by the factor $1 + \ell T/\alpha$. The rate constant of Eq. (A5.20) can now be "improved," i.e. made applicable to polyatomics subject to an effective potential, by the device of defining an effective threshold by a procedure similar to the one used above, with the result

$$k_0^{wc} \cong \frac{\omega}{Q_{vr}(1 + \ell T/\alpha)} \int_{E_0}^{\infty} N_{vr}(x) \exp[-x/(\ell T)] \, dx \qquad (A5.21)$$

where Q_{vr} and $N_{vr}(x)$ are now the vibrational–rotational partition function and density of states, respectively, of the fragmenting molecule. For application to a number of diatomics see [7, 9]. Table A5.1 gives an example of a calculation in which Eq. (A5.21) is applied to the decomposition of SO_2 at low pressure.

References

[1] B. Widom, *J. Chem. Phys.* **32**, 913 (1960).
[2] A. P. Penner and W. Forst, *J. Chem. Phys.* **67**, 5296 (1977).
[3] W. Forst and A. P. Penner, *J. Chem. Phys.* **72**, 1435 (1980).
[4] W. Forst, *J. Chem. Phys.* **80**, 2504 (1984).
[5] W. Forst, *Theory of Unimolecular Reactions* (Academic Press, New York, 1973), p. 192, Eq. (8.143).
[6] A. P. Penner, *Mol. Phys.* **36**, 1373 (1979).
[7] W. Forst, *J. Phys. Chem.* **84**, 3050 (1980).
[8] W. Forst, in *Advances in Chemical Kinetics and Dynamics*, Vol. 2B, J. R. Barker, Ed. (JAI Press, Greenwich, CT, 1995), pp. 470ff.
[9] W. Forst, *Int. J. Chem. Kinet.* **13**, 789 (1981).

Appendix 6
Gamma function $\Gamma(n) = (n - 1)!$

$$\Gamma\left(\frac{1}{2}\right) = \sqrt{\pi} = 1.7725$$

$$\Gamma\left(\frac{3}{2}\right) = \frac{1}{2}\sqrt{\pi} = 0.886\,23$$

$$\Gamma\left(\frac{5}{2}\right) = \frac{3}{4}\sqrt{\pi} = 1.3293$$

$$\Gamma\left(\frac{7}{2}\right) = (15/8)\sqrt{\pi} = 3.3234$$

$$\Gamma\left(\frac{9}{2}\right) = (105/16)\sqrt{\pi} = 11.6317$$

$$\Gamma\left(\frac{1}{4}\right) = -\frac{3}{4}! = 3.6256$$

$$\Gamma\left(\frac{3}{4}\right) = -\frac{1}{4}! = 1.2254$$

$$\Gamma\left(\frac{5}{4}\right) = \frac{1}{4}! = 0.9064$$

$$\Gamma\left(\frac{7}{4}\right) = \frac{3}{4}! = 0.919\,06$$

In general, $(-x)! = \dfrac{\pi x}{x! \sin(\pi x)}$

$$\Gamma\left(\frac{1}{3}\right) = -\frac{2}{3}! = 2.6789$$

$$\Gamma\left(\frac{2}{3}\right) = -\frac{1}{3}! = 1.3541$$

$$\Gamma\left(\frac{4}{3}\right) = \frac{1}{3}! = 0.892\,98$$

$$\Gamma\left(\frac{5}{3}\right) = \frac{2}{3}! = 0.902\,75$$

$$\Gamma\left(\frac{7}{3}\right) = \frac{4}{3}! = 1.190\,64$$

Reference

E. Jahnke and F. Emde, *Tables of Functions* (Dover, New York, 1945), p. 11.

Answers to selected problems

Chapter 1

Problem 1.2. From $k(t) = -(1/[M(t)])\,d[M(t)]/dt$ it follows that $[M(t)]/[M(0)] = \exp[-\int_0^t k(t)\,dt]$. The answer follows using the result of the previous problem.

Problem 1.3. Applying Stirling, absolute terms cancel out; then segregate log terms with $s - 1$ multiplier and apply $n - m \gg s - 1$.

Problem 1.4. $(n + s - 1)!/n! = (n + 1)(n + 2)\ldots(n + s - 1) \approx n^{s-1}$. Now $E = nh\nu$, so Eq. (1.1) yields $[E/(h\nu)]^{s-1}/(s - 1)!$. This is the *number* of states at E; to obtain the number *per unit energy*, i.e. the density, divide by $h\nu$. If the oscillators are non-degenerate with average frequency $\langle \nu \rangle$, E becomes simply the total energy and $(h\langle\nu\rangle)^s$ becomes $\prod_i^s h\nu_i$, using Eq. (1.18).

Problem 1.5. The expansion in Eq. (1.31) is

$$k_0 = \frac{\omega[1 - \exp(-x)]^s}{(s - 1)!}\left(\frac{(m + s - 1)!}{m!}\exp(-mx) + \frac{(m + s)!}{(m + 1)!}\exp[-(m + 1)x] + \cdots\right)$$

$$= \frac{\omega[1 - \exp(-x)]^s}{(s - 1)!}\frac{(m + s - 1)!}{m!}\exp(-mx)\left(1 + \frac{m + s}{m + 1}\exp(-x) + \cdots\right)$$

Now for $m \gg s$, we have, assuming that $x = h\nu/(kT)$ is small,

$$1 + \frac{m + s}{m + 1}\exp(-x) \approx 1 + \exp(-x) \approx \frac{1}{1 - \exp(-x)}$$

so that, after expanding the $[1 - \exp(-x)]^{s-1}$ term to first order, and using the result of Problem 1.4,

$$k_0 \approx \frac{\omega[1 - \exp(-x)]^{s-1}}{(s - 1)!}m^{s-1}\exp(-mx) = \frac{\omega(mx)^{s-1}}{(s - 1)!}\exp(-mx)$$

Chapter 2

Problem 2.1. Use the multiplier -2β; the condition for a constrained minimum is

$$\left(-2\beta|\alpha_{1i}|\frac{1}{2}\epsilon_i^{-1/2} + 1\right)\sum_i d\epsilon_i = 0 \tag{2.26}$$

which requires $1 - \beta|\alpha_{1i}|\epsilon_i^{-1/2} = 0$, so that $\epsilon_{i0} = \beta^2\alpha_{1i}^2$. Substituting back into the constraint yields $\beta = q_0/\Sigma_i\alpha_{1i}^2$ and $F_0 = \Sigma_i\epsilon_{i0} = q_0^2/\Sigma_i\alpha_{1i}^2$ follows.

Problem 2.2. The Boltzmann fraction at temperature T of any classical system with energy between ϵ and $\epsilon + d\epsilon$ is

$$f(\epsilon)\,d\epsilon = \frac{N(\epsilon)\exp[-\epsilon/(kT)]\,d\epsilon}{\int\limits_0^\infty N(\epsilon)\exp[-\epsilon/(kT)]\,d\epsilon}$$

where $N(\epsilon)$ is the density of states. For a harmonic oscillator of frequency $h\nu$ and quantum number n, $\epsilon = nh\nu$; if it behaves classically, this implies large n, which can therefore be considered to be a continuous variable, so that $d\epsilon = h\nu\,dn$. The density of states $N(\epsilon)$ is then given by $dn/d\epsilon = 1/(h\nu)$, i.e. a constant. The result for a single oscillator is then $f(\epsilon)\,d\epsilon = \exp[-\epsilon/(kT)]\,d\epsilon/kT$. For a collection of independent oscillators, each will contribute a similar factor, and Eq. (2.5) follows.

Problem 2.3. The following integrals are required:

$$\int\limits_0^\infty J_0(a\sqrt{t})\exp(-t)\,dt = \exp(-a^2/4) \qquad [1,\text{ p. }486,\ 11.4.29] \tag{2.28}$$

$$\int\limits_{-\infty}^{+\infty} \cos(ax)\exp(-x^2)\,dx = \sqrt{\pi}\exp(-a^2/4) \qquad [2,\text{ p. }236,\ 861.20] \tag{2.29}$$

$$\int\limits_{-\infty}^{+\infty} y^{-2}[1 - \exp(-y^2)]\,dy = 2\sqrt{\pi} \qquad [2,\text{ p. }231,\ 860.24] \tag{2.30}$$

[1] M. Abramowitz and I. A. Stegun, *Handbook of Mathematical Functions* (NBS Applied Mathematics Series 55, Washington, 1972).
[2] H. B. Dwight, *Tables of Integrals and Other Mathematical Data*, 4th Ed. (Macmillan, New York, 1965).

Since $E = \Sigma_i\epsilon_i$, integrate first $J_0(a_ix)\exp[-\epsilon_i/(kT)]\,d\epsilon_i/(kT)$ in Eq. (2.25) for each i in turn, using Eq. (2.28). The result is

$$\int\limits_0^\infty \prod_{i=1}^n J_0(a_ix)\exp\left(-\sum_i \epsilon_i/(kT)\right)d\epsilon_i = (kT)^n\exp\left(-\frac{x^2kT\sum_i\alpha_i^2}{4}\right) \tag{2.31}$$

The second term in the compound bracket of Eq. (2.25) is dealt with in the same way, which yields

$$\int_0^\infty \prod_{i=1}^n J_0\left(a_i\sqrt{x^2+\lambda_1^2 y^2}\right)\exp\left[-\sum_i \epsilon_i/(\ell T)\right]d\epsilon_i$$

$$= (\ell T)^n \exp\left[-\left(\frac{x^2\ell T\sum_i \alpha_i^2}{4}+\frac{y^2\ell T\sum_i \alpha_i^2\lambda_i^2}{4}\right)\right] \qquad (2.32)$$

where $\lambda_i = 2\pi\nu_i$. Integration with respect to x makes use of Eq. (2.29), and that with respect to x of Eq. (2.30), which leads ultimately to Eq. (2.10).

Problem 2.4. From Eq. (2.5) we have, by integration subject to $\Sigma_i\epsilon_i \le E$,

$$\int\cdots\int_{\Sigma_i\epsilon_i\le E}\prod_{i=1}^n d\epsilon_i = \int_0^E d\epsilon_1\int_0^{E-\epsilon_1}d\epsilon_2\cdots\int_0^{E-\epsilon_1-\ldots\epsilon_{n-1}}d\epsilon_n = \frac{E^n}{n!}$$

By differentiation we obtain the value of the integral for the *total* energy lying between E and $E+dE$:

$$\int\cdots\int_{E\le\Sigma_i\epsilon_i\le E+dE}\prod_{i=1}^n d\epsilon_i = \frac{E^{n-1}\,dE}{(n-1)!}$$

so F, the fraction of molecules with energies in the range $E \le \Sigma_i\epsilon_i \le E+dE$, is

$$F = \frac{1}{(n-1)!}\left(\frac{E}{\ell T}\right)^{n-1}\frac{\exp[-E/(\ell T)]}{\ell T}\,dE$$

Finally the fraction of molecules with energies $E \ge E_0$ is the integral of F, which, after evaluation by parts, yields

$$\int_{E_0}^\infty \frac{1}{(n-1)!}\left(\frac{E}{\ell T}\right)^{n-1}\frac{\exp[-E/(\ell T)]}{\ell T}\,dE = \exp[-E_0/(\ell T)]\sum_{i=0}^{n-1}\frac{1}{i!}\left(\frac{E_0}{\ell T}\right)^i$$

Equation (2.13) will then be recognized as ω multiplied by the expansion of the above series, starting with the last term.

Chapter 3

Problem 3.2. The time dependence of q is given by $q = A\sin(\omega t)$ (neglecting the phase factor), with $-A \le q \le +A$, $\omega^2 = k/m$ (k is the force constant, m is the mass). By inversion $t = \omega^{-1}\sin^{-1}(q/A)$, so the time necessary to pass from $-A$ to q_0 with *positive* momenta p is

$$t_+ = \omega^{-1}\sin^{-1}(q_0/A) - \omega^{-1}\sin^{-1}(-A/A) = \omega^{-1}\left\{\sin^{-1}\left[(E_0/E)^{1/2}\right]+\pi/2\right\}$$

$$= (1/4\nu) + [1/(2\pi\nu)]\cos^{-1}\left[(1-E_0/E)^{1/2}\right]$$

since $\sin^{-1}(-1) = -\frac{1}{2}\pi$, $\omega = 2\pi\nu$. The time necessary to pass over both positive and negative momenta is twice this. Cf. R. C. Baetzold and D. J. Wilson, *J. Phys. Chem.* **68**, 3141 (1964).

Problem 3.3. The K-rotor part of $E_r(J, K)$ can be summed analytically:

$$\sum_{-J}^{+J}(A - B)K^2 = \frac{1}{3}(A - B)J(J + 1)(2J + 1)$$

Since there are $2J + 1$ values of K for a given J, the K-rotor average is the above result divided by $2J + 1$. Then

$$\langle E_r(J, K)\rangle_K = J(J + 1)B + (K\text{-rotor average})$$

Problem 3.5. Take advantage of the fact that, in general, $N^*(y) = dG^*(y)/dy$; then integrate the numerator of $\langle \epsilon_t \rangle$ by parts, using $G^*(0) = 0$.

Chapter 4

Problem 4.1. Integration with respect to dq yields merely the linear dimension ℓ. Integration over positive momenta yields $p = +\sqrt{2mE}$, and integration over positive and negative momenta is twice this.

Problem 4.4. The beta function is

$$\mathcal{B}(z, y) = \int_0^1 t^{z-1}(1 - t)^{y-1}\, dt = \frac{\Gamma(z)\Gamma(y)}{\Gamma(z + y)}$$

so that

$$N_{vr}(E) = q_v q_r \frac{E^{v-1+r/2}}{\Gamma(v + r/2)}$$

Problem 4.5. Exploit the fact that the density for two free rotors is a constant.

Problem 4.6. The translational partition function for a particle is $q_t(\Bbbk T)^n$ ($n = \frac{3}{2}$). From Eq. (4.62) we have $x^* = (n + k)/E$. The stated result then follows from Eq. (4.69). By inversion $\mathcal{L}^{-1}\{x^{-n+k}\} = E^{n+k-1}/(n + k - 1)!$ (from tables or by taking the direct transform of the result).

Problem 4.7. Use $x^* = (n + k)/E$ throughout.

Problem 4.8. Equation (4.74) is a special case of a more general relation. Let

$$P(E, k) = \frac{\exp[\phi(x)]}{[2\pi \phi''(x)]^{1/2}}$$

where $\phi(x) = \ell_n Q + Ex - k\, \ell_n x$. Observe that

$$F' = \frac{dF}{dE} = \frac{dF}{dx}\left(\frac{dE}{dx}\right)^{-1}$$

$$\frac{dF}{dx} = \frac{\exp[\phi(x)]}{[2\pi \phi''(x)]^{1/2}}\left(\phi'(x) - \frac{\phi^{(3)}(x)}{2\phi''(x)}\right)$$

Now $\phi'(x) = (\partial \ell_n Q/\partial x) + E - k/x = 0$, from which $E = (k/x) - (\partial \ell_n Q/\partial x)$ and $dE/dx = -[(k/x^2) + (\partial^2 \ell_n Q/\partial x^2)] = -\phi''(x)$. On substituting for E in $\phi(x)$ and differentiating we have $\phi'(x) = -[x(\partial^2 \ln Q/\partial x^2) + k/x] = -x\phi''(x)$, so $\phi'(x)/\phi''(x) = -x$. Thus

$$F' = \frac{\exp[\phi(x)]}{[2\pi\phi''(x)]^{1/2}}\left(x + \frac{\phi^{(3)}(x)}{2[\phi''(x)]^2}\right)$$

This equation is valid for any x, and therefore also for $x = x^*$. Equation (4.74) is obtained in the special case $k = 1$.

Problem 4.9. $\partial/\partial x = -z(\partial/\partial z)$; $\partial^2/\partial x^2 = z(\partial/\partial z) + z^2(\partial^2/\partial z^2)$; but $z(\partial/\partial z) = 0$; then $\partial^3/\partial x^3 = (\partial/\partial x)(\partial^2/\partial x^2) = -z(\partial/\partial z)[z^2(\partial^2/\partial z^2)] = -2z^2(\partial^2/\partial z^2) - z^3(\partial^3/\partial z^3)$ etc. Then let $x \to x^*$, $z \to 0$.

Problem 4.11. Start with $E = nh\nu$, so that $E_z = \frac{1}{2}sh\nu$; multiply top & bottom by $(h\nu)^s$.

Problem 4.12. The total partition function is $Q = [1 - \exp(-x\omega)]^{-n}$, so, on writing $\mathcal{E} = \exp(-x\omega)$, we have from Eq. (4.62)

$$\frac{\partial \ln Q}{\partial x} = \frac{-n\omega\mathcal{E}}{1 - \mathcal{E}} = -E$$

Also

$$\frac{\partial^2 \ln Q}{\partial x^2} = \frac{n\omega^2\mathcal{E}}{(1 - \mathcal{E})^2}$$

Solving the first equation above for $\mathcal{E}$ yields x^*, from which the result for $N(E)$ follows.

Problem 4.15. (a) For $x \to 0$, we have $Q_v \approx [x^v \prod \omega_i]^{-1}$, so $x^* = (v + k)/E$; then $\phi''(x^*) = (v + k)/(x^*)^2$ and $Q(x^*) = [(x^*)^v \prod \omega_i]^{-1}$, which then leads to the result shown.

Problem 4.16. As shown in Problem 4.15, the classical partition function is $(\mathcal{k}T)^v/\prod \omega_i$. Applying the theorem of the zero shift gives

$$I_1(E, k) = \mathcal{L}^{-1}\left\{\frac{\exp(aE_z)\,Q(s)}{s^k}\right\}$$

$$= \frac{1}{\prod \omega_i}\mathcal{L}^{-1}\left\{\frac{\exp(aE_z)}{s^{v+k}}\right\} = \frac{(E + aE_z)^{v+k-1}}{(\prod \omega_i)\Gamma(v + k)}$$

The difference from Problem 4.15 is that here the vibrational ground state has energy $-aE_z$ but is nevertheless assigned energy E, so that the notorious undercount of the purely classical formula is corrected by increasing the energy in the numerator by aE_z.

Chapter 5

Problem 5.1. (b) The normalization constant $C(y)$ is given by

$$C(y) = \int_0^y \exp[-(y - x)/\alpha]\,dx + \int_y^\infty \exp[-(x - y)/\beta]\,dx$$

Problem 5.2. The diatomic $B(x)$ is $(\mathcal{k}T)^{-1}\exp[-x/(\mathcal{k}T)]$. Then $\langle x^n \rangle = (\mathcal{k}T)^{-1}\int_0^\infty x^n \times \exp[-x/(\mathcal{k}T)]\,dx = n!(\mathcal{k}T)^n$ and similarly $\langle x^{n-1} \rangle = (n - 1)!(\mathcal{k}T)^{n-1}$; the result given then follows.

Problem 5.3. (b)

$$\langle \Delta E(y) \rangle_{\text{down}} = \int_0^\infty (x - y) \exp[-(y - x)/\alpha] \, dx$$

Problems 5.5 and 5.6. Use rules of matrix–vector multiplication.

Problem 5.7. At $t = 0$

$$\frac{n(x_i, 0)}{[B(x_i)]^{1/2}} = \sum_k h_k f_k(x_i)$$

Multiply both sides by $f_j(x_i)$ and sum over i. Because the $f(x)$ are orthonormal, only $f_j(x_i) = f_k(x_i)$ survives.

Problem 5.9. Use $n(x, t) = n(x) \exp(-k_{\text{uni}} t)$.

Problem 5.11. Rewrite the steady-state form of the master equation (Eq. (5.39)) in the form

$$k(x) n(x) = \omega \int_0^\infty p(x, y) n(y) \, dy - (\omega - k_{\text{uni}}) n(x)$$

Integrate this equation with respect to x from E_0 to ∞, and consider that, as $\omega \to 0$, the threshold at E_0 acts as a perfect sink. Exploit the fact that $\int_{E_0}^\infty n(x) \, dx \approx 0$.

Problem 5.13. Use the classical expression $Q_v = [\mathit{k} T/(h\nu)]^v$ both for the reactant and for the transition state, assuming that $\nu \sim \nu^*$.

Problem 5.14. Use the high-energy form of $\langle \Delta E^2 \rangle$ from Problem 5.4.

Problem 5.16. The equivalent of $N(t)$ is obtained by summing the vector $n(x_i, t)$ of Eq. (5.26) over all x_i. Then

$$\bar{t} = \int_0^\infty N(t) \, dt = \int_0^\infty dt \sum_i \sum_k h_k \phi_k(x_i) \exp(\mu_k t) = -\sum_i \sum_k h_k \frac{\phi_k(x_i)}{\mu_k} \tag{a}$$

From Eq. (5.25) after multiplication by $[B(x_i)]^{1/2}$,

$$\frac{\phi_k(x_i)}{\mu_k} = \mathbb{J}^{-1} \phi_k(x_i) \tag{b}$$

When Eq. (b) is substituted into Eq. (a), $\sum_k h_k \phi_k(x_i)$ will be recognized as $\mathbf{N}_0$, the vector of initial (fractional) population.

Problem 5.17. Use classical expressions both for Q_v and for $N_v(x)$ (Appendix 1, Section A1.1).

Chapter 6

Problem 6.2. Use the substitution $(\epsilon_t - E_c)^{3/2} = y^3$. Since the Airy function in $p_1(\epsilon_t)$ decays to zero at energies below E_c (Fig. 6.7), the effective lower limit of the integral in Eq. (6.39) can be extended to $-\infty$. The integral is highly peaked and can be evaluated by the method of steepest descents. Cf. [20, 38], where $\langle P \rangle$ represents $\ell\,T\mathcal{9}_1(s)\exp(-E_c s)$ in the present notation.

To obtain a real integral by the method of steepest descents, let

$$I = \int\limits_0^\infty f(x)\,dx$$

where $f(x)$ is assumed to be a sharply peaked function. Let $\phi(x) = \ell n\, f(x)$, and find x^* as the solution of $\phi'(x) = 0$; then obtain the second derivative $\phi''(x^*)$ at x^*. Expand $\phi(x)$ in a Taylor series about x^*:

$$\phi(x) = \phi(x^*) + (x - x^*)\phi'(x^*) + \frac{1}{2}(x - x^*)^2\phi''(x^*)$$

Note that the second derivative will be negative since x^* refers to a *maximum*. I becomes (since $\phi'(x^*) = 0$)

$$I = \exp[\phi(x^*)]\int\limits_{x^*}^\infty \exp\left(-\frac{1}{2}(x - x^*)^2\,|\phi''(x^*)|\right)dx$$

Let $x - x^* = y$; then

$$I = \exp[\phi(x^*)]\int\limits_0^\infty \exp\left(-\frac{1}{2}y^2\,|\phi''(x^*)|\right)dx = \frac{1}{2}\exp[\phi(x^*)]\sqrt{\frac{2\pi}{|\phi''(x^*)|}}$$

For limits $-\infty$ to $+\infty$ the integral is twice the result.

Chapter 7

Problem 7.1. There are nine vibrational dof in CH_4, of which six have essentially the same frequencies as those in CH_3, and one is the reaction coordinate (C—H stretch). In the product there are three rotational dof for CH_3, and two dof of orbital motion. These correlate with three rotational dof of CH_4, plus two low-frequency modes in CH_4: these two are the transitional modes.

Chapter 8

Problem 8.1. We have $\hbar^2 L^2 = (\mu vb)^2 = 2\mu E_t b^2$, from which, by differentiation, $\hbar^2 L\,dL = 2\mu E_t b\,db$. Classically one integrates $\hbar^2 \int_0^L L\,dL = 2\mu E_t \int_0^{b_c} b\,db$. The stated discrete result for b_c follows on replacing $L\,dL \Rightarrow \frac{1}{2}(2L + 1)$ and using summation over L instead of integration (cf. Eq. (8.35)).

Problem 8.2. From Table 8.2, $\Gamma(J_r) = 2J_r$ for the case sphere + atom. Then

$$\Gamma(y, J) = \sum_0^{L_m}\sum_{|J-L|}^{J+L} 2J_r \qquad \text{for } L_m < J$$

$$\Gamma(y, J) = \sum_0^{J}\sum_{|J-L|}^{J+L} 2J_r + \sum_J^{L_m}\sum_{|J-L|}^{J+L} 2J_r \qquad \text{for } L_m > J$$

See also Problem 8.10.

Problem 8.3. For example, for $J_m > J$,

$$\Gamma(y, J) = \int_0^{L_-} dL \int_{J-L}^{J+L} dJ_r + \int_{L_-}^{L_+} dL \int_{|J-L|}^{J_r^*} dJ_r$$

etc. Use the result of Problem 8.8 for J_r^*.

Problem 8.4. Use the result of Problem 8.8 for J_r^*.

Problem 8.15. The result can be obtained directly without first obtaining $\Gamma(J_r)$, which has a complicated form. Let the rotational energies of the two linear particles be $B_1 J_1^2$ and $B_2 J_2^2$ and let $y = B_1 J_1^2 + B_2 J_2^2$. The conservation condition is $J_1 + J_2 + L = J$. Couple first $J_1 + L \rightarrow J_0$; this is equivalent to the case linear + atom, for which the semi-classical result from Eq. (8.36), using a vertical cut-off, is J_m^2 for $J_m < J_0$ and $2J_0 J_m - J_0^2$ for $J_m > J_0$. Next couple $J_0 + J_2 \rightarrow J$. For small J ($J \rightarrow 0$), $J_0 \approx J_2$; with this approximation the number of states is given by

$$\Gamma(y, J = 0) = \int_a^b J_m^2 \, dJ_2 + \int_0^a \left(2J_2 J_m - J_2^2\right) dJ_2$$

$$J_m < J_0 \qquad\qquad J_m > J_0$$

The integration limits are obtained as follows. If z is the rotational energy of the first linear fragment, $J_m = (z/B_1)^{1/2}$, so that $z = y - B_2 J_2^2$; thus, if $z = 0$, $y = B_2 J_2^2$; when $z = y$, $J_2 = 0$. Now, if $z = B_1 J_0^2$, $\approx B_1 J_2^2$, then $y = (B_1 + B_2) J_2^2$. So, for $J < J_0$, $0 \leq z \leq B_1 J_2^2$, or $b \geq J_2 \geq a$, where $a = [y/(B_1 + B_2)]^{1/2}$ and $b = (y/B_2)^{1/2}$. For $J_m > J_0$, $B_1 J_2^2 \leq z \leq y$, or $a \geq J_2 \geq 0$. The stated result then follows by evaluation of the above integrals.

Problem 8.19. (a) For $L^* < J$, the number of J-states at a given L is $2L$. The sum of states is then $\int_0^{L_m} 2L \, dL = L_m^2$; similarly for $L^* > J$, the number of J-states is $2J$, and the sum of states is $J^2 + 2J \int_J^{L_m} dL = J^2 + 2J(L_m - J)$.
(b) The number of L-states is $2J_r^*$ for $J_r^* < J$, and $2J$ for $J_r^* > J$; integrating with respect to J_r^* from 0 to J_m in the first case, and from J to J_m in the second, yields the two forms of Eq. (8.36).

Problem 8.20. As an illustration, consider the case of the E_t-distribution when $L^* > J$ and $J_r^* > J$ (cf. Fig. 8.14). The corresponding probability (before normalization) is the sum of the integrals

$$\int_0^{J_r^*-J} dL \int_{J-L}^{J+L} dJ_r + \int_{J_r^*-J}^{L^*} dL \int_{|J-L|}^{J_r^*} dJ_r$$

After some reduction the result is $2R_+(0) + R_-(L^*) - R_4 = R_2$ using Eq. (8.111).

Chapter 9

Problem 9.1. The characteristic determinant of the matrix in Eq. (9.2) is

$$
\begin{vmatrix}
a - \lambda & b & 0 \\
b & c - \lambda & 0 \\
0 & 0 & a + c - \lambda
\end{vmatrix}
$$

Expansion gives a cubic equation in λ, the roots of which are the eigenvalues.

Problem 9.2. Let the origin of a Cartesian system be at atom B and let r be the x-axis, with the y-axis perpendicular to r (Fig. 9.1). The atomic coordinates in this arbitrary frame of reference are then (masses m_A, m_B, and m_C)

	x	y	z
m_A	$-r$	0	0
m_B	0	0	0
m_C	$-r_e \cos \beta$	$r_e \sin \beta$	0

The coordinates of the center of gravity are then, with $s = m_A + m_B + m_C$ and $r_e = r_2 + r_3$ (the r's are by definition positive), $x_{cg} = \Sigma_i m_i x_i / s$, etc., i.e.

$$
x_{cg} = (-m_A r - m_C r_e \cos \beta)/s, \qquad y_{cg} = m_C r_e \sin \beta / s, \qquad z_{cg} = 0
$$

and the new atomic coordinates with respect to the center of gravity are (cf. N. Davidson, *Statistical Mechanics* (McGraw-Hill, New York, 1962), pp. 170ff.)

	x	y	z
m_A	$-(r + x_{cg})$	$-y_{cg}$	0
m_B	$-x_{cg}$	$-y_{cg}$	0
m_C	$-(x_{cg} + r_e \cos \beta)$	$(r_e \sin \beta) - y_{cg}$	0

These coordinates are then used to obtain elements of the moment-of-inertia tensor in Eq. (9.1), and from them the eigenvalue elements a, b, and c.

Problem 9.3. The probability of φ is $P(\varphi) \, d\varphi = \frac{1}{2} \sin \varphi \, d\varphi$ (see the text). By the cosine law, $J^2 = L^2 + J_1^2 - 2L J_1 \cos \varphi$, so, by differentiation with L and J_1 fixed, $\sin \varphi \, d\varphi = J \, dJ/(L J_1)$. The normalized probability of finding J, given specified L and J_1, is

$$
P(J) \, dJ = \frac{2J \, dJ}{(L + J_1)^2 - (L - J_1)^2}
$$

since $(L - J_1) \le J \le (L + J_1)$. The average J is then

$$
\langle J \rangle = \int_{L - J_1}^{L + J_1} J P(J) \, dJ = L + \frac{1}{3} \frac{J_1^2}{L}
$$

Problem 9.4. Start directly with the cosine-law result of Problem 9.3, and define $\langle J \rangle$ as the average over φ:

$$
\langle J \rangle = \frac{1}{\pi} \int_0^{\pi} \left(L^2 + J_1^2 - 2L J_1 \cos \varphi \right)^{1/2} d\varphi
$$

The stated result follows after transformation from φ to $\varphi/2$.

Problem 9.6. When $y_m(J)$ is written out explicitly using $b_6 = (54\mu^3 C_6)^{1/6}/\hbar$,

$$y_m(J) \cong \left(\frac{J}{b_6}\right)^3 = \frac{(J\hbar)^3}{(3\mu)^{3/2}(2C_6)^{1/2}}$$

Using r_{max} from Eq. (8.15) and $B_x \approx \hbar^2/(2\mu r_{max}^2)$, $L \approx J$, then

$$J^2 B_x \approx \frac{(J\hbar)^3}{(2\mu)^{3/2}(3C_6)^{1/2}}$$

Author index

Abel, B. 52
Abramowitz, M. 25, 99, 187, 259, 283
Alcaraz, C. 187
Allison, T. C. 187
Andersson, L. L. 52
Anicich, V. G. 259
Astholz, D. 166
Audier, H. 187
Augustin, S. D. 283

Baer, T. 52–3, 100, 187, 259
Baggott, J. 53
Barker, J. R. 14, 53, 166–7
Barker, R. A. 274
Bass, L. 205, 259
Bass, L. M. 259
Bauer, E. 187
Bell, R. P. 187
Benson, S. W. 14, 166
Berne, B. J. 166
Bernitt, D. E. 187
Beyer, T. 100
Bezel, I. 52
Bhattacharjee, R. C. 167
Bird, R. B. 166, 259, 292
Black, J. G. 14
Bloembergen, N. 14
Bonnet, L. 259
Booze, J. A. 53
Börjesson, L. E. B. 52
Borkovec, M. 166
Bowers, M. T. 187, 205, 259, 274
Bowman, C. T. 165
Bowman, J. M. 53
Bowman, R. M. 53
Boyd, R. K. 166
Brand, U. 166
Breshears, W. D. 188
Brown, F. B. 187
Brucker, G. A. 53
Bryant, J. T. 166
Buchler, U. 259

Bunker, D. L. 25, 205
Burleigh, D. C. 53

Caralp, F. 205
Carmeli, B. 167
Carr, R. W. 167
Carter, S. 292
Casassa, M. P. 188
Catanzarite, J. 259
Chan, S. C. 166
Chen, W. 53, 100
Chen, Y. 53
Chesnavich, W. J. 205, 259
Christianson, M. 99
Chuang, Y. Y. 100
Clarke, D. L. 166
Cohen, M. H. 167
Cohen-Tannoudji, C. 287
Collins, M. A. 52
Colwell, S. M. 187
Connor, J. N. L. 53
Current, J. H. 100
Curtiss, C. F. 166, 259, 292

Dantus, M. 53
Davidson, D. F. 165
Davidson, N. 308
Davidsson, J. 292
Davies, J. W. 167
Davis, M. J. 53
Davydov, A. S. 187
Dean, A. M. 167
Delos, J. B. 187
Demidovich, B. P. 167
Desouter-Lecomte, B. 188
Diau, E. W. 167
Diesen, R. W. 167
Dill, B. 167
DiRosa, M. D. 165
Diu, B. 287
Dove, J. E. 166
Droz-Georget, Th. 274

Dubinsky, I. A. 99
Dunbar, R. C. 168
Dutuit, O. 187
Dwight, H. B. 187

Eckart, C. 187
Eliason, M. A. 258
Emde, F. 299
Eng, R. A. 166
Eyring, H. 52, 99, 168, 187–8, 258

Farantos, S. C. 292
Feller, W. 167
Fisher, E. R. 187
Flagan, R. C. 100
Fletcher, F. J. 165
Forst, W. 52, 99–100, 165–7, 187–8, 204–5, 258–9, 274, 283, 297
Fowler, R. H. 100
Foy, B. R. 188
Frank, J. P. 53
Fröberg, C. E. 167
Fujimoto, T. 166

Gaynor, B. J. 166–7
Gebelein, G. 188
Gebert, A. 166
Gershinowitz, H. 3
Gilbert, R. G. 52, 99, 165–167, 187, 259
Gilderson, P. W. 165
Gill, E. K. 24, 187
Glasstone, S. 99, 187
Goddard, W. A. 100
Gol denberg, M. Ya. 25
Golden, D. M. 14
Gonzales, C. A. 187
Goodman, M. F. 14
Goos, E. 166
Gradshteyn, I. S. 99, 187, 259, 274
Gray, S. K. 53, 187
Green, N. J. B. 166, 167
Grice, M. E. 259
Gruebele, M. 53
Gush, H. P. 99
Gutman, D. 99
Gwinn, W. D. 99

Haas, Y. 167
Halpern, M. 99
Hamer, N. D. 14
Handy, N. C. 187
Hänggi, P. 166
Hanrahan, R. J. 14
Hanson, R. K. 165
Hartwigsen, C. 100
Hase, W. L. 52, 99, 100, 259
Hayward, B. J. 99
Henry, B. R. 99
Herbst, E. 259
Herschbach, D. R. 205
Herzberg, G. 53, 99, 283

Heydtmann, H. 167
Hippler, H. 165, 166
Hirschfelder, J. O. 166, 259
Hirst, D. M. 99
Hisatsune, I. C. 187
Hoare, M. 167
Hoare, M. R. 100
Holbrook, K. A. 52, 99, 166, 283
Horiuti, J. 205
Hsieh, T. C. 259
Huxley, P. 292
Hynes, J. T. 52

Illies, A. J. 187
Ionov, P. 52
Ionov, S. I. 53

Jagannath, H. 99
Jahnke, E. 299
Jaques, C. 53
Jarrold, M. F. 187
Jockusch, R. A. 167
Johnson, K. E. 53
Johnston, H. S. 187
Jolley, L. B. W. 259
Jordan, M. J. T. 99
Jortner, J. 188
Jung, K.-H. 167
Just, T. 167

Kac, M. 24
Kachiani, C. 166
Kaiser, E. W. 53, 100
Kang, S. H. 167
Karas, A. J. 52
Kassal, T. 166
Kassel, L. S. 14
Kaufman, H. 100
Keister, J. W. 187
Kemper, P. R. 259
Khundkar, L. R. 53, 259
Kiefer, J. H. 166
Kim, S. K. 167
Kimball, G. E. 99, 188
King, D. S. 188
King, K. D. 166–7
King, M. C. 3
King, S. C. 99, 205
Kirmse, B. 52
Klemperer, W. 259
Klippenstein, S. J. 259, 274
Klots, C. E. 100, 258–9
Knee, J. L. 259
Knyazev, V. D. 99, 167
Kolodner, P. 14
Kubo, R. 100
Kumaran, S. S. 166
Kuznetsov, N. M. 25

Laidler, K. J. 3, 24, 187–88
Lakshmi, A. 167

Laloë, F. 287
Landau, L. D. 273, 287
Laurie, V. W. 205, 292
Leblanc, J. F. 99, 205
Leung, A. 99
Levine, R. D. 52
Levy, D. H. 53
Leyh-Nihant, B. 188
Lifshitz, E. M. 273, 287
Light, J. C. 259
Lim, K. F. 166
Lin, M. C. 167
Lin, S. H. 99, 258
Lin, S. M. 258
Lin, Y. N. 165–6
Lindemann, F. A. 14, 166
Lindemann, L. 166
Locker, D. J. 165
Lorquet, A. J. 188
Lorquet, J. C. 53, 188
Louis, F. 187
Luther, K. 166
Lynch, G. C. 53

MacDonald, P. A. 187
Marchant, P. J. 166
Marcus, R. A. 99, 205, 259, 274
Marron, I. A. 167
McClurg, R. B. 100
McCluskey, R. J. 167
McCoy, A. B. 53
McGivern, W. S. 53
McMahon, T. B. 168
McEwan, M. J. 99, 259
McGraw, G. E. 187
Meisels, G. G. 259
Metsala, M. 24
Mielke, S. L. 53
Miller, W. H. 52, 100, 187–8,
    283
Moazzen-Ahmadi, N. 99
Montague, D. C. 53
Montroll, E. W. 166
Morse, P. M. 292
Murrell, J. N. 292

Nadler, I. 259
Nikitin, E. E. 187–8, 259
Nip, W. S. 166
Nitzan, A. 167
Noble, M. 259
Noid, D. W. 53
Nordholm, S. 14, 52, 99, 166–7
North, S. W. 53
Nyman, G. 292

Olschewski, H. A. 187
Olzmann, M. 259
Oppenheim, I. 166
Oref, I. 165–6
Osamura, Y. 187

Oum, K. 166
Ozier, I. 99

Pacey, P. D. 99, 165, 187, 205
Paech, K. 168
Pashtutski, A. 166
Pathria, R. K. 100
Pattengill, M. 205
Pauling, L. 205
Pavlou, S. P. 166
Pavlov, B. V. 24
Pechukas, P. 259
Peeters, J. 167
Penner, A. P. 297
Perona, M. J. 166
Petersson, G. A. 52
Pilling, M. J. 52, 166–7
Pitzer, K. S. 99, 283
Polach, J. 259
Powis, I. 53
Prásil, Z. 52, 100
Preston, R. K. 188
Price, D. 99
Price, W. D. 168
Pritchard, H. O. 53, 166–7, 188

Qin, Y. 187
Quack, M. 52, 100, 167

Rabinovitch, B. S. 14, 99, 100, 165–7
Rabinowitz, P. 167
Radhakrishnan, G. 259
Ralston, A. 167
Ramsperger, H. C. 3, 14, 52
Rankin, C. 259
Rapp, D. 100, 166, 187
Rayez, J. C. 52, 259
Reid, R. C. 292
Reid, S. A. 53
Reisler, H. 53, 259, 274
Rice, O. K. 3, 14, 52, 99
Rice, S. A. 53
Ridge, D. P. 274
Roberts, G. E. 99, 286
Robertson, S. H. 52, 99, 100, 166–7
Robinson, P. J. 3, 52, 99, 283
Rogers, P. 53
Rosenstock, H. M. 3
Ross, I. G. 187
Ross, J. 292
Rowland, F. S. 53
Ruijgrok, T. W. 100
Ryzhik, I. M. 100, 187, 259, 274

Sahm, D. K. 53
Samov, A. 274
Sanov, A. 53
Schaefer, H. F. 187
Schatz, G. C. 53
Schlier, C. G. 53
Schneider, F. W. 165

Schnier, P. D. 168
Schoenenberger, C. 167
Schranz, H. W. 14, 166–7
Schubert, V. 166
Schwarz, H. A. 14, 259
Schwarzer, D. 52
Schweinsberg, M. 53
Schwenke, D. W. 53
Setser, D. W. 167
Sewell, T. D. 52
Sharfin, W. 53
Sharp, J. R. 53
Shen, D. L. 53
Sherwood, T. K. 292
Shimanouchi, T. 99
Shin, H. 187
Shirk, J. S. 187
Shuler, K. E. 165–7
Shultz, M. J. 14
Sibert, E. L. 53
Skinner, G. B. 14
Slagle, I. R. 99
Slater, N. B. 3, 24–5
Smith, S. C. 52, 99, 100, 166–7, 259, 273
Snider, N. 168
Snider, N. S. 167
Solc, M. 14, 24
Solly, R. K. 14
Song, G. 99
Song, K. 259
Spicer, L. D. 166
St. Laurent, P. 100
Stearn, A. E. 187
Stegun, I. A. 24, 99, 187, 259, 283
Stein, S. E. 100
Stephenson, J. C. 188
Stone, J. 14
Strittmatter, E. F. 168
Su, T. 205, 259, 274
Sun, Q. 53
Sundaram, S. 166
Swinehart, D. F. 100

Talkner, P. 166
Tardy, D. C. 166–7
Teitelbaum, H. 166
Thiele, E. 14, 24–5
Thissen, R. 187
Thomas, H. 166
Thompson, D. L. 52, 187
Thölmann, D. 168
Tolman, R. C. 52, 99, 166, 293
Tonner, D. S. 167
Trenwith, A. N. 165
Troe, J. 14, 52, 100, 166–8, 186–7, 259, 273

Troude, V. 187
Truhlar, D. G. 52–3, 100, 187
Tsang, W. 166–7
Tucker, S. C. 187
Tully, J. C. 187
Tyler, S. C. 53

Uspensky, V. 259
Uzer, T. 53

Van Kampen, N. G. 166
Varandas, A. J. C. 292
Venkatesh, P. K. 167
Verboom, G. M. L. 259
Vereecken, L. 167

Wagner, A. F. 53
Wagner, A. G. 167
Wagner, H. G. 14, 186–7
Wahrhaftig, A. L. 3
Wallenstein, M. B. 3
Walter, J. 99, 168, 188
Wardlaw, D. M. 99, 100, 205, 294
Watkins, K. W. 165
Weiss, G. H. 166–7
Weiss, M. J. 259
Weston, R. E. 14, 259
Wharton, L. 53
Whitehead, R. 99
Whitten, G. Z. 99
Widder, D. V. 187
Widom, B. 167, 297
Wieder, G. M. 3
Wieters, W. 166
Wigner, E. P. 187
Williams, E. R. 168
Wilson, D. J. 24, 302
Wilson, E. B. 100
Wimalasena, J. H. 165
Witschel, W. 100
Wittig, C. 52–3, 259
Wong, W. H. 205

Yablonovitch, E. 14
Yahav, G. 167
Yamaguchi, Y. 187
Yarkony, D. R. 188
Yau, A. W. 167, 188
Yu, T. 167
Yuan, W. 166

Zahr, G. E. 188
Zaniewski, R. C. 168
Zewail, A. H. 53, 259
Zhao, M. 53
Zhu, L. 53, 99, 100
Zyrianov, M. 274

Subject index

Page numbers in boldface refer to more-thorough discussion

activated complex 28; *see also* transition-state
activation
    energy 124, **129**, 169
        as best estimate of critical energy 145
        effect of tunneling 180
        fall-off 129, 165
        high-/low-pressure difference 169
    enthalpy 130
    entropy 130
A-factor, *see* pre-exponential factor
Airy function 183
ammonia, inversion of 178
angular momentum 193, **215**
    compounding of 216
    conservation 193, 215
        effect of 251
    intrinsic, of fragments 215
    $J = 0$, significance of 226, 251
    orbital **212**, 260
        angular dependence 268
    vector addition of 218, 274
anharmonicity 39, 40, 68, 88
asymmetric top 282
Arrhenius rate constant 12, 124, 129, 144
average
    frequency 10
    lifetime **7**, 9
    of vector sum 274
    rate constant 18, **34**

Badger's rule 197
barrier
    absorbing 118, 137
    centrifugal 37, 43, 193, **212**, 215, 228, 249
    Eckart **172**
    hindering 191
    parabolic 175
        Laplace transform of 179
        tunneling probability 173
    rotational **43**

Bessel function 25
    modified 73
beta function **303**
Beyer–Swinehart algorithm 85
BIRD (black-body infrared-radiation dissociation) 163
Boltzmann
    equilibrium distribution 12, 108, 114, 115, 152, 155, 164
    fraction of oscillators 12
bottleneck, intramolecular 51
branching ratio 249
bulk properties **120**
butene ion, fragmentation channels 48, 248

canonical ensemble 25, **82**, 246
    fluctuation of energy 83
centrifugal correction factor **156**
    interpolation between high and low pressure 159
    variational 202
chemical activation **148**, 248
    distribution function 151, 170
    examples 152
    reaction efficiency 149
    steady-state sink method 150
    strong-collision solution 150
collision
    cross-section 113, 289
    efficiency **131**, 296
        as re-scaling 131
        as energy transferred per collision 132
    frequency **112**
    integral 289
    number Z 113, **289**
    probability of 10
    weak 131
collisional relaxation 106
complex bond breaking 37
conjugate momentum 27

314

conservation
  of constants of motion 212
  of probability 117
convolution **63**, 74, 176, 185, 227, 228, 238, 250
  for angular-momentum-conserved vib–rot sum
    228
  for internal–"external" states 250
  integrals
    for vib–rot sum of states 89
    interpolation for arbitrary J 238
    reduction of **237**
Coriolis coupling 38, 50
counting of states 35, 84
critical
  configuration **28**
  energy 17, 43
    as perfect sink 125
  impact parameter 113, 260, 289
  oscillator 4, 6, 10
    rate of decay 14
  threshold 6
cross-section versus sum of states 210
cumulative reaction probability 36
cut-off, in L–J_r space
  horizontal 218, 221, 227, 260
  vertical 218, 221, 226

Dawson's integral 95, **282**
de-activation by collision 12
de Broglie wavelength 209, 210
decomposition, probability of 7
degrees of freedom
  external 26, 209
  internal 26
  pertinent 38
  semi-classical **276**
delta function 32, 255, **287**
density of states 39, **57**
  as derivative of $\Gamma(y, J)$ 262
  classical approximation 15
  "exact" 58
  in J_r–L space 254
  of free rotors **64**, 279
  of hindered rotors **69**
  second integral of 62
  thermodynamic version 82
detailed balance 19, 108, 293
deviation from equilibrium, correction factor for 165,
    295
diffusion tensor 133
Dirichlet's integral 277
dissociation, probability of 5, 11
down-collision 111
drift vector 133

effective
  mass, in tunneling 173
  number of oscillators 14
  potential 42, 193, **212**, 265, **267**
    angle- and distance-dependent 267
    angle-averaged 269

for polyatomic fragments 272
  Lennard-Jones 44, **288**, **291**
  maximum of 213
  Morse 193, **290**
  zero-point-corrected **196**
  threshold 170, **296**
eigenvalue equation 116, 169
eigenvalue–eigenfunction expansion **115**,
    293
electronic degeneracy 86, 246
elliptic integral 74, 275
energy diffusion 134
energy levels, broadening of 34
ensemble average 29
entrance channel 41, 265
equilibrium constant 126, 145, 202, 246
error function 101, 237, 240, 281, **282**
ethane, fragmentation of 41, 42, 106, 126
ethylene ion 248
exact enumeration of states 59, 67, 85
exit channel 41, **265**
exponential
  decay 119
  time dependence 7
  transition probability 108, 132, 294
external
  energy 211, **216**
    conservation of 216
  rotations 41
  states **211**
    density of 211, 253
    sum of 211, 219, **250**

fall-off 14, 22, 128, 132
  in recombination reactions 128
  pressure domain of 127
femtosecond chemistry 50
fluctuation in rate constant 35
flux through critical surface 36, **208**
  minimization of **189**
Fokker–Planck equation **133**
force constants 290, 291
formaldehyde
  anharmonic count 88
  tunneling in 177
fragments
  approximation for non-linear symmetry
    229
  distribution of energy **249**
    internal–"external" 250
    translational–rotational **252**
FT-ICR (Fourier-transform ion cyclotron resonance)
    163

gamma
  distribution 22, 23
  function 298
generalized coordinate 27

HeI$_2$ 51
HgI$_2$ 50

hindered rotors **69**
  one-dimensional **72**, 192
    classical 73
    quantum 77
  two-dimensional 69, 191, 205
HNCO 270

ICN 50
impact parameter 113, 260
incomplete gamma function 23
incubation time **119**, 296
  in multiphoton dissociation 139
initial transient 122
intramolecular vibrational redistribution (IVR) 29, 40,
      51
inverse transform **285**
  zero-shifted 144, 145
inversion integral **78**
ion cyclotron 163
ion–molecule reactions **247**
isomerization 37
isotope effect in tunneling 174

J-rotor 66, 201, **281**
J-shifting approximation **46**, 91, 196
J-states, counting of **218**
J_r, solution for 261
J_r–L space, integration in **223**

K-rotor 38, 42, 50, **66**, 93, 193, 195, 243, 269,
      281
  active 93, 245
  J-restriction in **93**
Kac formula 25
Kassel rate constant 9
kinetic energy release 49, 253

Landau–Zener formula 183
Laplace transform 11, **61**, **78**, 143, **284**
laser experiments 49
Lindemann scheme 12, 164
L-states, counting of **219**

master equation **107**, **112**, 164
  as integral equation 293
  discrete form of **114**
  eigenfunction–eigenvector expansion
      115
  expansion for weak collisions 133
  matrix equivalent 114
  non-collisional input **147**, 149
  similarity transformation 115
  solution
    analytic, at low-pressure limit **293**
    formal **115**, 116
    numerical **136**
      Gauss–Seidel method 142
      power method 143
      reduced matrix approach 139
  steady state **122**
    low-pressure limit 125

symmetrized 114, 115
transient in 119
two-dimensional
  reduction of 153
  strong-collision version 155
Maxwell–Boltzmann distribution 289
mean first passage time (MFPT) **136**
  higher moments of 139
methane
  harmonic count in 59
  fragmentation 70, 247, 248, 268, **272**
methyl isocyanide, isomerization of 106, 132
methyl-radical recombination 126, 261
microcanonical ensemble **29**, **40**, 54, 83
microscopic reversibility 36, 208; *see also* detailed
      balance
modes
  conserved 191
  transitional 70, **190**
    as hindered rotors 191
molecular parameters, source of 56
moment of inertia 266
  orbital **268**
    angle-dependent 269
    for polyatomic fragments 272
  principal **266**
    approximation for 269
  tensor 266
    eigenvalues of 266, 274
Morse oscillator **68**, 87
multichannel reactions **160**
  rotational effects 162
  strong-collision solution 161

NCNO 245, 268
NO_2 50, 267
non-adiabatic transition 39, **181**
  crossing point in 183
  Laplace transform of **185**
non-random dissociation **21**
non-statistical distribution 258
number of states, defined 57

oscillator
  anharmonic 68, **87**
    coupled 88
  classical 276
    Hamiltonian of 53
  degenerate 5
  harmonic 67, **85**
    with cut-off 4, 17, 54
  quantum 39, **85**

partition function **57**, 87, 195, 240
  electronic 246
  J-conserved **240**
    interpolated 241
  Laplace transform of **61**
  r-dependent 194
  r-interpolated 194
    inversion of 195

rotational **278**
  for spherical top 278
  for symmetric top 101, **281**
translational 280
Pauling's rule 197
PEPICO 48, 49, 248
phase point 27, 29, 31
phase space **27**
  critical surface 40
  dividing surface in 28
  surface integral 31
  trajectory 27, 30
  volume integral 31, 53, **276**
phase-space theory 212
  rate constant by **243**
photoisomerization 106
Pitzer rotor 73, 75
population density
  steady-state 118
  time-dependent 112
potential
  angle-dependent 268, **270**
    in ion–molecule reactions 271
  angular dependence **269**
  attractive part of 212, 288
  central 42, **207**, **265**
  Eckart 172
  Extended Rydberg 194, **290**
  hindering 70, 72
  Lennard-Jones 214, **288**
    inflection point of 291
  Morse 193, **290**
  non-central **265**
  parabolic 45, 175
  parameters 214
  rotational, *see* rotational energy
  Stockmayer 289
potential energy surface 26
pre-exponential factor 8, 14, 19, 124, 130, 145
  as entropy of activation 130
  experimental values of 124
  in non-adiabatic transition 186
  temperature dependent 145
probability $P(E, z, J)$ **251**
  for diatomic and atom 263
    classical version 263

QRRK **4**
quantization 32
quantized systems **83**
quantum states
  counting of 32
  number of 57
  per unit energy 33, 57

random dissociation, probability of 23
randomization 8
  failure of 51
rate constant (*see also* thermal rate constant)
  classical 31
  effect of tunneling 176

low-pressure limit **118**
maximum possible 36
microcanonical **30**, 33
  by phase-space theory 243
  variational equivalent 190
pressure-dependent 13, 19, **122**
recombination 125, 246
semi-classical 32
specific-energy 33
time-dependent 15, 117
weak-collision, low-pressure 294
reaction
  coordinate 167, 18, 27, 30, 40, 194, 265, 269
    as bond extension 272
    as center-of-mass distance 268
    separable 40, 172
    translational energy in 31
  equilibrium 113
  path 30
    degeneracy 36
  Type 1 **37**, **43**, 126, **189**, **265**
    barrier for 44, 297
    bond-fission **207**
      variational treatment 189
    centrifugal correction for **156**
  Types 2 and 3 **37**, 45
    centrifugal correction factor 157
    low-pressure 158
reactive system
  loss of probability in 117
  temporal evolution of 122
recombination reactions 126, 128
  temperature dependence 203, 247
relaxation time 120
  in shock tube 295
reversible isomerization 117
rotational
  constant 42
    effective 230, 272
    for orbital motion 268
    geometric average 229
    reduced 230, 272
    spherically averaged 225, 229
  energy 42, 218
  potential 212
  sum of states 64, **277**
    general formula 279
rotations
  external 41, **216**
  internal 38
RRK 1, 4
  rate constant 9
  thermal reaction **11**
RRKM
  failure of 51
  rate constant **33**
  trajectory 30

saddle point 27, 79
scattering matrix 36
separatrix 52

shock tube 106, 120, 295
simple bond breaking 37
Slater theory
    demise 20
    distribution
        of gaps 21
        of lifetimes 24
    energized molecule 17
    model 16
    specific dissociation probability 18
    trajectory 30
    up-zeros 18, 21
space-fixed axis degeneracy 218
spherical
    fragments **256**
        density of external states 264
    rotor 91
    top 47, **278**
        reactant 244
spin–orbit interaction 183
statistical
    factor 37
    limit 246
statistics after dissociation 211
steady state **117**
    distribution 19, 118
        strong-collision version **126**
    in reactive system 117
    population 123
        as Gauss–Seidel vector 142
    rate constant 118
        as inverse of MFPT 138
steepest-descents method **79**
    approximaton for sum/density of states 80
    as smooth-function approximation 86
    aspects of numerical implementation 97
    first-order approximation 80
    for harmonic oscillators 85
        and free rotors **89**
        and hindered rotors **91**
        and K-rotor **95**
    for Morse oscillators **87**
    for partition function as sum **86**
    general routine for $G_{vr}(E, J)$ **95**
    graphical representation of integrand 79
    logarithmic formulation 82
    real integral by 306
    second-order approximation 81
    thermodynamic aspect 82
Stirling's approximation 15, 102
strong-collision assumption 13, 19, **126**
sum of states $G(E)$ 32, 35, **57**
    anharmonic correction 88
    exact enumeration 85
    for rotations **277**
    for translation 280
    Laplace transform of 62
    r-dependent 189, **196**
sum of states $\Gamma(y, J)$
    approximation for arbitrary J 236
    for complex cases **229**

for internal–"external" states 242
for linear and atom 220, **260**
for sphere and atom 222, **260**
for sphere and linear **232**
    high–low-J **235**
for sphere and symmetric top **234**
for two linear fragments 262
for two spherical tops **230**
    high–low-J 232
general case **222**
    high-J 226
    low-J 225
interpolation for 237
minimum energy y_m 227
semi-classical 220, 222
supercollisions 109
survival probability 7, 23
switching function 191, 197
    adjustable parameter 203
    Gaussian 198
    hyperbolic tangent 198
symmetric matrix 116
    characteristic equation 138
symmetric top 38, 42, 66, 265, **280**
    average rotational energy 54
    oblate 281
    prolate 281
    separable, symmetry number of 66
    symmetry considerations 282
symmetry number 94, 219, **277**

thermal ensemble 18
thermal rate constant 12, 123
    as Laplace transform **143**
        case of dissociation 144
        case of recombination 145
        of cross-section 210
    as lowest eigenvalue 118, **123**
    decline at high pressure 128
    effect of tunneling 178
    example of Laplace inversion 147
    fall-off **127**, 128
    high-pressure **124**, 201
        Arrhenius form 124
        in non-adiabatic transition 186
        thermodynamic formulation 130
        transition-state-theory result 123,
            156
    J-resolved 200, 246
    low-pressure 13, 20, **125**, 127, 169, 201
        as rate of collisional activation 125
        diffusion model 134
        eigenvalue equation for 125
        strong-collision **126**
        weak-collision 170, **294**
    rotational average 155, 159, 201
    strong-collision 126, 169
    temperature dependence **128**
thermal reactions
    activation energy defined 128
    examples of 105

reaction order 127
rotational effects 153
thermal system **105**
  collisional energy transfer in 106
  model of 106
Thermochemical Kinetics 131
threshold rate 36
time-lag between activation and reaction 12
transition kernel 114
transition probability **107**
  double exponential 109
  exponential 108, 294
    moments of 168
  factorized, vib–rot 153
  moments of **111**, 132, 133
  non-adiabatic **182**
    decay of 183
  normalization of 107
  off-set Gaussian 108, 110
  quasi-diatomic 293
  rotational, strong-collision form 154
  Schwartz–Slawsky–Herzfeld (SSH) 110
  strong-collision 110, 168, 169
  symmetrization of 114
transition state **28**, 30, 31, 32, **35**, 40, 50, 192, 194
  as local minimum in number of states 190
  as prolate top 38
  as separated fragments **208**, 211
  at extremum of effective potential **213**
  forward/reverse flux in 209
  loose 130, 199, **207**, 211, 247
  sum of states, J-dependent 211
  tight 130, 247
  traditional 207
  Type 1 130
  variational 199
  zeroth-order approximation 41
transition-state switching 199
transitional modes **190**
  in CH_4 fragmentation 70
  zero-point energy 196

translational–rotational distribution
  linear–atom 255
  sphere–atom 255
  sphere–sphere **256**
transport matrix 114, 123
  partitioning of 119, 125
    in reduced matrix approach 141
tunneling 39, **171**
  effect on reaction threshold 177
  experimental evidence for **177**
  imaginary frequency 173
  probability 173
    Laplace transform of 179
    WKB formula 176
Type-1 reactant 37, 38, 43, 126, **265**
  as prolate symmetric top 193

uncertainty principle 32
up-collision 111
  average energy transferred in 111

van der Waals complex 51
variational procedure **189**
  refinement of 273
vibrational
  barrier 44
  partition function 85, 87, 276
  potential 42
vibration–rotation
  density/sum of states for specified J **90**
  energy transfer **297**
  interaction 38
  partition function 297
VTST (Variational Transition-State Theory)
    192

zero-point energy 32, **197**
zero-pressure kinetics **163**
  Boltzmann limit 164
  master equation in 165
  "truncated" Boltzmann 164